FIRST DAWN OF ASH

REBIRTH

BOOK 3

AMY RENFROE

ALSO BY AMY RENFROE

ROMANCE NOVELS

Remember Paris

Emerald Promise

Forbidden in Dubai

His Eyes on the Sparrow

Kokomo Drive

DETECTIVE SERIES

Ashcroft and the Case of the Broken Timepiece

The 13th Valve

VAMPIRE TRILOGY

A Thousand Years to Love You

A Thousand Years of Blood

A Thousand Years of Fire

APOCALYPSE SERIES

First Dawn of Ash — The Impact

First Dawn of Ash — Exile

First Dawn of Ash — Rebirth

First Dawn of Ash — Ascent

ACKNOWLEDGMENTS

I would like to express my heartfelt gratitude to my beta readers, ***Sarah Sheppard*** and ***Esther White***. Your unwavering support, insightful feedback, and encouragement throughout the writing of this book have been invaluable. Thank you for your dedication and for believing in my vision. Your contributions have significantly enriched this work.

And to ***Joe—Joey***—my son, my steady-hearted hero, the soul behind the character who carries this story on his back. You've lived through storms that would've swallowed lesser men, and you still stand with grit, humor, and that quiet, stubborn strength that refuses to quit. This trilogy has a heartbeat because of you. Thank you for letting me borrow your courage, your fire, and your name. You'll always be the hero, on the page and off.

Formatted with Vellum

CONTENTS

INTRODUCTION

FLEET LOG — UNOFFICIAL
Recorded by: Joe Alvarez
Time-stamp lost to weather interference
(Entry archived pre-Rebirth)

If my voice sounds rough, it's because my hands are shaking so hard the recorder keeps slipping, and I'm not starting this over again. The storm already took enough out of us—I'm not giving it my words, too.

I don't even know what time it is. The sky's been black for hours, or maybe minutes. Hard to tell when all you can see is water being thrown at you sideways, like the ocean's trying to beat the world into something new.

The hurricane hit harder than the models suggested… not that we trust models anymore. Everything's different since the flood. Weather patterns move like they're angry. Like the planet's got a temper now. And tonight? It came for us. All of us.

Most of the fleet was clustered tight—Savannah at the center, Halcyon diving, surfacing, diving again as she tried to keep her systems dry. The tugboats were barely holding a line, getting tossed like toys. Every wave slammed us hard enough to rattle the bulkheads. But I didn't care about any of that.

I cared about Mercy-2.

Because that's where Mom and Dad were.

The storm wall hit like a freight train. One second we were riding swell, the next we were staring into a rotating wall of water taller than Mercy-2's entire superstructure. The first time I lost sight of them, my stomach just... fell. Like gravity let go.

That's when the fear started. Not the usual fear—the kind that sits in your chest and waits. This was the kind that crawls up your throat and digs in like claws. The kind you try to swallow but it sticks. The kind that whispers, This is it. This is the one you lose.

I'm not proud of it, but I yelled at the ocean. Full volume. Over the wind. Over Manny shouting telemetry I couldn't hear. Over the alarms that wouldn't shut up. Over Valery telling me to step back before a wave knocked me straight off the deck. I yelled at something ancient and merciless like it owed me an apology.

Because my parents were out there.

And I wasn't losing them. Not after everything we'd already survived.

The deck pitched hard to port, and I slammed into the railing, barely keeping my footing. The waves weren't crashing—they were exploding. Every one of them hit like the sea was punching the ship with both fists. Spray stung like gravel. I kept wiping my eyes, but it didn't matter. Rain, saltwater, tears—they all felt the same in a storm like that.

I had my radio clutched in one hand so tight my knuckles hurt. Every few seconds I'd lift it to my mouth, try again:

"Mercy-2, respond. Mercy-2, this is Savannah. Mom—Dad—do you copy?"

Nothing. Just static. Wind. A low groan somewhere deep in the hull.

The silence chewed at me. Made every second stretch into something unbearable. I'm trained to stay calm, to think through crisis, to prioritize systems, crew, structure… but training doesn't mean a damn thing when your parents are on a ship that keeps disappearing behind mountains of water.

Manny came up beside me, shouting something, pointing at the bow. I couldn't hear him at first—just the storm roaring down the lines. Then a flash of lightning tore the sky open and—

There she was.

Mercy-2.

For one bright, awful second, I saw the whole ship—bow high, stern buried in foam, her lights flickering like haunted eyes. She looked like she was fighting for her life, like she was climbing a mountain she didn't choose to climb.

I heard myself make a sound I'd never made before. Something low. Something broken. Something I hope nobody else on the deck heard.

The wave dropped. Mercy-2 vanished again.

I leaned forward over the railing like I could will her back into existence. My fingers were shaking so bad I nearly dropped the radio. A cold certainty crept up my spine—a whisper saying that was the last time you'll ever see them. I slammed the thought out of my head so fast I felt dizzy.

Valery grabbed the front of my jacket. "Joe! Joe—look at me!"

I did. Barely.

"We can't help them if we go overboard," she said, her voice sharp, clear, cutting through the storm like a knife. Her hair was soaked, plastered to her cheek, and her eyes—God, her eyes were steady. "You stay on your feet. You hear me?"

I nodded. Or I think I did. Hard to tell what my body was doing half the time.

Another hit. A big one. The deck shuddered under my boots, metal groaning like the whole ship was praying. Water sloshed across the bow, lifting half the crew off their feet. The floodlights flickered. We lost internal power for a second. Just a second—but long enough to feel like the storm swallowed us whole.

Every time I tried to log coordinates, the pen slipped. Every time I tried to check the sonar overlay, the screen fogged. My hands wouldn't stop shaking. I kept picturing them—Mom holding onto the rails, Dad shouting survival orders over the wind, both of them trying to look braver than they felt.

Not knowing if I could get to them.

Not knowing if I was too late.

I tried the radio again. And again. And again. My voice cracked on the third attempt. By the fifth, I could barely breathe. I don't know what broke me first—the fear or the helplessness.

Another lightning flash—this one closer, brighter. I could've reached out and touched the sky. The thunder rolled over the fleet like something alive.

And in that brief electric flare—

Mercy-2 surfaced again.

But wrong.

Listing hard. Deck half-submerged. Something gone from her silhouette. Something missing. Something broken.

I don't know if the scream I heard was mine or the wind.

Everything inside me went cold. Not the chill of rain—not the bite of ocean spray. This was deeper. Sharper. Like ice running through my veins, freezing every memory I had of them.

And then she vanished again.

That was the moment I made a promise. Not out loud. Not with words. Something bone-deep.

If the storm took them, I'd go after them.
If the sea fought me, I'd fight back.
If the world was ending again—I'd go down swinging.

Because they're my parents.
Because I love them.
Because I wasn't losing them—not now, not in the dark, not in the middle of an ocean that thinks it can take everything from us.

I don't know how this log will end.
I don't know how that night will be remembered—if at all.

But this? This moment?
Frozen in saltwater and fear and lightning?

This is where the fight started.
The real fight.

End of log.

CHAPTER 1
THE WAKE

The ping didn't stop when the clocks rolled past noon.

It followed them through the ship, four seconds apart, until no one could pretend it was random noise.

By the time the Savannah finished her first tight arc to southwest, the bridge had the signal pinned in a narrow cone: low-band buoy, not fleet standard, not drift. Weak, but steady.

"Two-three-two off the bow," Morales said. "Range less than a klick. It's sitting in the water and yelling at us."

Raines gave the order without a speech. "Put a skiff over the side. Tag it, don't run it down."

Deck crews moved without needing details. Winch cables whined. A skiff slapped into the gray water and peeled off toward the bearing, spray breaking white off its bow.

On the bridge, the ping kept going.

Ping.

Charlie stood near the aft bulkhead with Lenny at his feet, life vest still buckled, eyes red but clear. He didn't speak. Every time the sound came, his fingers tightened in the dog's fur.

Joe stayed by the forward glass, watching the distant skiff bump up and down the swell. His hands were steady. His face wasn't.

Valery sat at the console, headphones tilted to one ear, eyes locked on the rangefinder. "Closing. Two hundred meters... one-fifty... one hundred..."

"Visual?" Raines asked.

The watch officer lifted the binoculars. "Float on the port side of the skiff—small cylinder, white top, looks taped as hell. They've got it."

On the open channel, the skiff pilot's voice crackled thin: "Savannah, we have contact. One jury-rigged buoy, repeat, one buoy. Markings nonstandard. Bringing it back now."

Ping.

The sound grew louder as they homed in, then cut off mid-cycle when the skiff crew killed the tiny transmitter. The silence hit harder than the noise.

On deck, the skiff thumped the hull. Hands reached down and passed the thing up—a scratched biohazard canister wrapped in mesh and tape, scarred and battered from days at sea. Someone had lashed a bent antenna to the side with med tape. It looked like trash.

No one treated it like trash.

"Clear a table," Morales ordered. She took the canister herself to the midships bench. Joe, Valery, Manny, and Charlie followed. Lenny trotted behind them, nails clicking on the wet deck.

Up close, it looked worse. The gasket bulged where it had been overtightened. Inside the mesh, a narrow med pouch sat sealed in plastic.

Valery cut the tape carefully. She eased the pouch free, wiped the salt off, and held it to the light. A small folded scrap of paper sat inside, ink smeared by condensation.

Nobody breathed while she opened it.

Seven blocky letters filled the page.

BREATHE.

Written hard enough to groove the paper. The ink blurred at the edges where someone had exhaled on it before sealing the pouch. It smelled faintly of metal, antiseptic, and a ship that wasn't ready to die.

Valery's voice trembled: "That's... that's Donnie's block print."

Charlie stepped forward, voice shaking but steady. "I told you. He sealed the med lab. He sent this. They're still under there with air."

Manny scrubbed a hand over his face. "Could've been sent before—"

Charlie cut him off with a harsh shake of his head. "No. He said if you can't talk, you send a sign. That's what this is. He made a lung out of that room. He would've. He showed me how."

Morales looked at Joe.

Joe took the note. The grooves, the angles, the weight of each letter —pure Donnie. The kind of writing meant to last through oil smudges and saltwater and years of hard work.

He folded it along the old crease and tucked it into his chest pocket. It crinkled there, faint as a second heartbeat.

"We're not burying them twice," he said. His voice was rough but clear. "We find the wreck. We find the pocket. We bring them out. Or we prove they're gone. But we don't leave them guessing."

Raines nodded once. "Then we hunt."

The Gulf lay quiet—too quiet for truth. After storms, the sea pretends to forget, smoothing its wrinkles, humming a false lullaby. But the Savannah's crew knew better. Quiet was just noise that hadn't remembered how to scream yet.

The fleet fanned out over the coordinates where Mercy-2 had gone under.

One ship, two tugs, four skiffs, two helicopters, and a submarine formed a grim halo above the dead water. From above, it might have looked like patience. From the decks, it felt like punishment.

Joe Alvarez hadn't slept in forty hours. He'd closed his eyes, but all he saw was the hatch—the glass circle filling with water, his mother's hand pressed against it, Donnie's silhouette braced against the failing frame. The jar sliding beneath the waves at dawn. The buoy. The word.

BREATHE.

That kind of memory doesn't fade. It syncs itself to your pulse.

He stood at the forward console on the bridge. Morales ran command beside him, voice flat, clipped, precise—fatigue honed into discipline.

"Hydrophones deployed. Dragging at one knot. Sweep set sub-forty hertz. Depth limit, four hundred meters."

Joe nodded. "Copy."

"Valery?" Morales called.

Valery sat cross-legged at her station, headphones pressed tight. Her monitor cast her in blue. She'd been filtering noise since before sunrise—whale song, static, the sea's endless arguments with itself.

"Channels one and two clean," she said. "Three still has static. Might be the tugs."

"Or wishful thinking," Morales muttered.

Valery didn't smile. "I'll take wishful."

The bridge felt like a silent chapel. No one talked unless needed. The ship hummed beneath them like a slow, mechanical heartbeat.

Outside: gray sea, gray sky, gray horizon. The world looked scraped bare.

"Comm check, all vessels," Morales ordered.

Voices answered, ragged but steady.

"Begin search pattern alpha," she said. "Five-mile spacing. Report anything irregular."

Joe adjusted the gain. Ghost lines marched across the screens—sonar's version of memories.

"Drag depth steady at one hundred meters," Morales said.

"Copy," Joe answered, hand brushing the paper in his pocket again.

Valery murmured, "Something under the base noise… low… buried…"

"Sea talking to itself," Morales replied.

Valery shook her head. "No. Patterned."

She twisted the filter dial. A faint pulse shimmered across the spectrogram—barely-there, steady, stubborn.

"What frequency?" Joe asked.

"Seventy-two point three. Consistent."

"Mechanical?" Morales offered.

"Too clean."

The pulse repeated—three beats, pause, two beats.

Joe leaned close. "That's not chance."

Morales exhaled. "You think it's them."

Joe didn't answer.

He didn't need to.

The bridge went still except for the quiet click of relays and the slow tick of the wall clock. Outside, gulls circled the waves—rare this far out, which made the moment feel heavier.

Joe reached for the comm. "All ships, hold current heading. Decrease speed to half-knot. Prepare for narrow-band sweep."

Morales hesitated. "You think it's—"

"I don't think anything yet," he said softly. "But I want to listen before I start believing."

He adjusted the dials. The pulse sharpened—three beats, pause, two more. Then silence.

Valery's eyes widened. "That's not random."

Joe nodded once, throat tight. "No. It's not."

Another long breath passed through the bridge. Hope was a dangerous thing—sharp as a blade, sweet as poison.

Outside the windows, the sea rolled under a thin sheet of gray light. The fleet drifted into a cautious halo, engines throttled low, the whole world holding its breath with them.

The sound came again. Sure. Steady.

Three beats.

Pause.

Two more.

Morales stared hard at the display. "Well, I'll be damned."

Joe didn't smile. He never did when something mattered too much. His jaw set, eyes fixed on the spectrogram as if he could drag meaning out of it by force.

"It's a start," he said. "Mark coordinates."

Outside, the faint sun sliced across the water. The Savannah held position, listening to a voice barely louder than memory.

No one said it aloud, but the truth sat in the room like an extra heartbeat:

Someone was calling back.

The next stretch of time became math—relentless, patient, merciless math.

Morales spread the search grid overlay across the entire bridge table, a glowing lattice of lines and angles. Five miles squared. Broken into loops. Systematic. Surgical.

The Savannah drifted through the first leg, dragging her hydrophones through gray water like silver ears listening for ghosts.

Joe stood beside the console, eyes pinned to the waveform. Every thirty-nine seconds, the pulse came.

"Signal-to-noise ratio's rising," Valery murmured. She traced the pattern in the air with one hand. "If the signal's strengthening, either we're closer or the pocket's shifting upward."

Morales snorted. "Or the ocean's finally playing fair."

Joe shook his head. "Ocean doesn't play fair."

Valery pressed on. "Still—it's clean. Cleaner than this morning."

Morales gave her a long look. "Cleaner doesn't mean alive."

Valery didn't blink. "Doesn't mean dead either."

The air thickened again. Every crew member on the bridge felt the tension—not hope exactly, something older and heavier.

Outside, the hull groaned gently beneath them. Calm seas were liars.

"Triangulating," Valery said, fingers flying. "Combining hydrophone arrays."

A blue cone unfurled across the holo-map.

Another from the Halcyon.

A third from the tugs.

The three translucent shapes intersected over a crooked diamond of sea—maybe a mile long, half as wide.

"There," Morales said. "Echo range. Two-point-one nautical miles."

"Depth?" Joe asked, moving closer.

Valery checked the numbers. "Fluctuates—one-seventy to two-twenty meters."

"That's breathing depth variation," he said without thinking.

Morales shot him a sharp look. "Ships don't breathe, Alvarez."

"No," Joe said quietly. "People do."

Lenny padded into the bridge at that moment, nails tapping across the deck. He made a slow circle around Joe's legs, then lay down near the window, head up, ears twitching.

Every time the pulse hit, his tail flicked once.

Valery noticed first. "He hears it."

Joe swallowed. "He used to sleep under Dad's workbench. He always heard things the rest of us missed."

The next signal hit harder—cleaner, brighter. A double waveform.

Morales frowned. "Reflection?"

Valery shook her head. "Delay's wrong. Too clean. Too synced."

Joe leaned in. "Two sources."

Valery nodded, voice hushed. "Two heartbeats."

Joe closed his eyes for half a second. "Mom and Dad."

Morales didn't correct him this time. She didn't look away from the screen. "Send coordinates to the captain. Raines needs to see this."

Valery transmitted. Her hand shook once.

Ten seconds of static.

Then Raines's voice: steady, gravel-edged.

"Thermal imaging active. Surface clear. Scanning subsurface density."

Valery held her breath.

Suddenly: "Localized pocket. One-fifty meters down. Temperature delta four degrees over ambient."

Valery's breath hitched. "That's air."

"Could be trapped methane," Raines said. "But it reads like an air pocket."

Joe pressed his palm to the console. "Then we found them."

Silence swept the bridge—not disbelief, not fear.

Reverence.

Morales finally moved. "Prep a dive plan. Controlled descent.

Valery whispered, "And if we're too late?"

Joe answered before Morales could.

"We won't be."

The next pulse rolled through the hull—soft, stubborn, alive.

Joe touched the glass, staring at the patch of gray sea that hid everything.

"They're still talking," he murmured.

Morales nodded once, crisp and certain. "Then we'd better listen."

Time lost its shape after that.

Hours smeared together—gray water, dim lights, tired eyes memorizing waveforms until the sea felt like it was staring back.

Rain drifted in thin needles. Deck plates shone slick. The crew moved like shadows.

Hope became less a feeling and more a ritual.

Every few hours they recalibrated: hydrophones cleaned, sensors reset, logs updated. The same numbers marched down the lists.

Valery didn't leave her station except to wash her face and curse under her breath. Her headphones became part of her skull. She lived in the rise and fall of sonar tones. In half-dozing moments, Morales caught her whispering: "thirty-nine seconds… thirty-nine… still there."

Morales logged everything—every flicker, every ghost-reading, every almost-signal. Her handwriting was tight, razor-clean.

"02:50 — static return

03:40 — false positive

05:04 — thermal ripple

06:33 — unknown harmonic, discarded"

She never wrote the word hope. Not once.

Below deck, Manny and Guido swapped hydrophone relays again. No talking, no jokes. The work itself became a superstition: if they kept the gear alive, maybe it would keep the people alive.

Joe walked the outer deck in the cold dark, murmuring equations like prayers—current drift, displacement, pressure equalization.

Guido called down from above, coffee in hand: "Talk nice to her, hermano. She might talk back."

Joe didn't look up. "I'm trying."

Lenny slept beneath the main console. Every ping made him lift his head, whine once, then drop it again—as if even he was pacing his hope.

Then—near morning—

Valery whispered, "Morales… listen."

Morales leaned in. "What do you hear?"

Valery stripped the top band off the filter. Static peeled away.

Something moved beneath it.

A hum. Soft. Slow. Alive.

Joe was at her side in seconds. "Play it again."

She did.

Morales locked onto the coordinates. "Track scatter. Drift. Where's it coming from?"

"East," Valery said. "Depth one-six-zero."

"Temperature?" Morales demanded.

"Four degrees above ambient. Localized. Not spreading."

Morales froze.

"Holding heat?"

Joe exhaled. "Something's breathing down there."

Valery hit transmit. "Savannah to Raines—we have thermal contact."

Raines came back instantly. "Copy. Marking. Stand by for drone."

Joe's fingers tightened on the console until his knuckles went white. He stared at the glowing patch of sea.

Morales whispered, "All night of static."

Valery didn't look away. Tears streaked her cheeks. "And then it sang."

Joe touched the screen, voice barely air.

"Hold on. We hear you."

The Gulf was a sheet of pewter that morning, so flat it looked artificial, like someone had ironed the world. The Savannah's engines whispered against the current, thrusters nudging her along the contour of the seafloor. Everyone on the bridge had gone quiet again—that strained, knife-edge silence that only happens when fear and math are trying to agree on something.

Morales stood at the mapping console, fingers light on the keys, watching the drone's telemetry crawl across her screen. Valery hovered behind her, headset on, eyes darting between the sonar trace and the live camera feed. Joe, Guido, and Manny waited near the bulkhead like men outside a hospital room—ready for anything, able to do nothing.

Charlie leaned in the doorway with Lenny in his arms. "I told you they're alive," he said quietly. "There's air in that med-bay. Donnie would've sealed it and pressurized it. He showed me how it could be done." He set Lenny down and waited.

"Depth one-six-zero meters," Valery said. "Drone One stable. Visibility… two meters at best."

Morales adjusted the gain. "Stay low. Move east by ten. If we clip the debris field, we lose line of sight."

"Copy," Valery murmured.

The drone nosed into the gloom. Its lights speared floating rust flakes, drifting bits of Mercy-2's bones, and a dangling IV bag swaying like a dead jellyfish. A twisted ladder. A torn stretcher. The ghost fingerprints of collapse.

"There," Joe said suddenly. "Pause. Back five degrees."

The image slid backwards. A bulge in the seafloor appeared—curved, symmetrical, too intact for a collapse. The plating held its shape like it was stubborn, not broken.

Morales leaned close. "That's not debris. That's structure."

"Pocket," Joe said. "Intact."

Valery's hands flew. "Running gas scan."

Numbers scrolled: rising CO_2, falling O_2, stable nitrogen.

Morales exhaled. "Something's breathing in there."

The bridge stilled again. Even breathing felt loud. Guido crossed himself. Morales didn't look away.

The drone drifted closer, camera lights sweeping what had once been the med-bay bulkhead. Mercy-2's medical insignia—faded, cracked—still clung to the metal. Beneath it, where the plating had buckled but not broken, tiny bubbles leaked upward, curling like breath.

"Zoom," Morales ordered.

The image tightened. Three cylindrical canisters were bolted into the seam—reactor scrubbers, rerouted with medical tubing. Old emergency O_2 bottles patched into the return line. Coils from a cooling unit strapped along the vent path. Improvised. Brilliant. Insane.

Morales's voice dropped. "He built a pressure-stable life-support loop out of scrap."

Valery's throat caught. "He bought them time."

The drone edged closer. A panel lay half-submerged in silt. Faded white paint. Indented letters.

MED BAY – 2

And beneath the buckled edge, something impossible flickered.

Light.

Weak. Flickering. But alive.

Morales whispered, "Auxiliary cell. They kept a battery running."

Joe stepped closer to the monitor, breath fogging the glass. "Dad rerouted power. Or Mom did."

Valery nodded. "She knew that system. She could've kept it stable."

For a long moment, they didn't speak. The drone's lights glowed against the med-bay's torn metal like a candle behind stained glass.

Morales snapped out of it. "Signal the captain. We have visual confirmation."

"Copy," Valery said. Her voice cracked on the second syllable.

Raines responded within seconds, gravel-low but controlled: "Copy. Marking coordinates. Hold position. Prepare for dive briefing."

Joe stared at the screen as if he could will the metal to open. His pulse hammered in his neck. He didn't blink.

Morales lowered her voice. "All night of noise…"

"And then this," Valery said. Tears ran, unnoticed.

A faint rhythmic blink pulsed beneath the plating. Not steady enough to be a system. Not random enough to be chance.

Morales leaned in. "Is that—"

Joe's voice broke in, quiet and certain. "Heartbeat or light. Doesn't matter. It means they're still there."

Behind him, Charlie whispered, "I told you."

Joe put a hand on the console, steadying himself. "Prep a controlled dive.

Valery answered softly. "Already on it."

Morales didn't look away from the screen. "Then let's bring them home."

CHAPTER 2
BREATHLESS

The first sound wasn't an explosion. It was a low, heavy groan rolling through the Mercy-2's ribs. The deck flexed under Donnie's boots. Red strobes kicked alive. Alarms layered until they stopped sounding like warnings and started sounding like weather.

"Pressure wave," Donnie said, then louder: "Brace!"

Steel shrieked—rivets complaining, a bulkhead throwing bolts. Water found a door somewhere forward and made it history. Donnie's hand hit the med-lab hatch wheel.

"Amy!"

"I'm here," she called, stepping into view from the far side of the room—hair braided tight, eyes steady in the horn-light. "¿Qué pasa?"

"Crush load off the port keel. We have seconds. Inside—now."

He spun the wheel and shouldered the hatch. Trays rattled. A clamp skittered to his boot. Amy grabbed the crash cart and planted her feet. "Who's hurt?"

"Everybody if we don't seal." His voice was calm in the way that made her move faster.

He ripped the emergency kit off the bulkhead—polymer tubes, foam canisters, flex-collars—and dumped it on the counter. "We're making a lung."

"Tell me what to hold."

"Vent first—collar it," he said, pointing. "Anything that leaks is the enemy until we've got pressure under control."

She jumped to the counter and cinched the vent throat with a metal clamp. He ran a glittering bead of sealant around the hatch frame, thumb smoothing it into the corners. "Foam."

Amy slapped a canister into his palm. He punched the valve; cream swelled into seams and stiffened hot. The ship groaned again. The floor canted three degrees. Dust sifted from the ceiling grid.

"Back inside," he ordered, snagging the corridor valve with one hand.

"Donnie—"

"Please, Amy."

She moved. He cranked the valve. Black water shouldered down the corridor with a cold roar. The hatch frame shuddered once, then steadied as the pressure on both sides learned each other's number.

"Equalizing," he said. "Good girl."

"Your hand," she said, catching a thin cut across his knuckles. She wrapped it fast with gauze. "You bleed worse than you admit."

"Everything does. That's why we carry tape."

He tapped the bulkhead, listening. Tight. He pried the service panel above the med-fridge and exposed a coil nest. "We'll need this." He kept his eyes on the angles and lengths, already stealing heat for later.

Amy yanked the respirators and an old O_2 bottle from a cabinet. She cracked the valve. A soft hiss answered. "Not empty."

"Believe the hiss, not the sticker," he said.

The ship took a long diagonal shove. Two ceiling tiles snapped. A fracture chased across the porthole like lightning.

"Prep's over," Donnie said. "We commit."

He spun the hatch dogs home and checked the gauge: delta sliding toward green. Sealant gleamed like new weld. The corridor outside was a black wall stitched with bubbles.

"What now?" Amy asked, breathing through her nose—three in, hold, three out—the way she'd taught Joe under hurricane blankets.

"Make a room that buys us hours," Donnie said. "Soda-lime pulls CO_2. The ventilator cart has canisters—we rip them, build a loop, and bleed O_2 back in." He pointed at the fridge coils. "Run them as a condenser to tighten the air. It's ugly, but it'll work."

"Like canning peaches," she said.

"Exactly. Jars just bite harder."

He gutted the ventilator, fed scrubber canisters into a crude manifold, and tied the coil through with medical tubing and tape. He patched the half-full bottle into the return line. Gas hissed. The rig answered with a slow, steady sigh.

"She's holding," Amy said, eyes on the seams.

"She's holding," he echoed.

Darkness rolled through as the generator coughed and died. The emergency torch painted the lab in weak amber. The air took on a sweet metal taste. The lower half of the room filled to their waists; the upper stayed dry.

Donnie flipped open his beat-up ledger and penciled numbers.

Air volume: ~37 m^3

Leak rate: 0.04 /min (guess)

Scrubber life: 8 hrs optimum

"If we stay calm," he said, "we've got nine good hours, maybe twelve if we nap like cats. If we panic—six."

"And if it takes longer?"

"Then we talk quieter." He squeezed her hand. "We're not dead until the math says so."

They sat shoulder to shoulder on the counter, legs tucked up. The med-lab stopped groaning and settled into a listening quiet. The home-built system ticked and sighed. For a while they just breathed.

Then the air thinned more than he liked.

"We need a messenger," Donnie said, pulling an empty biohazard canister, gasket, mesh, waterproof tape. He sealed a note in a med pouch—**BREATHE**—plus a microchip from a dead transponder. He built a ridiculous little buoy and wired it to the waste vent riser.

Amy braced the joint while he set a CO_2 extinguisher at the base. "Will it go?"

"It'll go." He cracked the valve. A muffled thump ran up the duct. The canister kicked twice and vanished. Bubbles streamed into dark.

They watched until the glow outside the porthole trembled and went out.

"So now we wait," Amy said.

"Now we live," he said, squeezing her fingers.

They breathed with the machine and each other until time lost its edges. Outside, the ship's noise faded to the slow heartbeat of steel cooling under pressure. Inside, their jury-rigged lung whispered them toward dawn.

Dawn broke clear enough to work. Wind was light, swell low, visibility decent. The fleet held a tight ring over the coordinates. The Savannah ran point. The Halcyon tracked sonar and standby support. The House, two tugs, four skiffs, and two helicopters held their posts.

On the Savannah's bridge, Morales stood at the chart table with a marker in her hand and the dive window written in block letters: 0520–0610. She didn't look away from the board when she spoke. "Final checks. We follow procedure. We don't rush. Confirm teams."

"Dive team ready," Joe said. He was already in his suit, helmet latched to his harness, gloves clipped. The exothermic torch rig sat in a cradle beside him.

"Cutter Two ready," Guido said over the intercom from the launch deck. "Torch primed, spare rods loaded."

"ROV hot," Valery reported from the comms console. "Camera clear. Manipulator working. Micro-vent routine pre-programmed. I'll run vent pulses on your call."

"Telemetry and recording up," Perez added. "Hydrophones clean. Thermal overlay steady. Live feed to the bridge and launch deck."

"Decompression prepared," Dave said from medical. "Chamber at pressure, scrubbers fresh, O_2 and IVs set. We can take two plus one tender immediately."

Morales capped the marker and finally looked up. "Good. Keep chatter tight. Call only what matters. If you're about to say more than six words, cut it to three." She turned to Joe. "You find the seam. You keep the heat moving. You do not pop it."

"Understood," Joe said.

The bridge door opened and Charlie stepped in, holding Lenny against his chest. He had on a life vest that looked too big and a face that looked too old for eleven. He stopped just inside the threshold like he knew better than to run on a bridge.

Morales's eyes softened a fraction. "You can stay on the bulkhead, kid. No questions during the descent."

Charlie nodded hard. "Yes, ma'am."

Joe crossed the room and crouched to him. "You stay with Morales," he said. "You listen. If it gets loud, you cover Lenny's ears."

Charlie swallowed. "Bring them back."

"I will," Joe said. He meant it the way men mean promises they can't afford to break.

Charlie shifted Lenny to one arm and wiped his face with the back of his hand. He didn't cry. He just held the dog tight and stared at the live-feed monitor like it was the only window that mattered.

Morales checked the clock. "Zero-five-one-eight. Helm, hold station. Launch deck, stand by."

"Standing by," Guido answered.

Joe pulled his helmet on and sealed the ring. The suit hissed once as the regulator settled. Valery ran the comm check. "Joe, give me count."

"Five by five," he said.

"Guido?"

"Five by five."

"Bridge copies," Morales said. "Dive team to the lift."

Joe jogged down the ladder to the launch deck. Air was cool. Deck plates were dry. The crane boom hung square over the water. Two deckhands clipped his harness to the lift line and tugged twice to confirm. Guido stepped into his cradle and checked his torch once more.

Manny stepped up, hand on Joe's shoulder. "I'll be your safety. If anything twitches wrong, I'm on the line."

"Copy," Joe said. He bumped Manny's fist with a gloved knuckle.

Morales's voice came over their helmets, steady and even. "Now hear this: this is a controlled breach. We treat that room like a lung. We vent before we cut. We cut thin before we open. If pressure spikes, you stop. No heroics."

"Copy," the team said together.

Valery's voice followed. "ROV moving to Seam Alpha. Light on the mark. I'll hold five meters off the plating to avoid backscatter."

Perez added, "Thermal pocket stable—delta four over ambient. Depth one-six-zero meters. No cross-traffic."

"Launch," Morales ordered.

The lift took weight and swung them gently over the water. Joe looked once at the rail. Valery had a hand on the comms board. Perez leaned in over her shoulder. Morales stood tall, hands on the table edge. Charlie stood in the doorway, Lenny crushed to his chest, eyes locked on Joe's helmet. Joe gave him a small nod. Charlie nodded back, jaw set.

"Lower," Morales said.

The winch paid out line. Cold water rose to Joe's boots, then his waist, then his chest. The sea closed over his visor with a flat slap, and the world switched to the sound of breath and the slow thrum of pumps. Light faded in even steps. Numbers replaced sky.

"Twenty meters," Valery called. "ROV ahead and left. Follow my light."

Joe adjusted trim and followed the white cone in the green. Silt drifted like dust. The wreck came up out of the dark—a collapsed outline and a hundred angles. Panels bent. Beams torn. The med-bay section sat under a slab like a box under a table.

"Target at your one o'clock," Valery said. "ROV light on Seam Alpha. You're three meters off."

"Copy. Closing," Joe said. He reached the plating and set his glove along the seam. He felt the faint tremor through the suit—a thin vibration, not noise, not current. Pressure on one side, the Gulf on the other.

Guido eased in beside him. "Torch One ready."

"Hold torches," Morales said. "We vent first. Valery?"

"Micro-vent ready." The ROV's arm pinched the small needle valve they'd set the night before. "Cracking one degree."

A thin stream of bubbles slid along the seam and drifted up slow. No surge. No shake.

"Stable," Valery reported. "Internal pressure reading 1.8 atmospheres and easing. Delta now point-three."

"Proceed to thinning cut on the top seam only," Morales said. "Keep heat low. Move. Stop. Vent. Move. Stop. Vent. No puncture."

"Copy," Joe said. He brought the torch to life. The water flashed around the flame and went clear. He set the tip to the edge and walked the heat along the line, slow and even. The metal softened in a thin band. He kept the bead narrow, drew breath by the count, and watched his temp gauge.

"Heat steady," Valery said. "Good pace. You're at eight centimeters."

"Vent," Morales ordered.

"Micro-vent," Valery confirmed. The bubbles washed Joe's wrist. The seam stayed quiet.

"Continue," Morales said.

Guido took the parallel side. They moved together—two steady lines, no rush. Every twenty centimeters, Valery feathered the vent. Perez kept the numbers rolling in a flat voice: "CO_2 dropping point-one… O_2 holding at 12.9… thermal steady… structure creep negative."

Charlie's voice broke in from the bridge mic before Morales could stop him. "Is my dad still breathing in there?"

Morales put a hand over the bridge talk and kept her voice calm. "Charlie, you keep listening. We've got them."

There was a small sound, Lenny's soft whine. Then Charlie went quiet again.

"Fifteen centimeters," Valery called. "Vent ready."

"Vent," Morales said.

Joe lifted the torch and let the bubbles run along his glove again. Warm. Real. He put the torch back and advanced another span.

A beam shifted above with a dull groan. Silt clouded. Visibility dropped by half.

"Hold," Morales snapped. "All stop."

They froze. The silt drifted past. The beam settled without collapsing.

"Stable," Valery said after three long breaths. "Proceed."

They moved again.

"Partial O_2 internal at 12.4," Perez reported. "CO_2 2.7. You're buying time."

"Noted," Morales said. "No deeper than coin thin, Alvarez. If you see blistering, you stop."

"Copy."

Joe kept the heat walking. Elbow steady. Shoulder tight. Breath even. He didn't talk. He didn't look away. He thought only about distance, temperature, and the faint tremor under his palm.

"Ten centimeters to corner," Valery said. "Vent ready."

"Vent," Morales said.

"Micro-vent," Valery confirmed.

Bubbles slipped up again, warm against Joe's sleeve. He pictured his mother's hand on the other side of the metal and forced the thought away so he didn't change his pressure. He kept the line true.

Morales checked the time. "Zero-five-thirty-one. You are on schedule. Keep it boring."

Guido gave a short laugh in Joe's ear. "Boring is my favorite kind."

"Five centimeters," Valery said. "Three... two..."

"Hold," Morales said. "Vent."

"Micro-vent active," Valery answered. "Delta now point-two. Interior easing."

Morales drew a slow breath the team could hear. "Good. Begin second pass. Same depth. Same pace. No penetrations."

"Copy," Joe said. He reset his grip, steadied his arm, and started the next run. Above them the Gulf stayed quiet. On the bridge, a boy held a dog and watched a screen, and the clock rolled toward the minute when a door would finally open.

The wreck loomed larger the deeper they worked. Floodlight beams sliced through the murk, touching broken frames and torn plating. The med-bay hatch lay half buried beneath twisted debris, the faded letters barely visible through the film of silt.

"ROV repositioning," Valery said. "Holding at your nine o'clock. Light steady."

"Copy," Joe answered. He steadied himself with one hand against the hull and watched the thin seam glow under the torch. "Temperature holding 120 Celsius. No spatter."

"CO_2 2.8," Perez called. "O_2 twelve even."

Morales's voice was flat but calm. "You're right where you need to be. Keep the cut thin."

Guido's torch moved in parallel, the two arcs of light crossing like a zipper being opened in slow motion. Every few seconds, Joe paused to let Valery feather the vent. Each vent came with a string of small bubbles, rising lazy and warm. The sound in his headset wasn't noise anymore; it was breathing.

"Pressure differential down to point-one-eight," Valery said. "Good response. The pocket's relaxing."

Morales exhaled quietly into the mic. "That's what we want. Gentle. We let her breathe before we ask her to talk."

Up on the Savannah, the bridge was silent except for the hum of electronics. Charlie sat on the floor next to Lenny with his knees pulled to his chest. He didn't take his eyes off the feed. Morales kept glancing at him but didn't send him away. The boy's face was pale, focused, the way Donnie used to look when fixing an engine.

Valery's voice filled the room from the speaker. "Micro-vent engaged. Bubbles consistent. Seam stable."

"Looks good down here," Joe said. "Another ten centimeters and we hit the corner."

"Roger that," Morales answered.

Charlie whispered to Lenny, barely loud enough for the mic to pick up. "You hear that? He's close." The dog's tail thumped once against the deck.

"Hold for pressure," Valery said. "Reading shift on inside gauge."

Joe froze mid-cut. "Talk to me."

Perez read the numbers. "Interior pressure 1.7 and steady. False alarm."

Morales nodded to herself on the bridge. "Proceed."

Joe's torch slid forward again, precise and slow. Guido followed half a meter behind. The ROV drifted over the hull, its light brushing across what looked like makeshift tubing and canisters fixed to the bulkhead inside.

Valery's tone was analytical. "Confirming the scrubber array we mapped yesterday. Reaction heat up two degrees—still cycling."

"Copy," Morales said. "Configuration unchanged. Proceed by the plan."

Guido let out a low whistle through the comm. "That's your old man, hermano. Smart as sin."

Joe's chest tightened. "Yeah. That's him."

"Vent again," Morales ordered. "Ten-second pulse."

The ROV adjusted its valve. More bubbles rose, the water shimmering around them. Joe lifted his torch and felt the warm slip of air against his forearm.

"Stable," Valery said.

Morales nodded, satisfied. "Then we keep working. No surprises. Nice and slow."

On the bridge, Charlie gripped the railing now instead of the floor. His knuckles were white. "Come on, Dad," he whispered.

Perez glanced at him. "They've got it under control, kid."

Charlie didn't look away from the screen.

Below, Joe paused long enough to catch his breath. The seam ran half the length of the hatch now, glowing a dull red where the metal had softened. Guido floated close, his torchlight flickering against Joe's visor.

"You good?" Guido asked.

"Arm's dead, but yeah."

"Then let's finish it."

"Copy."

"Partial O_2 twelve even," Perez said. "CO_2 2.9. Still rising."

Valery adjusted her filters. "Getting interference—picking up faint vibration. Could be inside movement."

Morales frowned. "Define 'movement.'"

"Amplitude consistent with internal knock pattern. Same 3-and-2 sequence logged yesterday—stronger signal."

Joe pressed one gloved hand flat to the plate. The thump came again—three short, pause, two. "Confirmed," he said quietly. "Same signal."

Charlie's breath caught. "That's them."

Lenny barked once, sharp, startling everyone on the bridge. Morales smiled faintly. "Witness confirmed."

She turned back to the comm. "All teams, listen up. We've got continuous contact inside the pocket. Proceed by the numbers. No mistakes. You hear me?"

Every voice came back steady and sure: "Copy."

"Alright," Morales said. "We open that door."

Joe steadied the torch. Guido raised his. Valery lined up the ROV light on the hatch seam. Perez logged the time.

"Begin third pass," Morales ordered. "Top seam only. Keep it clean. Let's bring them home."

Joe's torch hissed back to life, blue-white flame curling through green water. The seam brightened again, turning from dark steel to dull copper, then to glowing orange. He moved in slow increments—two inches of melt, then stop. "Vent," he called.

"Micro-vent," Valery replied. The ROV turned the small valve a quarter twist. A line of bubbles slid past his faceplate, warm and steady. "Delta point-one," she added. "Holding stable."

"Copy," Joe said, lowering the torch again.

Guido's voice came through, calm but hoarse. "Right side clear. I'm halfway down Seam Bravo."

"Looks good," Morales said from the bridge. "You're both clean. No overheat, no drift."

Perez's voice followed. "Partial O_2 inside is eleven-point-nine. CO_2 climbing to three percent."

Morales turned toward Valery. "How long before the air's unbreathable?"

Valery didn't look away from her screen. "At this rate—twenty minutes, maybe less."

"Then we buy them twenty-one," Morales said. "Keep working."

On the bridge, Charlie stood perfectly still, watching the feed. Lenny rested his chin on the boy's knee, eyes fixed on the same screen. The blue glow from the monitor painted both their faces pale. Charlie whispered, "They're talking back." Morales heard him but didn't correct him. It wasn't wrong.

Below, the wreck shifted with a sound like metal remembering to breathe. Silt poured from a crack above, turning the water opaque for a few seconds. Joe ducked his head and waited for it to settle.

"Structural stable?" Morales asked.

"Stable," Guido said. "Beam movement minimal."

"Copy. Keep it slow."

Joe adjusted his angle. His left arm burned from holding the torch steady so long. "We're through ninety percent of the top seam."

"Then prep for the corner cut," Morales ordered. "But don't puncture until I give it."

Valery leaned into her mic. "Pressure difference point-zero-eight. You're close enough to taste it, Joe."

He smiled faintly inside the helmet. "That's not on my diet plan."

Charlie let out a shaky laugh from the bridge. Morales glanced at him. "He's fine," she said, more for the crew than for herself.

Guido's torch sputtered once, then flared back to full. "Torch One stable again. Just a short bubble in the line."

"Keep your focus," Morales said. "If you lose the arc, call it."

"Copy."

Perez looked up from his monitor. "Thermal up one degree on the inside wall."

Valery leaned closer. "That's not bleed from the torch—it's internal. Same heat bloom as before. Stronger."

Joe put his glove to the metal again. He waited. Three slow knocks. "Still steady," he said. "They're holding."

Guido's voice broke. "Dios mío... still fighting."

"Copy that," Morales said, tone clipped but full. "Let's finish it."

Valery triggered another vent pulse. Bubbles streamed out, each one carrying a breath from inside the med-bay. "Pressure equalized to 1.9. Stable."

"Proceed to final pass," Morales said. "Keep it shallow."

Joe traced the last few centimeters, torch cutting through steel that gave up too easy after holding on too long. The glow faded as he lifted off.

"Top seam complete," he said. "No rupture."

Guido floated beside him, panting slightly. "Same here."

"Copy. Prepare to open the micro-vent full for ten seconds," Morales said. "Valery, give me count."

Valery's fingers hovered over the controls. "Ready."

Morales took a breath. "Go."

Valery turned the valve. The ROV's camera caught a rush of silver bubbles that flooded upward like quick rain. The sound through the comm was a deep sigh—the ship exhaling after two days.

Joe reached out with his hand and let the bubbles hit his palm. "Warm," he said quietly.

"Means they're alive," Valery answered.

The bridge stayed silent for a few seconds. Even the hum of machinery seemed to fade. Morales finally spoke. "Good work. Maintain hold. Valery, scan internal gas levels again."

"CO_2 still dropping," she said. "O_2 stabilizing at eleven point five. That's not possible unless they're scrubbing manually."

Joe's hand tightened on the hull. "He's keeping it running."

"Then we don't waste another minute," Morales said. "Guido, prep pry bar. Joe, get ready to support the hatch."

"Copy," they said together.

"Valery," Morales continued, "slow vent another quarter turn. Let her breathe again before we open her lungs."

The ROV moved its arm. More bubbles. More warmth. Then the seam gave a small shudder, like the ship itself was agreeing. Joe felt it vibrate against his chest plate.

"That's our cue," Morales said. "Begin breach on my mark. Guido, Joe—you lift together. Valery, hold vent line steady. Perez, watch pressure."

"Ready," Joe said.

"Ready," Guido echoed.

"On three," Morales said. "One. Two. Three."

Joe and Guido pulled together. The seam flexed, then split with a sound like a sigh. A slow exhale of air escaped into the sea, turning the water into a stream of glittering bubbles.

"Pressure equalized," Valery said. "Delta zero-point-zero. We're clear."

Morales's voice finally cracked with relief. "Then open the door."

The hatch gave way with a groan that shook through both men's arms. A stream of bubbles poured out, flashing white in the ROV's lights. For a heartbeat, Joe thought the ocean would claim the pocket, that it would all rush out and collapse—but it didn't. The air was warm and held its ground.

"Pressure holding steady," Valery said through the comm. "Two-point-one internal, two-point-one external. You're clear to breach."

"Copy," Morales replied. "Proceed."

Guido wedged the pry bar under the seam and levered it up. The hatch rose slow, reluctant, until a strip of light slid through the silt. Joe steadied the door while Guido worked it open the rest of the way.

"Lights on," Morales ordered.

Joe aimed his floodlight inside. The beam cut through drifting dust and debris. The med-bay appeared, a half-flooded room frozen in time. Metal shelves twisted. Tools floated. Sheets of paper drifted like leaves in dirty snow.

Then he saw them.

Amy lay against the far wall, half-submerged, her hair spreading like dark threads through the water. One arm clutched a sealed med log to her chest. Donnie sat slumped beside the pressure valve, hand still on the pipe he had built, his knuckles white even now. His chest moved—slow, shallow, but real.

"Contact," Joe said, his voice cracking in his helmet. "Two survivors. Visual confirmed."

The bridge erupted. Valery gasped out a sound halfway between a laugh and a sob. Perez swore softly in disbelief. Morales didn't speak; she just pressed her hand flat on the console.

"Say it again," Valery demanded.

"They're alive," Joe smiled and said. "Both of them."

Morales's voice steadied. "Copy. Begin extraction."

Guido was already unclipping the rescue hoods. "Got your back, hermano. Let's bring them home."

Joe moved inside first. The air was thick, heavy with condensation and heat. His boots hit solid deck. His floodlight swept over broken monitors and floating instruments. He knelt beside his mother, checked for a pulse. Faint, but there. He turned his head toward Donnie. "Dad?"

The old man's eyelids fluttered. His lips parted, letting out a shallow breath. Joe almost tore off his own helmet. "He's breathing!"

"Vitals confirmed," Valery said, voice trembling. "O_2 inside the pocket is down to nine percent. You've got to move."

"Copy."

Guido moved in beside him, anchoring the extraction harness. "Strap your father first."

Joe looped the straps over Donnie's chest, tightening the clips. He reached for Amy next. Her eyes opened halfway. "Joey…"

"Yeah, Ma. I'm here."

She smiled weakly, her voice rasping against dry air. "Took you long enough."

Joe laughed, the sound small and broken. "We got a little lost."

Guido worked fast, securing both harnesses. "Ready for lift," he said.

"Move," Morales ordered. "Structure's unstable. Get them clear."

The walls groaned again. A section of ceiling beam bent with a long metallic whine. Silt poured down, blinding them for a few seconds.

"Pressure drop," Valery said. "You've got a leak starting on the east bulkhead."

"Copy," Joe said. He hooked his arm under Donnie's shoulders, guiding him toward the hatch. "Come on, old man, we're going home."

Amy tried to stand, but her legs gave out. Joe caught her under one arm. "Easy, Ma. I've got you."

Guido covered the rear, scanning the dark corners with his light. "Hurry it up," he muttered.

Valery's voice came over the channel again, faster now. "Internal O_2 at eight-point-three. CO_2 climbing. You're out of air, Joe."

Morales's voice followed. "Ascend now. No pauses. Guido, you lead the way."

Joe didn't waste another word. He tightened his grip, one arm on each of them, and kicked toward the hatch. The harness cables tugged against his chest as the lift line took their weight.

The ship moaned—a sound that made everyone on the bridge flinch.

"Partial collapse detected," Perez shouted. "You've got thirty seconds!"

"Copy!" Guido yelled back. "We're moving!"

The med-bay shook behind them. Water surged through the open hatch as if the ocean had finally found its way in.

"Go!" Morales barked.

Joe pushed off with everything left in him. The world turned into bubbles and light and the muffled roar of collapsing steel. He didn't look back. He couldn't.

The ROV camera followed, catching the final image of the med-bay folding in on itself. The walls twisted inward, the ceiling bent, and the last of the air escaped in one great sigh before the whole section disappeared in a cloud of silt.

"Surface team, prepare for recovery," Morales ordered. "They're coming up hot."

Joe's breath came in short bursts. His helmet display showed his tank dipping low, but all he could see was his parents. Amy's head rested against his shoulder; Donnie's mask hissed faintly with each shallow breath.

Valery's voice steadied, a lifeline through static. "Fifteen meters, Joe. You're clear of debris. Keep your angle shallow."

"Copy."

Guido turned back and held out a gloved hand, guiding him toward the light bleeding down through the water. "Almost there!"

The wreck below vanished into shadow. The ocean above looked bright for the first time in days.

"Ten meters," Valery said. "Almost there."

Joe kicked once more, dragging his parents with him, heart pounding, lungs burning.

"Surface in sight," Guido shouted. "Bring them home!"

They broke the surface in a spray of white foam and light. Joe gasped into the headset, half choking on relief. Guido's voice was raw with laughter and exhaustion. "They're up! Both of them—alive!"

"Copy," Morales said from the bridge, voice cracking. "Surface team, move now!"

The Savannah's deck exploded with motion. Crew ran to the rails, throwing lines and slinging the rescue rig over the side. The crane motors groaned as the harness cables tightened. Valery leaned over the railing, shouting through her headset. "I've got eyes! They're right below port! Get them up slow!"

Two divers splashed into the water and guided the rescue frame toward Joe. His limbs felt like lead, but he didn't let go of either parent. Donnie's chest still rose and fell inside the mask; Amy's fingers twitched against his suit. "Hold on," Joe whispered. "You're safe now."

The divers clipped their lines in. The crane lifted, water streaming off them in heavy curtains. Guido steadied the frame, shouting for clearance. "Go slow! Keep them level!"

The harness swung up and over the rail. Deckhands reached out, grabbing for the ropes, pulling the survivors in with practiced hands. Amy and Donnie were laid side by side on stretchers. The hiss of oxygen and shouted orders filled the air.

Morales jogged across the deck, headset still around her neck, clipboard in hand. "Get them to decompression, now!" she barked. "Pulse ox on both. IV ready. Let's not lose them after all this!"

Valery climbed up from the ladder, dripping, eyes glassy with tears. "They're breathing," she said, her voice cracking. "They're really breathing."

Joe pulled off his helmet. His hair plastered to his face, his eyes bloodshot. He stumbled to the rail, gripping it with one hand while catching his breath. For a second, he couldn't move. He just stared at them—his parents—breathing.

A small sound broke through the noise behind him.

Charlie was running down the deck barefoot, Lenny at his heels. The dog barked nonstop, high and frantic. Charlie's voice cut through the air. "Joe!"

Joe turned just as Charlie reached him. The boy's eyes were wide and wet. "I told you!" he said, breathless. "I told you they were still alive!

Joe dropped to one knee, catching him in a hug so tight the boy squeaked. "You were right," Joe said, voice rough. "You were right, kid."

Charlie wiped his eyes with the back of his hand. "Can I see them?"

"Yeah," Joe said softly. "Come on."

They walked together toward the med team. Lenny followed close, whining low, tail wagging weakly. Amy and Donnie lay under thermal blankets, their skin pale but warm. Oxygen masks fogged with each breath. A medic pressed a stethoscope to Amy's chest and nodded. "Strong rhythm," she said. "She's fighting."

Guido leaned against a bulkhead, shaking his head and smiling through tears. "We just stole two people from the sea, hermano."

Morales stood nearby, arms crossed, eyes rimmed red. "Not stole," she said quietly. "Took them back."

Valery helped Joe out of his suit, hands quick and steady. "You're shaking."

"I'm fine," he said, voice hoarse. "Just... tired."

"You should sit."

"Not yet."

Charlie knelt beside Amy, taking her hand. "Mom?" he whispered. Her eyelids flickered, and she turned her head slightly. "Charlie..." Her voice was barely air.

"I'm here," he said. "I told them you'd be okay."

A ghost of a smile touched her lips. "You always... know."

Donnie's fingers twitched beside her. Joe leaned over, pressing his gloved hand to his father's shoulder. "Hey, old man. We made it."

A weak exhale answered him. "About... time," Donnie rasped, his eyes still closed.

Morales wiped a hand across her face. "Get them inside. Chamber's ready."

The medics lifted the stretchers carefully. Joe followed, Charlie close behind with Lenny trotting between them. As they passed the rail, the sun broke through the clouds, spreading gold across the deck. It glinted off the wet steel, the cranes, the dripping suits—a world washed clean.

When they reached the decompression chamber, Morales turned to the crew. "That's it, people. Pressure seal, two hours minimum. I want full vitals logged every five."

"Yes, ma'am," the medic said.

The chamber door sealed with a heavy clunk. Through the small window, Amy and Donnie lay side by side, faint movement under the blankets.

Valery stood next to Joe, her hand brushing his arm. "They'll make it," she said quietly.

Joe stared through the glass, eyes unfocused. "They're tougher than the ocean."

Charlie looked up at him. "Does that mean we're all together again?"

Joe's throat tightened. "Yeah, kid. We are."

Lenny let out one soft bark, as if agreeing.

Morales logged the final report on her clipboard. "Ten-forty-two," she said softly. "Two recovered alive."

For a long moment, no one moved. The only sounds were the soft hum of the chamber and the endless breathing of the sea around them.

Then Valery exhaled, the first true breath she'd taken in hours. "We did it," she whispered.

Joe didn't answer. He just nodded once, eyes on the small window, watching the faint rise and fall of two chests that had refused to stop.

Behind them, the waves rolled against the hull—steady, patient, alive.

CHAPTER 3
RECOVERY

The decompression alarm began as a low hum—the kind that vibrated in the deck before it reached the air. It was the sound of waiting ending. Inside the observation bay aboard the Halcyon, the crew froze in place. No chatter, no orders—only the faint hiss of oxygen and the slow blink of green lights marking pressure equalization. Morales stood at the console, her fingers white around the gauge handle. "Surface equilibrium reached," she murmured. "Ready when you are."

Joe didn't wait. He hit the release lever, and the hatch bolts unlocked one by one, metal sighing like lungs learning to breathe again. A soft cloud of condensation spilled from the seam, curling through the air. Then the door cracked open with a hollow pop and a rush of cool air that smelled of antiseptic, rust, and hope.

He stepped forward, Charlie right beside him, Lenny padding close on small, steady paws. The boy clutched the fabric of Joe's sleeve, eyes wide behind tear-streaked lashes. The dog's ears were forward, every muscle quivering with anticipation. Together they crossed the threshold into the chamber—three shadows cut from the same piece of love and fear.

Inside, the light was soft gold—the color between storm and sunrise. Amy sat half-upright against the padded wall, pale but breathing. Donnie lay beside her under a sheet of thermal foil, chest rising in fragile rhythm. The sight hit Joe like a blow. For a heartbeat he couldn't move.

Then Lenny broke from his side, barking once—high and bright—before launching himself into Amy's lap. Wires tangled, foil crinkled, and Amy gasped out a sound that wasn't quite laughter, wasn't quite a sob. Her arms folded around the little dog like she'd been waiting a lifetime to feel something warm. "Mi pequeño," she whispered, burying her face in his fur. "Mi niño bueno."

Charlie couldn't hold still anymore. He ran the last few steps and fell into her arms. "I knew it!" he cried, voice cracking. "I knew you were alive! I told them, Mom—I told them Dad would seal the med-lab. Nobody believed me, but I knew!"

Amy looked up at him through tears, her voice trembling. "Oh, sweetheart… I knew you'd be waiting." She pulled him close, wrapping her arm around his shoulders.

And that was when Joe finally moved. The strength went out of him all at once—days of fear collapsing under the weight of joy. He dropped to his knees beside them and folded both of them into his arms—his mother's trembling body, his kid brother's shaking shoulders, the warmth of the dog pressed between them. He held them all as if he could fuse them back into the world by sheer force.

Amy's fingers found his face, tracing his jaw the way she used to when he was small. "Joey," she whispered, voice cracked and paper-thin, "you came for us."

He nodded, forehead against hers. "Always."

Donnie's voice came rough and warm, gravel turned gentle. "Knew you'd come raising hell."

Joe let out a wet laugh. "Guess I learned from the best."

Donnie managed a tired grin. "Damn right."

Behind them, the hatchway filled with the quiet presence of the others—Morales with her clipboard, Valery biting her lip to keep from crying, Manny standing in the back with grease still streaked on his sleeves. Even Guido, gruff as iron, wiped his eyes and muttered something in Spanish that sounded suspiciously like a prayer.

Morales cleared her throat, blinking hard. "Vitals holding. Oxygen within range. Let's move them to med for full evaluation."

Joe didn't look up. "Give them a minute."

And the room obeyed. No one moved, no one spoke. The ship hummed beneath them—a deep, living sound. Outside, the Savannah floated steady over what used to be the Gulf, now a gray ocean swallowing the world.

Amy held Lenny close, one hand tangled in his fur, the other gripping Joe's sleeve. "You came," she said again, softer now, a prayer more than words.

"I never stopped," he said.

Donnie gave a low chuckle. "He's stubborn, that one," he murmured, pride threading through the rasp.

Valery's voice floated from the doorway, trembling but warm. "Welcome home."

Joe pressed his forehead to Amy's once more. "Rest now, Mom," he whispered. "We'll take it from here."

Lenny gave one small bark—not of alarm, but of affirmation. The crew smiled through their tears. For the first time in too many days, the air didn't taste like loss.

They were all home again.

The med bay had never felt so holy.

Not because of prayer—though Morales whispered a few—but because every sound in that room meant life still dared to continue. Machines hummed in low harmony, monitors blinked steady green, and the air smelled of salt, antiseptic, and second chances.

Amy sat propped up on a cot, her skin still pale but threaded with color. Her hair clung damp against her neck. Every breath was a fragile victory. Donnie lay in the next bed, wrapped in thermal foil, his arm tethered to an IV line, his other hand draped over his chest like he was daring anyone to argue that he'd survived.

Joe stood between them, one hand gripping the edge of Donnie's cot, the other clenched at his side. He hadn't slept in two days, and the exhaustion had settled in deep, turning his voice low and gravelly. He looked like a man who'd been punched by relief and didn't know how to stay upright afterward.

Morales checked the monitor again. "Oxygen holding steady. CO_2 down. Lactic levels improving." She glanced toward Joe. "They're stable, but fragile. They need rest and fluids."

Joe nodded without looking away from them. "Whatever they need, they get."

Valery moved quietly around the room, setting out warm packs, checking seals, adjusting the oxygen mask over Amy's face. When she brushed a strand of hair from Amy's forehead, Amy smiled faintly. "Feels like heaven," she murmured through the mask.

Valery smiled back. "You're a tough one."

In the corner, Charlie sat cross-legged on the floor, Lenny curled against his legs, the dog's muzzle resting on the boy's knee. Every few seconds Charlie's eyes darted toward the monitors as though he could will the numbers to stay steady. He whispered something under his breath, maybe counting heartbeats, maybe praying.

Manny came in carrying two canisters of medical-grade water and a stack of clean towels over his arm. "Morales said you'd need these." His voice was rough but warm, that same mechanic's steadiness that had held half the fleet together.

Joe gave him a tired grin. "You're a damn miracle worker, you know that?"

"Not compared to your old man," Manny said, setting the supplies down. He looked at Donnie—the man pale and stubborn enough to outthink the ocean—and shook his head. "Two days down there? Hell, I've seen engines give up faster than that."

Donnie's eyelids fluttered open. His voice rasped like rusted wire. "That's 'cause engines don't have wives naggin' 'em to stay awake."

Amy rolled her eyes affectionately. "He means that lovingly."

Morales smirked, scribbling notes on her tablet. "I'll document it as 'positive motivation.'"

The room laughed, soft and brief—the kind of laugh that sounded like air being returned to lungs.

Joe dragged a chair closer to Donnie's bed. "How did you do it, Dad? How'd you stay alive that long?"

Donnie took a long breath, slow and shallow. "Pressure's a bully, son. It only wins if you let it. So I made friends with it instead."

Morales frowned but leaned in, curious. "Made friends how?"

"I stopped fightin' the water," Donnie said, voice steadying. "Flooded the lower corridor to equalize pressure, sealed the outer hatch, and trapped the air where it mattered most—the med lab."

Amy's voice was thin but firm. "He cursed the ocean, and I prayed at it. Between us, it listened."

Manny chuckled quietly. "Figures. One of you argues with physics, the other argues with heaven."

"Worked, didn't it?" Donnie said, his smile weak but proud.

Charlie climbed to his feet and stepped closer, his small hands gripping the railing beside his father's cot. "You really did it, Dad," he whispered, awe softening the word. "You saved her."

Donnie opened one eye, looking at the boy. "Saved both of us, kiddo. Can't lose your Mom—she's the only one who remembers where I put my tools."

Amy squeezed his hand. "And don't you forget it."

The monitors beeped in steady rhythm, a metronome of defiance. Valery adjusted the tubing on the oxygen mask, and Morales checked vitals one more time before stepping back. "Both of you are stabilizing faster than I'd hoped," she said. "You might actually make me believe in miracles."

"Not miracles," Donnie murmured. "Math, duct tape, and stubbornness."

Joe laughed softly. "That's our family motto."

Manny clapped Joe on the shoulder. "And you wonder why I stick around. Somebody's gotta keep you people from outsmarting yourselves."

Morales exhaled, her voice dropping quieter. "You two should rest. We'll keep a watch."

Charlie reached for his Mom's hand, his voice soft but certain. "We're all together again."

For a moment, nobody spoke. The only sound was the rhythmic hum of the Halcyon's engines, steady as a heartbeat beneath them.

And in that quiet, every person in the room believed—just for a little while—that maybe the world could begin again.

• • •

The Savannah's briefing room wasn't built for moments like this. It was meant for orders, not miracles. Yet here they sat—Amy wrapped in a blanket, Donnie pale but upright, Joe at their side, and the crew arrayed around the long metal table like witnesses to something sacred. The hum of the ship's turbines filled the silences no one wanted to break.

Captain Raines stood near the viewport, hat tucked under one arm. The light from the ocean shimmered faintly across his weathered face, cutting the edges of his authority into something gentler. "You two gave this fleet a story we'll be telling until the sea runs out of salt," he said quietly.

Amy gave a tired smile. "Let's just hope no one ever has to repeat it."

Raines inclined his head. "That's the plan."

Morales opened her logbook and looked up. "Captain, before we move forward, we need to document exactly how they survived. We still don't have a full picture."

Donnie leaned back, his voice gravel but steady. "Well, Doc, it's not much of a picture—more like a jigsaw I glued together underwater."

Manny chuckled from where he leaned against the wall. "Hell of a jigsaw, old man."

"Damn right," Donnie muttered, cracking a faint grin. "When the hull started folding, I knew the med-lab had the thickest plating. Bio-containment reinforcement. It was our only shot. So I flooded the corridor outside, sealed the pressure doors, and equalized it."

Morales blinked, impressed. "You used counter pressure?"

"Exactly," Donnie said. "You can't beat the ocean head-on, so you trick it into thinking it already won. Keeps her calm."

Amy nodded softly. "He told me if the sea wanted to crush us, it'd have to do it gently."

A ripple of soft laughter moved through the room. Even Raines's mouth twitched.

Valery leaned forward, elbows on the table. "And the oxygen? You were down there over two days. No system should have lasted that long."

Donnie rubbed his jaw. "Medical ventilator scrubbers, tied into the refrigeration coils. Condensed moisture gave us water. Recycled air. Not perfect, but enough to breathe slow."

Morales shook her head, astonished. "You made a life-support system out of trauma foam and a fridge."

Amy smirked. "Donnie always said anything can be fixed with tape, wire, or spite."

Manny laughed. "Add math to that, and you've got a religion."

Charlie, sitting quietly beside Amy, looked up from petting Lenny. "I told them you'd pressurize it," he said proudly. "I knew you'd find a way."

Amy smiled, reaching to brush his hair back. "You always believed in us, sweetheart."

Donnie's voice softened. "He's got a mechanic's heart already. Don't ever lose that, boy."

Raines finally stepped closer, resting his hands on the table. "What you built down there wasn't just survival. It's proof we can adapt. The next time any ship goes under, we'll know how to fight back."

Joe's throat tightened. "You turned a death trap into a blueprint."

Donnie shrugged modestly. "Just bought us time, son. You're the one who found us."

Joe met his father's eyes, pride and love sitting heavy in the silence. "We found each other."

Amy looked around the table. "So what now, Captain? What's next?"

Raines glanced toward Morales, who handed him her tablet. A faint static sound filled the room—a rhythmic beeping pattern, low and persistent.

"Twenty minutes ago," Morales said, "this started coming through the comm arrays. We isolated the frequency. It's repeating every seven seconds."

Valery's eyes narrowed. "That's not a natural ping."

"No," Raines agreed. "It's a signal. Too faint to identify yet—but it's coming from north of our current position."

Joe frowned. "North? That's inland."

"Used to be," Morales corrected softly. "Now it's open sea."

Raines looked out the viewport again, toward the horizon where sky met water. "We've charted it," he said. "If the source is still transmitting, it's not wreckage. It's alive."

The room fell quiet. Even Lenny stopped panting.

Amy gripped Donnie's hand, her voice barely a whisper. "You think it's people?"

Raines didn't answer immediately. "That's what we're going to find out," he said at last. "But we'll need every ship running and every person steady." He looked at Joe. "That means you, Alvarez. You're leading the recon with Morales and Valery. Manny's on engines."

Joe nodded, pulse quickening. "Aye, Captain."

Raines turned toward the door. "We move at dawn."

The lights dimmed slightly as the turbines shifted power flow, and outside, the ocean stretched endless and unknowable.

Charlie squeezed Amy's arm. "Maybe there are others, Mom."

Amy looked at him, her face lit by that soft, trembling hope that hurts to hold. "Maybe there are," she whispered.

And for the first time since the fire fell from the sky, hope didn't sound impossible.

The Savannah's bridge settled into that narrow band of quiet where work lives: no panic, no chatter, just the click of toggles, the whisper of pencil on chart, and the heartbeat-hum of turbines shifting load. Outside, night thinned to pewter. The sea wore a skin of silk and ruin.

Raines stood centerline with his hat tucked under one arm, the picture of a man who'd outlasted storms by glaring at them. "Bring it up," he said.

Morales patched the feed to the overhead. A thin tone ticked into the speakers—seven seconds apart, steady as a metronome that refused to die.

Valery's hands moved fast over the comm array. "VHF carrier with a dirty harmonic. Someone's squeezed a beacon out of scrap."

Joe braced on the rail behind her, eyes on the spectrum. "Bearing?"

"Initial hit puts it 006 to 010," Valery said, "but multi path is messing with us. We're getting bounce off structures."

"Structures?" Joe asked.

"High ones," Morales said, already marking on the paper chart. "Skyscraper tops. Towers. Anything tall that kept its head when the world lost its body."

Raines glanced at the chart—the drowned mouth of the Mississippi behind them, the ghost of I-55 a dotted memory under gray water. "We're north of the old Louisiana line by sixty nautical. Plot a cautious intercept—heading 008, speed five knots." He turned to the helm. "Make your revolutions gentle. We're not waking any ghosts today."

"Aye," the helmsman said. The deck answered with a shiver as the Savannah leaned into the new course.

Morales keyed the fleet net. "All vessels, this is Savannah Actual. Formation Delta—stretched column. Halcyon hold astern two cables, tugs Marlin and Kite on our quarters. Skiffs secured. Helo One and Two pre-flight cold; crews on standby."

"Halcyon copies," came the reply, a low rumble of steel and discipline.

Joe tapped the engine-room line. "Manny, how's my heart down there?"

Manny's voice came up warm and wry under the machinery: "Purring like a cat that won the lottery. I can give you six knots if Captain Stare-Down insists."

"Five's fine," Raines said, not looking at the speaker. "We'll spend speed when we can buy certainty."

"Copy," Manny said. "Holding five, tuning for quiet. I'll squeeze another percent efficiency out of turbine two if I steal a bolt and a blessing."

"Steal both," Joe said. "I'll sign after."

The tone pinged again—seven seconds, then seven more. It did not waver.

Valery slid a new filter across the signal. "Pulling the harmonic apart… there." A faint tail revealed itself in the spectrograph, a ragged thread at the edge of audibility. "It's riding echo from metal and glass. That's city skeleton."

Morales drew a second bearing line. "Cross with last sweep… mark. If I'm not lying, that puts our source north-northeast, past whatever's left of Jackson and east of Little Rock's old skyline. High ground should start somewhere beyond the Ouachitas—if the peaks are still honest."

Raines nodded once. "Honest mountains. I wouldn't mind meeting those again."

"You and me both," Joe said.

Charlie had slipped onto the bridge like a rumor and took up his spot near Amy's blanketed shoulder where she sat in a jumpseat, color back but strength still miserly. Lenny lay flat at her feet, chin on paws, the alertness of a small creature who had Shouldered Important Duties and planned to keep them. Charlie listened to the tick of the tone with solemn eyes, counting under his breath, lips moving—one… two… three…

"Seven," he whispered with each arrival, like a spell that worked only if you respected it.

Amy brushed his hair back. "Steady, mijo."

Raines flicked his gaze to the port glass. The horizon had started showing teeth—dark prongs stabbing up through haze. At first they looked like broken waves. Then the light caught edges. Windows. Corners. The bones of a skyline.

"Dead city ahead," the helmsman murmured.

Valery magnified the forward cam. Stumps of towers rose from the plain of water, some sheared clean, some leaning like drunks who'd fallen and decided the floor would do. Debris rafts eddied between them. A billboard's top quarter poked up and offered a sun-bleached promise no one could read.

Morales tightened formation. "All ships, steer one-zero-zero meters port of verticals. Watch for snags. Depth sounders to active."

Ping. Seven seconds. Ping.

"It's stronger," Valery said. "Multi path aligning. We're close to a direct line."

Joe looked at the chart again, at the gray expanse that used to be county names and highways and human habit. He felt the weird vertigo that came with navigating over drowned maps. "If the beacon sits high, it could be on a tower roof. Or a radio mast still peeking above."

"Or a water tank," Manny offered over the squawk box. "Farm tanks used to be stubborn."

"Or a church steeple," Amy said softly. "They were built to be seen."

Raines weighed each possibility in a single nod. "We'll treat every tall thing as both sanctuary and trap."

"Copy," Morales said, already drawing approach arcs around three tallest silhouettes that stabbed ahead like punctuation.

Heard first from the bow lookout: "Debris field two hundred meters. Plastic and wood. Possible bodies." The last word hung like a held breath.

"Helm, starboard two degrees," Raines said. "We won't write new gravestones with our wake."

They eased past. The Marlin's small diesel coughed and steadied, holding the outer curve like a shepherd. The Kite slid neatly in behind, her rubber fenders kissed by the swell. Above, Helo One's crew pulled covers and ran hands across a rotor like a priest blessing an old bell.

"Give me ranges," Raines said.

Valery read the numbers. "Primary spike at six-point-eight nautical, altitude unknown. Secondary ghosts farther north. If the primary's man-made, it's patient."

"Patient is good," Manny said. "Means it's not on fire."

Morales slid the nav pencil under her thumb and made one last mark. "Captain, we'll be skirting the rumored high ground if we keep this line. If there's anything left of the Ouachitas or the far edge of Ozarks, we'll see shadow at dawn."

Raines glanced to the east, where night had already started to bruise into light. "Dawn's keen on secrets. Let's see what she gives up."

Ping.

The tone cut sharper now, bright enough that even Lenny lifted his head. Charlie counted "seven," and this time couldn't help the breath that followed it—half laugh, half prayer.

Joe leaned to the engine-room mic. "Manny, hold us at five. If the line narrows, I'll ask for six. Keep turbine two quiet; I need our noise floor down."

"You'll have it," Manny said. "I tuned her like a church mouse." A beat, then softer: "Tell Amy I'm saving the bolt she dropped on my deck. Lucky thing."

Amy smiled toward the speaker. "Hold onto it, Manny. We may need all the luck left."

Raines straightened, the room coming to a fine point around his voice. "All hands, listen up. We've got a live signal north of our track—unknown friend, unknown temper. Savannah leads. Halcyon covers our back. Tugs hold our hips. Helicopters stay leashed until I say otherwise." He slid his hat back on and let the brim find its old, certain angle. "We move quiet. We move kind. We move like we're about to meet neighbors."

Valery's eyes stayed on the scope. "Captain—new data. The pulse just doubled—two quick pips, then seven-second gap. That's deliberate."

"Code?" Joe asked.

"Not one I know. But it's a mind, not a storm."

The bridge exhaled as one organism.

"Then bring us to them," Raines said, softer now, as if he didn't want to scare the sound away. "North by a whisper. Keep her steady."

The Savannah slid through the ruin with the patience of a pilgrim. Behind her, Halcyon kept station like a broad-shouldered brother. The little tugs padded at their flanks. Above, the helicopters crouched under canvas, waiting for the sky to earn them.

Ahead, the skyline thinned, and beyond its broken teeth the horizon hinted at a darker line—something not water, not sky. Maybe the first ridge. Maybe nothing at all.

Ping. Pip-pip. Seven breaths of hope. And the fleet went to meet it.

Morning came pale and shivering, a thin mist rising from the drowned world. The Savannah moved slow and sure, her hull carving through the silver surface like a blade through breath. Behind her, the Halcyon and two tugs followed in perfect rhythm—four ships bound together by stubbornness and the shared illusion that direction still meant destiny.

The signal still pulsed through the comm system. Pip. Pip. Seven seconds. It had become part of the ship's heartbeat.

Joe stood on the bridge beside Captain Raines, his hands wrapped around a mug gone cold hours ago. He hadn't slept, but he didn't feel tired—not in a human way. More like a machine that hadn't yet remembered to rest.

"Bearing's holding true," Valery said from the console, her eyes reflecting the green glow of the sonar feed. "Signal strength increasing. We're inside ten nautical now."

Morales glanced over her shoulder. "Frequency's clean. It's not weather interference. Somebody built that transmitter to last."

Manny's voice came through the comm, thick with engine-room reverb. "Turbine two's running smooth. If you want six knots, I can give you seven, but she'll grumble about it."

"Five will do," Raines answered. His tone was low but full—like thunder considering whether to return. "Let's not announce ourselves before we see what's waiting."

Amy sat on a bench by the port bulkhead, wrapped in a gray blanket. Her color was better now, though exhaustion clung to her like salt. Lenny rested beside her feet, nose twitching each time the ping came through the speakers. Charlie sat cross-legged at her side, keeping count of the pings on his fingers. He'd been doing it for hours.

"Whoever's out there," he said quietly, "they're patient."

Amy smiled faintly, brushing her fingers through his hair. "Patience means they're alive."

Joe turned from the console and looked out over the water. The horizon was changing. Where dawn met sea, faint humps began to rise —uneven, ragged, but solid. His pulse kicked. "Captain, visual contact," he said, pointing. "Starboard bow. Bearing zero-one-two."

Raines moved closer to the glass. The peaks were faint but there—the tops of drowned mountains pushing their shoulders through the ocean's skin.

"Land," Morales whispered. The word came out like disbelief wrapped in prayer.

Raines didn't smile, but his voice softened. "Mountains," he said. "That means the world hasn't forgotten how to stand."

Valery cross-checked radar. "Two ridge lines visible. Between them, the signal's origin." She hesitated. "Elevation reads above sea level—

barely. If the transmitter's there, it's sitting on the last dry ground for hundreds of miles."

Amy leaned forward, gripping the edge of her seat. "Then someone made it."

"Or something," Raines said, his eyes narrowing. "We won't assume until we see."

Manny's voice rumbled through the speaker again. "I'll bring power up for a hover approach. If you want to send a drone, I can warm up auxiliary feed."

"Negative," Raines said. "We go in careful, not blind." He turned toward Joe. "Alvarez, prep a small recon team. You, Morales, and Valery. Manny runs comm from below. We'll need eyes on whatever's calling us."

Joe nodded. "Copy that." He glanced toward his mother. "You'll stay here. No arguments."

Amy opened her mouth—then saw his face and sighed. "Fine," she said. "But take food, water, and a rope. Don't you dare come back empty."

Charlie piped up, "And take Lenny! He can smell danger."

Lenny wagged his tail as if volunteering.

Joe laughed softly. "Tempting, but he's guarding you two." He knelt, pressing his hand to the dog's head. "You hear me, Len? Watch them."

Lenny gave one quiet bark that sounded like understood.

By the time Joe and Morales suited up, the Savannah had slowed to a crawl. The engines whispered; even the sea seemed to hush. Above them, the sun pushed through haze—not bright, but determined.

"Forward visual confirmed," Valery said through the headset. "That's no wreck. That's structure."

Joe adjusted his scope. Through the mist, something loomed—a tower rising from the waves, its sides draped in seaweed and rust, antenna still intact. And on its summit, a makeshift mast blinked a faint red light.

"Signal's strongest there," Morales said. "Reading solar power trickle. Somebody maintained that."

Raines's voice crackled over comms. "Recon skiff ready on lower deck. You've got two hours before we lose the tide. Make it count."

Joe turned toward Morales. "You heard him."

She nodded once, setting her jaw. "Let's go meet whoever's been whispering to the sea."

Below, Manny's engines purred to life, the tugs adjusting their lines to brace the big ship's drift. On the bridge, Amy and Charlie stood by the viewport, watching the skiff lower into the glittering water. Lenny pressed his paws to the glass, whining once—not fear, just recognition.

Amy laid a hand on the glass, whispering under her breath, "Find us something worth believing in, Joey."

Outside, the skiff's engine coughed, caught, and steadied. Joe gripped the tiller, Morales beside him, Valery at the radio. The sea ahead shimmered with reflection—a path of light leading toward the blinking red beacon.

The Savannah's crew watched in silence as the small craft moved into the mist. The signal pinged again—louder now, more certain.

Ping. Ping. Seven seconds.

Then, faintly, the second tone returned—two quick pips, deliberate, human.

Joe smiled beneath his breath. "They're answering."

And with that, the fleet turned north, toward the broken peaks and whatever still lived beyond the water.

CHAPTER 4
NORTHBOUND ECHO

Day 462 Post-Fire

Approx. fifteen months since the Rain of Fire

Location: 31° N, 91° W — north of the drowned Louisiana corridor

Course: 008° toward inland ridge / signal bearing confirmed

The sound came again — soft, deliberate, and human.

Ping. Pip-pip. Seven seconds.

Morales stood over the sonar console, her hand hovering above the gain dial as if afraid she might scare the signal away. This wasn't the random chatter of echo returns — this pattern breathed. She turned to Raines.

"It's not oceanic. Origin's elevated. Rough estimate puts it sixty-five meters above the surface."

Raines leaned over the chart, tracing the faint green line northward. The drowned Mississippi basin sprawled below them — a graveyard of cities sleeping under a gray mirror.

"So whatever's calling us isn't floating."

"Land-based," Morales confirmed. "A transmitter anchored to something solid — probably steel. Maybe radio tower, maybe an old repeater mast. I'm filtering multi path now."

The bridge went quiet except for the hum of engines and distant machinery. Dawn smeared pink across the water — soft, muted, exhausted.

Valery adjusted the spectral overlay. "Harmonic is clean. Battery-powered, maybe solar trickle. Whoever set it up knew what they were doing."

Raines's jaw worked once. "Fifteen months of silence… then this."

His voice didn't lift, but something in his eyes thawed.

"The world still has a pulse."

Amy sat against the bulkhead in a gray ship blanket, Lenny at her feet and Charlie pressed close beside her. The boy counted the interval with quiet precision.

"Seven seconds again."

Amy brushed a hand through his hair. "Seven's a good number, mi hijo. Means we're supposed to hear it."

Joe stepped onto the bridge — pale with fatigue but carrying a spark of something dangerous: hope.

"What do we have?"

"Land-based signal," Morales repeated. "Bearing zero-zero-eight. Estimated six to eight nautical miles inland from our current waterline."

Raines tapped the map where drowned Louisiana gave way to the shadow of Arkansas.

"That's the inland shelf. If anything survived the rise, it'll be there."

"Ouachitas range," Morales murmured, studying contour ghosts. "If the ridges stayed above water, there could be dry pockets — plateaus."

Joe's heart kicked. "First solid ground in over a year."

Raines turned to him slowly.

"Aye — but we approach like guests, not conquerors."

He keyed the intercom.

"All vessels, this is Savannah Actual. New heading zero-zero-eight. Speed five knots. Maintain Delta formation. Halcyon astern two cables; tugs Marlin and Kite on quarters. Helo-1 and -2 cold standby. We're going north."

The Halcyon answered steady and low.

Beneath their feet, the Savannah shifted — not fast, but with deliberate confidence — the way an old warhorse turns toward the smell of land.

Manny's voice crackled through the comm, warm with that sleepless engine-room humor.

"Main turbines trimmed for lean burn. I can give you five knots all day without complaining. Six if you bribe her with coffee grounds."

Joe cracked a tired smile. "We'll keep her happy, Manny. Save the caffeine for miracles."

"Copy that," Manny replied. "Tuning turbine two to whisper mode. She likes sneaking more than sprinting."

Charlie grinned. "Manny talks about the ship like she's alive."

"She is," Joe murmured. "She's our lungs."

The ping came again — closer now, sharper.

Valery's eyes widened. "Signal strength up twenty percent. Closing distance."

Morales zoomed the chart, bearing line crossing dark elevation beneath mist.

"That's terrain," she whispered. "If the transmitter's there, it's sitting above sea level."

Amy sat forward — muscles tense beneath the blanket.

"Then someone made it. Someone survived."

Raines stared ahead, voice low and almost reverent.

"Maybe the world's been waiting for us to remember it."

Outside, sunlight finally cleared the horizon — streaking gold across dark water.

The fleet followed single-file — Savannah first, Halcyon behind her like a steady shadow, the tugs trailing like loyal dogs.

Above them, gulls circled — one, then two — where no gulls should be.

Morales noticed without looking from her console.

"Even the birds know something's changing."

Raines nodded once.

"Land leaves a scent on the wind. You'll feel it when it reaches us."

The sonar pulsed again — steady, relentless.

Joe and Valery exchanged a glance.

"That rhythm's intentional," he said.

Raines folded his hands behind his back.

"Then whoever sent it already knows we're coming."

And the Savannah pressed forward — chasing the first human voice they'd heard in a year.

The Savannah leaned into her new heading like an old mare turning toward home. The water ahead darkened to slate, ripples small and tight under the early wind. Morales stood at the helm beside Raines, tracking the fleet's formation on the radar overlay — the Halcyon holding steady astern, the tugs Marlin and Kite sweeping the edges like faithful sheepdogs.

"Formation steady," she reported. "Fuel consumption holding. Running clean on both turbines."

Raines nodded.

"Keep it that way. We need enough reserves for a full retreat if those mountains are nothing but ghosts."

Below deck, the temperature shifted — warmer, louder — the heartbeat of machinery surrounding steel and heat. The engine bay flickered with reflections from gauges and pressure valves, every surface humming with restrained power.

Manny stood beside turbine two, wiping grease from his wrist. The readout hovered near perfection — 98% efficiency.

"That's my girl," he muttered, tapping the panel like a shoulder.

Footsteps clanged along the upper catwalk.

Ashley Konkel crossed without looking down — sleeves rolled, clipboard tucked tight, focus sharp as a blade. She checked bolts, vent housings, pressure seals. Work, not conversation. Movement, not memory.

Manny froze only for a heartbeat — then let her pass.

She didn't see him.

He didn't call her name.

Some things didn't break — they bent and waited for eyes brave enough to look back.

He swallowed hard and turned back to the console.

"Some things fix slower than engines," he whispered, and the turbine purred as if it understood.

Up on the bridge, Morales watched the efficiency shift.

"Engine noise reduced twelve decibels. She's running quiet now."

"Tell Manny," Joe replied softly. "He's the only one who can make steel breathe like a lullaby."

Valery smirked, eyes on her sonar return. "Let's hope that lullaby doesn't wake anything sleeping under us."

Amy and Charlie had moved to the forward observation window. Charlie perched with his knees tucked up, counting aloud again — steady, rhythmic.

"Still seven."

Amy stroked Lenny's ears. The dog lay sprawled across both their feet, tail tapping with each interval.

"He thinks it's another heartbeat," Amy murmured. "Maybe it is."

Joe turned slightly.

"Heartbeat of land. Or of someone calling across the dark."

Outside, the water shifted color — from cold ocean blue to the murkier green-gray of river sediment and history rising. Bits of the old world drifted past: a traffic light head, a plastic playground swing, the rusted roof of a submerged car.

Ghosts without graves.

"Depth change," Morales said, tapping her display. "Dropping from forty meters to twenty-five. Shelf rising."

Raines leaned forward.

"That's the earth climbing back toward us. Watch the shallows and submerged structures."

The Halcyon mirrored their course, her engines thrumming low behind them. One of the helicopters rocked gently against crosswind, rotor blades shifting like a bird testing forgotten wings.

"Fuel report," Raines called.

Morales read it off. "Savannah at seventy-one percent reserves, Halcyon at sixty-four, tugs just under fifty. We can sustain five knots for nine more days before rationing."

"That's enough."

"We learn what waits north before we burn what's left."

Joe stiffened suddenly. "Captain — interference spike. Mid-band."

Valery leaned over her console. "That's not sonar bounce. Atmospheric scatter — reflection off something solid."

"Land," Morales whispered — barely audible.

Closer.

Raines straightened.

"Five knots steady. Full eyes. Full silence."

Charlie pressed his forehead against the glass — breath fogging the pane. A faint smudge hovered on the horizon — something between mist and memory.

"Is that it?" he asked.

Amy watched the same haze, her voice a whisper trembling between fear and awe.

"Maybe, sweetheart. Maybe that's what home looks like now."

Lenny whined — soft, hopeful — and the bridge let out a quiet ripple of exhausted laughter.

The ship continued — slicing through tide and time.

The sonar pulsed again — steady, patient.

Ping. Pip-pip.

Seven seconds.

And the world whispered back like something waking.

By midday, the light had turned a tired gold — the kind that made the world look older than it was. The Savannah pushed through still water, her bow cutting quiet ripples that rolled against half-submerged towers and the ribs of old bridges. What used to be Arkansas glimmered beneath the surface — an underwater skeleton of glass, asphalt, and memory.

Valery sat at the radar console, eyes shifting between sonar returns and the faint blue arc of an old interstate invisible beneath them. "Echo scatter increasing," she said. "Getting bounce from vertical surfaces ahead — lots of them."

Raines leaned closer. "Define lots."

"Too many to count. Probably rooftops… maybe skyscrapers."

Morales adjusted the gain. The ship speakers hummed and then steadied — the pulse now louder, clearer.

Ping. Pip-pip.

"Signal strength's up thirty percent," she said.

"Not just louder — closer. Intentional."

Outside, the water reflected shards of light like broken glass. Shadowy outlines of buildings rose beneath the surface — some tilted, some standing stubbornly upright as if refusing to acknowledge the sea.

A billboard jutted crookedly from the water — unreadable except for one word clinging to the metal in peeling paint:

HOPE.

Amy watched from the observation bench, one hand resting on Lenny's warm back. "It feels like driving through a graveyard," she whispered.

Joe nodded. "Yeah... but these are tombstones that forgot how to sink."

Raines didn't disagree. "Keep eyes sharp. Anything vertical could be hiding trouble underneath."

The comm crackled.

"Engine room to bridge," Manny said. "Turbine two, stable running lean. I can give you a burst if you need to dodge a skyscraper."

Joe sighed. "Let's try not to make demolition part of the mission."

Manny chuckled. "Then tell the ocean to move her furniture."

A ripple of laughter drifted across the bridge — the kind that wasn't joy, but release.

Ahead, the flooded world thickened — more debris, more structure, more ghosts.

An old radio tower leaned out of the water at a forty-degree tilt, barnacles crusted around its rusted ribs. The beacon pulse came from somewhere beyond it — unseen but insistent.

Valery fine-tuned the scope. "Origin reading zero-zero-nine. Distance: four point three nautical miles. Height: still over sixty meters. Whatever's broadcasting isn't drifting."

"Meaning," Morales said softly, "there's land underneath it."

Amy sucked in a shaky breath. "Land."

Charlie leaned forward and pointed. "Mom — look!"

Just visible beneath the rippling surface, a line of rooftops stretched ahead. In the distance: the faded curve of a water tower. A name clung to its rusted belly:

MENA.

Morales updated the chart. "Edge of the old Ouachitas range."

Raines folded his arms. "If that beacon's anchored there, then something — or someone — stayed alive."

Valery's console chimed.

"Captain," she said quietly, "I'm detecting a secondary signal. Faint. Human voice modulation. Low-band. Riding the beacon."

Joe moved fast. "Play it."

She tapped the command.

Static.

Then — broken words, buried in distortion:

"...still... north... ridge... alive..."

The bridge froze — no breath, no movement.

Charlie whispered,

"That's a person."

Amy swallowed. "Someone's waiting."

Raines's voice was controlled — but reverent.

"Or warning."

No one spoke.

Valery checked the harmonic again. "Not automated. Not repeated code. That was someone pressing transmit."

Morales nodded once — jaw tight.

"Survivors."

Raines stood still a moment longer — weighing hope against caution.

Then:

"No weapons unless fired upon. Maintain five knots. Eyes open."

Outside, the Halcyon mirrored their wake — silent, steady. The tugs drifted closer like sheepdogs guarding a fragile herd.

Every vessel in the formation moved with one shared breath.

Ahead, a lookout shouted from the bow:

"Structure dead ahead! Fifty meters high! Antenna mast — blinking red!"

Valery pulled the feed onto the bridge display.

A tower rose through mist — tall, skeletal, crowned with rust and salt-streaked lights. The beacon flashed its rhythm:

Ping. Pip-pip.

Seven seconds.

Morales confirmed coordinates. "That's our transmitter."

Raines stared at the blinking light.

"We hold distance before dark," he said quietly.

"No sense greeting strangers when shadows do the talking."

Amy stared at the tower — eyes wet, breath trembling.

Charlie squeezed her hand.

Joe's voice softened.

"We'll see it up close tomorrow."

The ship drifted in deliberate silence toward the blinking red heartbeat of the world.

Ping. Pip-pip.

Seven seconds.

And no one — not on any deck — moved until the sea itself seemed to exhale.

The light outside dimmed toward bronze as evening settled over the fleet. The ocean grew eerily still — the kind of stillness that made sailors check instruments twice just to be certain the world hadn't stopped moving.

Raines stood near the navigation console, hands clasped behind his back, posture rigid from hours of tension. Through the viewport, the red pulse of the beacon blinked steadily — rhythmic, patient.

Ping. Pip-pip.

Seven seconds.

Morales adjusted stabilizers and scanned the instruments. "All ships holding formation. Current steady. No drift."

"Good," Raines said.

"We keep position until dawn. No mistakes in the dark."

Valery leaned over her monitors. "Signal modulation unchanged. Whoever's transmitting hasn't adjusted power or frequency despite our approach. Could mean they don't know we're here."

"Or they're waiting," Joe said quietly.

From the bench near the bulkhead, Amy lifted her head. "If they've survived this long… they've learned to be careful."

Charlie sat cross-legged beside her, Lenny's head resting in his lap. He wasn't counting anymore — just listening.

"If they have land," Charlie said softly, "they have food… right?"

Amy brushed his hair aside. "Maybe, sweetheart. Maybe they even have trees."

The bridge fell silent — reverent — at that word.

Trees.

Raines turned toward Morales.

"Plot projected terrain rise. I want safe approach lanes mapped before morning."

"Already on it."

Morales pulled up hybrid overlays — old topographic scans layered with real-time depth maps. Mountains, once measured in miles above sea level, were now measured in meters above drowning.

"The Ouachitas used to hit two thousand feet," she said. "Even after the flood… something had to remain."

"Maybe enough for settlement," Joe murmured.

"Or at least a start," Raines added. "If we find firm ground, we test it before anyone steps ashore. Soil, radiation, pathogens. Nothing rushed."

Amy raised her hand slightly. "Captain… when it's time, I can help test the soil. Plants, contamination signs… rot signatures. I know what earth is supposed to smell like."

Raines met her gaze and nodded once.

"You'll get that chance, Amy. But rest now — we'll need clear minds when the sun rises."

She gave a tired, crooked smile. "You sound like my husband."

From the comm panel, Donnie's voice chimed in — warm, worn, amused. "I heard that. Careful, Captain — I charge royalties for using my tone."

Raines allowed the smallest smirk. "Heal up, Chief. I'll need your hands before long."

"Soon as Morales stops hovering over me like a broody hen."

Morales snorted.

"Keep talking and I'll duct-tape you to the bed and label you 'fragile.'"

Soft laughter rolled through the bridge — tired but real.

When it faded, Raines refocused.

"Night watch rotation starts now. No unnecessary lights. No outside movement. We wait."

Morales keyed commands. "Formation locked. Drift anchors deployed. Halcyon and tugs confirm standard hold."

Valery typed rapidly, preparing payloads. "Drones prepped for first-light reconnaissance."

Joe stepped toward the viewport — eyes fixed on the blinking red light ahead.

"Captain," he asked quietly, "what do you think we'll find?"

Raines didn't look away from the window.

"Doesn't matter what I think. We'll find whatever's left — and make it enough."

The bridge lights dimmed as the ship shifted into night cycle. Instruments glowed blue and amber — the heartbeat of the Savannah.

Amy stood and nudged Charlie gently. "Come on, love. Dinner first. Then bed. Tomorrow's a big day."

Charlie hesitated — eyes on the distant tower.

"Will we see land when we wake up?"

Joe ruffled his hair. "That's the plan."

Charlie grinned — small and fierce. "Then I'm waking up early."

Lenny barked once — sharp, decisive — like a stamp of agreement.

The door closed behind them, leaving the bridge quieter — solemn.

Morales updated her screen.

"Course locked. Systems steady. Fleet holding."

Raines exhaled slowly.

"Then we wait for morning."

Outside, the wind shifted — carrying something faint, unfamiliar.

Not salt.

Not rust.

Earth.

Joe lifted his head.

"You smell that?"

Valery nodded — eyes wide and soft with disbelief.

"Yeah," she whispered.

"The world hasn't died. It's just… waiting."

Night on the open sea was never truly dark anymore. The sky still carried that faint stain from the old fires — a reddish cast that turned the stars into bruises and made the horizon glow like something half-remembered. The fleet drifted under it, engines idled, hulls creaking in a rhythm that sounded almost like breathing.

Joe couldn't sleep. None of them really did anymore.

He stood at the bow beside Valery, the metal rail cold beneath his palms. Behind them, the Savannah's lights burned low and warm — a tiny constellation drifting through a flooded graveyard of a world.

The wind carried warmth now — faint, impossible — a smell of not ocean, but something older.

Dirt.

Green.

Alive.

Joe inhaled slowly.

"You ever think," Valery murmured, eyes still on the horizon, "that the world isn't ending anymore… just changing into something new?"

Joe kept his gaze forward.

"I think it ended already," he said. "We're just walking through the echoes."

She huffed a soft laugh — not amused, but understanding.

"Then let's hope the echoes are friendly."

The red beacon blinked far ahead — steady, unwavering.

Ping. Pip-pip.

Seven seconds.

But now — standing on the bow, nothing between them and the sound except water and the unknown — Joe felt intention behind it.

Not just a signal.

A call.

The bridge doors slid open behind them. Morales stepped out, coat zipped high against the wind.

"Raines wants you both inside."

Joe and Valery followed her back into the bridge.

The room was quiet — not the absence-of-noise kind, but the held-breath waiting kind.

Raines stood at the view window, hands braced against the sill, posture statue-still.

Valery sat — fingers already flying across the controls.

"What happened?"

Morales didn't answer — she pointed.

Valery's monitor pulsed with a new waveform.

"Signal just changed," Morales said.

Joe leaned in.

"How?"

Valery filtered noise, sharpened modulation, and pushed it through external speakers.

The beacon pulse now sounded like this:

Ping. Pip-pip. Pip.

A new beat. Same rhythm — but human-made.

Joe stared.

"That's deliberate."

Valery nodded once.

"Second ping offset by two milliseconds. Someone pressed transmit manually."

Static crackled — then a voice broke through, faint, but real.

"...north ridge... not alone... still here..."

Amy's voice crackled over the internal comm, breathless:

"Joe. Was that...?"

"Yes," he said. "A real voice."

Donnie's voice followed — rough, steady, full of exhausted disbelief.

"Then we go find them."

Raines turned from the window — expression carved from steel and something dangerously close to hope.

"We move at dawn."

Joe nodded — pulse hammering.

Valery's eyes gleamed with quiet wonder.

"That was human," she whispered. "Human. After all this time."

Raines pointed toward the misted silhouette of mountains barely visible ahead.

"Morales — keep us steady. No one fires unless fired upon. Halcyon and tugs mirror formation at first light. No aircraft until we confirm landing stability."

"Yes, sir."

A bark echoed sharply from the stairwell.

Lenny.

Amy laughed through the comm — fragile, emotional.

"Even the dog knows it's time."

Raines's expression softened — only slightly.

"Good instincts. That one."

Outside, the sky shifted — black softening to charcoal, then ash, then the faintest blush of amber.

The first sunrise in months where direction felt like a promise — not a threat.

Joe rested a hand on the cold metal railing.

"We've chased ghosts long enough," he murmured. "Now we see who's left to answer."

The beacon flared — bright red through the fog.

Ping. Pip-pip.

Pip.

Seven seconds.

A ship didn't breathe.

But everyone aboard the Savannah did — together.

For the first time since the Rain of Fire.

Because finally — unmistakably — the sea was speaking back.

CHAPTER 5
CAPTAIN'S BURDEN

Fleet Log – SAVANNAH, Day 466 Post-Fire

Position: 33° N, 92° W

Course: 008° toward inland ridge / beacon signal active

Health Advisory: Command medical deferment pending

———

Captain Raines stared at the log terminal until the letters bled into one another. His fingers trembled on the keys. The screen's glow painted him in pallor, hollowing the lines of his face. The sea rolled soft outside the viewport, but he felt every shift in his bones — the ache that never left since the Gulf turned from home to graveyard.

He drew a slow breath and continued typing.

"Signal persists. Seven-second pattern, human frequency confirmed. Direction consistent with inland transmission. Fuel reserves steady. Morale… improved."

He paused, letting the hum of the ship fill the silence. It sounded like breath through an old lung — weary, determined. He rubbed the side of his neck where the stiffness had turned to heat days ago. Morales had noticed the flush. She'd told him to rest. He'd smiled, promised he would, then didn't.

Rest was for men who could afford tomorrow.

Raines leaned back and reached for the flask inside the drawer. He didn't drink anymore — not since the world burned — but tonight the dull fire helped the trembling. He tipped it once, enough for a swallow, and let it coat his throat like steel turned liquid.

His hand drifted back to the log. He typed slower now, choosing every word as if the ocean itself might judge him for lying.

"Operational command to remain under current leadership until incapacitation. Successor identified in contingency record C-12. Authorization code RAINES-DELTA."

He hesitated, the cursor blinking like a heartbeat. The next words felt heavier than any order he'd given.

"Designated successor: Chief Engineer Joseph Alvarez. Competence in command, mechanical integrity, and moral steadiness beyond measure. Should this ship lose me, it will not lose its course."

He stared at the sentence until it stopped feeling like betrayal. Joe wasn't military — not by training, not by the book — but he was built for crisis, and crisis was all the world had left. Every man under Raines's old Navy would've understood the truth: sometimes survival didn't salute rank.

He saved the log, encrypted it, and sat back in the dim light. The ship groaned around him, steel and age sharing the same sigh. Outside, the sea stretched flat as slate, and the faint ping… pip-pip… seven seconds came through the comm speaker like a clock that refused to die.

Raines tilted his head toward it. "You keep calling," he murmured. "We're coming. Just let me last long enough to see you."

The door hissed open. Morales stood there, tablet in hand, dark circles under her eyes. "Captain," she said softly, "you should be in quarters. You look—"

"Like I've earned my wrinkles?" he offered.

"Like hell warmed over," she corrected.

He gave a ghost of a grin. "Good. Means I still look like a captain."

She crossed her arms. "Doctor's orders say no bridge duty until that fever drops."

"Doctor's orders don't steer ships," he said, eyes back on the horizon. "We're within twenty miles of the ridge. I'll rest when I see the top of it."

"Sir—"

He cut her off gently. "You'll have to get used to not calling me that soon enough."

Morales frowned. "What do you mean?"

Raines didn't answer. He just glanced at the small file now sealed on his terminal and then out the viewport again. The sky was starting to turn from silver to bruise, dawn pressing up from beneath the water's edge.

"You ever hear the sea hum before sunrise?" he asked quietly. "It's like the world remembers its old voice. Reminds me of when I believed we could still fix it."

Morales's tone softened. "We still can."

He smiled faintly. "Maybe you can. Me…" He flexed his hand, the tremor refusing to stop. "My time's almost spent."

The comm crackled — Manny's voice, rich with static and grease. "Bridge, this is engine room. We're holding five knots steady. If Captain Stare-Down's awake, tell him turbine two's singing in key for once."

Raines managed a dry chuckle. "Copy that, Manny. Keep her humming."

Morales gave him a long look. "I'll be back in ten minutes. If you're still here, I'm dragging you out by your collar."

He didn't argue, didn't promise. He just watched her leave and listened to the sea.

The ping came again. Seven seconds. Then again.

He whispered to the sound, "Hold the line, Alvarez. You're next."

And somewhere deep below, the engines pulsed — steady, obedient, alive.

The bridge lights were low, the color of old copper before dawn. Rain misted the glass, faint and steady, beading down the viewport in tired lines. The world outside was pale and endless — gray sea beneath gray sky — but the sound within the ship was never still.

The ping came every seven seconds. Sometimes two quick notes first, sometimes only one. To anyone else it was static; to Joe Alvarez, it was a pulse.

He stood near the plotting table, both hands flat on the steel. The maps were long gone to ruin, replaced by digital fragments and hand-drawn overlays that showed what little the world remembered of itself. North. Ridge. Land. Hope.

Behind him, Captain Raines coughed. It wasn't the usual smoker's rasp anymore — it had weight, the kind that came from deep inside. Morales had tried to order him off the bridge twice already. Joe was about to make it three.

"Captain," Joe said quietly, not looking up from the chart. "You need to lie down. You're burning up."

"I've burned before," Raines said, voice a gravel scrape that might once have been laughter.

"Not like this." Joe turned, jaw tight. "You're shaking. You shouldn't even be standing."

Raines gripped the helm rail with both hands. His knuckles were the color of bone. "I've stood in worse places, Alvarez. You think I'll leave this bridge when the horizon's whispering back at us?"

Joe exhaled hard, patience straining thin. "You wrote me into your logs, didn't you?"

That got Raines's attention. His eyes — faded steel under brine and exhaustion — flicked toward Joe. "Who told you that?"

"Nobody had to," Joe said. "You've been grooming me for months without saying it. You've been letting me handle briefings, fleet mechanics, navigation. I can read the writing, Captain. You're planning for me to take command."

Raines said nothing.

Joe stepped closer. "If that's true, then it starts now. You need rest. The fleet doesn't need a dead man steering it."

For a moment, Raines's expression softened — not anger, just the sadness of someone who'd already made peace with what the other man refused to accept. "The fleet needs more than mechanics and guts, son. It needs a heartbeat. Until I stop hearing that ping, I'm not done."

Joe's hand hit the table harder than he meant. "You think being stubborn's the same thing as being strong? We're twenty miles from a signal that could be human, and if you collapse before we reach it, who does that help? Morales is ready. Valery's ready. Hell, I'm ready. Don't make us watch you die standing."

That stopped Raines cold. The words hung in the air like the echo after thunder.

Outside, the ping sounded again — faint, patient, steady.

Raines turned his head slowly toward the sound. "You know," he said quietly, "when I was your age, I thought strength meant staying upright no matter what. The ocean taught me different. Sometimes the best thing you can do for your crew is not fall until they've learned how to stand without you."

Joe's jaw flexed. "And when they already have?"

"Then you give them time to realize it."

Morales entered then, tablet clutched in one hand, her hair damp from the mist on deck. "Captain, your temperature's at one-oh-three," she said. "You can't keep this up."

Raines didn't turn. "Morales, if I step away, this ship loses its center. You know what happens when a wheel loses its hub?"

Morales narrowed her eyes. "It wobbles. Then it breaks."

"Exactly," he said. "That's why I stay until we reach land."

Joe started forward. "You're not making sense anymore—"

"Enough!" The word tore from Raines's throat like it had claws. Then the strength left him as quickly as it came. His hand slipped from the helm rail. His shoulders trembled once, barely visible, but Joe saw it.

Raines steadied himself with effort. "Alvarez, if I fall, you step up. Until then, you stand down. Understood?"

Joe didn't answer.

Raines turned, his gaze sharp enough to cut through fog. "Understood?"

Joe clenched his jaw. "Aye, sir."

Morales frowned but didn't push it. The captain had that look again — the one that said he'd already made peace with his ghosts. She exchanged a worried glance with Joe as she turned toward the med console.

When she was gone, Raines leaned close enough for only Joe to hear. "When I go, don't let them turn the fleet into a funeral procession. Keep her moving north. Find that signal. Find whoever's still breathing out there."

Joe swallowed. "You're not going anywhere."

Raines smiled faintly. "I'm already halfway gone, son. I just need to hand you the map before I vanish."

The ping came again. Two pips, seven seconds, soft as breath.

Joe's eyes went to the radar where the pulse glowed faint green. "Then we'll find them, Captain. I swear it."

"Good," Raines said, his voice cracking under the weight of both fever and relief. "Then I can die with a little peace."

Joe turned toward him sharply. "Don't talk like that."

Raines's grin flickered like a dying light. "Not talking. Just... logging my last order."

He walked away before Joe could speak — slow, deliberate steps toward the viewport where the sea glowed pale in the distance. He stood there, framed by dawn light, a captain still holding his post as the world turned beneath him.

Morales's voice echoed from the comm: "Fleet course stable. Beacon signal holding steady. Ten miles to contact."

Raines closed his eyes, whispering so softly even Joe barely caught it: "Then let me see the shore."

The engine room was its own weather: hot, metallic, and humming with a pulse that never matched the sea. Pipes ticked as they shed heat. Bearings whispered. The big shafts turned like patient planets, and every rotation wrote another line of scripture on steel.

Manny stood between Turbine Two and the reduction gear, one palm on the casing as if he could read the ship by Braille. Sweat ran into his collar and tasted like brine and grit. Above his head the deck shivered softly—helm corrections traveling down the bones of the Savannah. North. North. Hold five knots. Maybe six, if the old man's will beat the math.

"Tell me something good," Perez said, shouldering through a tangle of lines with a diagnostic slate hugged to his chest. He wore ear cups crooked—half on, half off—so he could hear both the machine and the man.

Manny didn't bother with good. "What'd the accelerometers say on your midnight sweep?"

Perez tilted the tablet so Manny could see. The graphs glowed sickly green. "The shaft looks normal at five knots," he said. "But check out what happens around five-point-six." He zoomed in. A thin ridge rose along the line — small, almost invisible, the kind of flaw only someone who'd spent years listening for trouble would even notice. "The propeller blades are starting to react to it. It's not cavitation yet, but the water isn't clean. All the junk out there is creating tiny swirling pockets."

"City bones," Manny muttered. "Glass, metal dust, all the world's trash." He touched the thrust bearing and felt the heat even through his gloves. "If we push the speed past six knots, those swirls hit the perfect rhythm. The vibrations build. We'll crack the shaft if we're not careful."

Perez's expression tightened. "Check the thermal cams."

They moved to the monitor wall. Manny flipped through views until Turbine Two's housing came up in heat colors — hot orange sliding toward an angry, overheated yellow along the front side.

Perez whistled under his breath. "She's running hot in the upper band."

"That's not just heavy load," Manny said. "Look here." He traced a pale line on the screen, a brighter seam that pulsed in time with the ship. "The casing is flexing against its mounts every time we change course. We're carrying torque the bearings don't want."

"Why now?" Perez asked. "We've pushed her harder before."

"Not across a field of collapsed skyscrapers." Manny pointed upward, meaning the drowned city they were moving through. "All that debris changes how the water moves."

The port intercom crackled. Valery's steady voice drifted through the static: "Bridge to engineering—what's our safe speed if we need it?"

Perez and Manny shared a quick look. Manny hit the button. "Engineering here. We can hold five knots clean. Anything over six risks vibrations we don't like. Recommend staying at the current speed until we're out of the city rubble."

"Copy," Valery said. She didn't ask the Captain for more. She never did. But Manny heard the question already moving up the ladder.

Perez set the slate down and ducked under a handrail, vanishing behind the alternator. "Give me a minute. I want to see the old sin."

"What sin?" Manny called.

Perez's voice came back tinny. "The one you commit when you tell yourself you'll only override safety once."

Manny frowned. He popped an access panel and slid out the interlock tray. Three toggles wore fresh scratches, and one bore the smudge of a thumbprint he didn't like seeing. "Who lifted the thermal trip yesterday?"

Perez reappeared with a face like a bad diagnosis. He held up a breaker module with a makeshift shim—a shaved-down brass washer—that defeated a spring meant to send them to idle when case temps spiked. "Found your confession."

Manny swore. "Which watch?"

"Quick fix, neat work. Looks like the Bridge wanted it temporary and it just ended up staying that way." Perez clicked the module back into place but kept the thin spacer in his hand like something rotten. "If the Captain pushes us to six knots, this little cheat will let him. And if the vibrations spike into the danger zone, we'll still be spinning at failure temperature because the safety trip will think everything's just warm summer weather."

Manny didn't get angry often. When he did, it arrived cold. "Pull the shim. Restore the trip."

Perez hesitated. "You sure? He'll feel the drop if it hits."

"That's the point." Manny took the washer, flipped it once, and pocketed it like a coin he didn't plan to spend. "Machines protect fools and captains the same way."

Perez cracked a tired grin. "You calling the Captain a fool?"

"I'm calling him dying," Manny said, too blunt to take back. He stared at Turbine Two, at the orange bloom creeping like a sunrise they didn't want. "And good men get stubborn when the map runs out."

Above them the hull thrummed. A slight starboard correction rippled through the braces—Morales dancing the ship around something sharp. The vibration register kissed the red line for a heartbeat and slipped back.

Perez pointed. "That's the band I'm scared of. You feel that? Whole frame tried to sing."

Manny felt it in his teeth, his knees, even in his old scars. "She's telling us the speed where she'll give out."

Perez opened another panel. "I'll shift the RPMs a little, make the vibrations uneven. It'll sound rough, but it'll keep the ship safer." He tapped in a series of commands. The readouts jumped, then settled slightly off-balance. The whole room's hum changed—less smooth, more shaky, like a choir slipping off key just to protect its voice.

Manny nodded. "Better."

The overhead speaker clicked again, this time the Captain's voice, worn and iron. "Engineering, give me five-point-eight for two minutes. We've got cross current and tight verticals. I want the bow clear."

Perez froze, eyes on Manny.

Manny keyed the mic. "Engineering copies. Recommending five-point-six maximum for safe vibration envelope. Two minutes at five-eight risks shaft resonance. Strongly advise against."

Silence. Then Raines again, softer but cutting. "Five-point-eight. Two minutes. That's an order."

Manny closed his eyes once. When he opened them, he spoke without wobble. "Aye."

He leaned over the throttle block. "Bring her up slow," he told Perez. "No leaps."

Perez eased the throttle forward. The big turbines climbed toward the set RPM like giants rising to full height. The deck seemed to dim under their boots—Manny couldn't tell if it was real or just his nerves. He kept one hand on the turbine casing and the other near the manual shutdown switch, like someone deciding whether to open a door to trouble.

The vibration line on the graph climbed. Thin. Sharp. Hungry. Manny watched it rise, listened to the bolts around him, and felt the ship speak in a way the officers up on the bridge would never hear.

"Thirty seconds," Perez said, eyes on the timer.

The ping came through the bones of the ship even down here—two quick sparks, seven seconds of hush. Human, Manny thought. Or human enough. He pictured Amy upstairs with the boy and the dog;

he pictured Joe holding a future like a welding rod bright enough to stare at. The Captain wanted to buy them a faster answer. The ship wanted to keep them alive long enough to hear it.

The ridge rose one more rung.

Manny hit the intercom, voice flat. "Bridge, Engineering. You've got ten seconds at five-eight before I dump us to safe. We're flirting with a crack."

No reply.

"Nine," Perez said under his breath. "Eight."

The deck thrummed in a way Manny hated—too clean, too pretty, like glass before it sings apart. He slid his thumb onto the kill.

"Seven."

A shadow crossed the thermal bloom—cooler air from a vent Morales must've opened to ease casing temps. Smart. Not enough.

"Six."

Manny pictured the washer in his pocket. He pictured Valery's scope, Amy's bandaged hands, the boy's solemn counting. He pictured the Captain, grey-eyed and finished, trying to will a continent back into being.

"Five."

"Dumping," Manny said, and killed ten percent.

The turbines moaned, dropped, settled. The ridge fell off the crest and rattled itself into harmless.

Perez let out a breath he hadn't noticed he'd been keeping hostage. "We live to sin another minute."

The intercom hissed. Raines didn't shout. He didn't have to. "Engineering, report."

Manny keyed in. "Machinery safe. Harmonics exceeded red band. I cut to preserve shaft integrity. We'll give you all the speed we can without breaking the ship, Captain."

Silence stretched. Then: "Understood."

The channel clicked dead.

Perez leaned his head against the warm steel. "You just saved his command and pissed him off in the same breath."

Manny shrugged, shoulders heavy. "He can be mad at me when we reach land."

Perez glanced toward the ladder well, where bright white light slanted down from a world that still believed in dawn. "If we reach land."

"We will," Manny said, and he believed it because anything less would be treason. He wiped a black smear from his knuckles onto a rag and tucked the brass washer deeper into his pocket. "Let's go tell Joe the ship's truth before pride writes us a new gravestone."

They moved together through the heat and the thrum, past machines large enough to be mistaken for gods. Above them, the ping kept time: pip-pip—seven seconds—hope.

The storm that took Raines didn't start with thunder. It started with silence.

The kind that creeps into a ship when everyone feels something's wrong but nobody says it.

By late afternoon, the Savannah had crossed into what used to be Arkansas — or the ghost of it. The flooded plain shimmered under the gray sky, skyscraper tips and mountain bones poking through like broken ribs. The beacon still pinged steady through the comm every seven seconds, two short tones first. Hope was close enough to taste.

But Captain Raines hadn't left the bridge all day.

Joe had been pacing between the helm and the navigation desk for hours, his nerves strung taut as the ship's antenna wire. Morales was on comm rotation, Valery at sonar, and Manny had just come up from the engine room, wiping grease from his arms.

The Captain stood near the viewport, his reflection washed thin in the glass. His hat hung crooked in his hand. His eyes had gone glassy, his lips pale.

"Captain," Joe said quietly. "You haven't eaten since dawn."

"Appetite's for men with choices," Raines muttered, not turning.

"Then at least sit down," Morales urged, glancing at Valery, who gave a small nod. "Sir, your color's wrong."

"I've looked worse." He coughed into his sleeve. The sound was wet and low, dragging too long.

Manny leaned on the rail beside Joe. "Boss," he said softly, "turbine two's holding, but she won't take much push. Same for you, sir. Machines and men both need cool-downs."

Raines gave a half grin. "Always the poet, Manuel."

"Engineer," Manny said, serious now. "And the ship's talking to us. You're not hearing her anymore."

Raines turned — slow, like his body was made of rusted hinges. His gaze swept the bridge, touching each of them like a farewell disguised as inspection.

"This ship," he said, voice shaking but hard at the edges, "has carried what's left of a species across an ocean that forgot mercy. You think I'll step down before we find a damn place to stand?"

"Sir," Valery said gently, "you've already found it. It's us. All of us. You kept us together."

He blinked at that, like she'd said something he couldn't quite hold onto. Then the deck shifted under a crosscurrent, and he caught the rail too late. His knees buckled.

"Captain!" Morales lunged forward, tablet clattering to the deck.

Joe caught him before he hit the floor. The weight of the man felt all wrong — too light, bones against fabric. He lowered him onto the mat near the helm while Morales checked his pulse.

"Faint, rapid," she said. "He's crashing."

Manny knelt beside her, hand steady on Raines's shoulder. "He's burning up," he said. "Fever's breaking his strength."

Raines's eyes flickered open, cloudy but still aware. He tried to speak, but only a whisper came. "Steady course… north…"

Joe leaned closer. "We've got it, Captain. You rest."

Raines grabbed his sleeve with surprising force. "No," he rasped. "You… you steer."

"I am—"

"Not the ship." The words came jagged, forced between breaths. "The fleet."

Joe froze.

Raines coughed again, a deep, tearing sound. Morales pressed a cloth to his lips. It came away red.

Valery's voice trembled. "We need to get him to med bay now."

Raines shook his head weakly. "Too late. Bridge is my bed."

Manny swallowed hard, looking away. He'd seen death before — on decks, in storms, in the eyes of those who didn't make it out of the fire. But not this one. Not the man who'd kept them alive long enough to start hoping again.

Morales squeezed Raines's wrist gently. "Sir, listen to me—"

"Quiet," he whispered. His gaze found Joe's, sharp again for one last spark. "You're command now. Log C-12… confirms it."

Joe's throat locked. "No. Not like this."

"Exactly like this," Raines said. His hand slipped, but Joe caught it. "Don't let this ship die standing still."

Outside, the ping came again — pip-pip, seven seconds — louder now, clearer, as if the world itself were answering.

Raines's eyes followed the sound. His breath steadied for a moment, and he smiled faintly. "Hear that? Someone's waiting."

Joe squeezed his hand. "Then we'll find them, Captain. I swear it."

Raines's eyes stayed open, watching the horizon. The hum of the engines seemed to fade beneath the sound of the sea — soft, rhythmic, almost kind.

Then the Captain exhaled once, slow and long. The next breath didn't come.

Morales looked up, voice breaking. "He's gone."

Valery turned away, pressing a hand to her mouth. Manny bowed his head. No one spoke.

Joe stayed kneeling beside him, hand still around the old man's wrist, unwilling to let go. "Rest easy, Captain," he whispered. "We'll keep her steady."

Outside, dawn finally broke across the horizon — faint light spilling over the drowned world like a benediction. The beacon still pinged through the comm.

Pip. Pip. Seven seconds.

This time, it sounded almost like goodbye.

The bridge stayed quiet for a long time after Captain Raines's heart stopped.

Nobody spoke. Nobody moved. The hum of the Savannah's turbines became the only sound left, soft and patient, like the ship itself was holding its breath. The smell of salt and oil lingered in the air — the scent of endurance, of a world still turning whether it wanted to or not.

Morales sat back on her heels, staring at the empty space that used to belong to him. She pressed her stethoscope against his chest one last time, as if disbelief could change physics. It didn't.

Valery turned off the comm. Her hand hovered over the switch longer than necessary. The steady ping of the northern beacon filled the silence again — pip-pip, seven seconds — indifferent to the fact that a good man had just died chasing it.

Manny leaned on the railing beside the helm, his head bowed. The lines in his face deepened, the grief sitting heavy in his jaw. "He went standing," he said finally, voice low. "Same way he lived."

Joe didn't answer. He was still crouched beside Raines, one hand over the old man's shoulder, the other resting against the cold deck. The captain's hat had rolled a few feet away; Joe picked it up, brushed the brim with his thumb, then set it across the man's chest.

Morales swallowed hard. "We should move him to med bay."

"Not yet," Joe said quietly. His voice had no tremor in it — only weight. "He'd want one last watch on his bridge."

Nobody argued.

The deck beneath them shuddered as the autopilot kept its slow northern climb, guiding the ship through the still water. Dawn light cut through the mist, painting the windows in soft gold and copper. The whole room seemed to glow — an unintentional memorial, light from a sun that didn't know who it had just lost.

Valery finally spoke. "The fleet will need to know."

Joe nodded, standing slowly. His movements were deliberate, the kind of slow that comes from exhaustion and grief knotted together. "Send it fleet-wide," he said. "All channels."

Valery hesitated. "You want me to—"

"I'll say it," Joe interrupted.

He stepped to the comm, his reflection a ghost in the dark glass, and keyed the broadcast line. The ship's intercoms hummed alive, connecting the fleet — the Halcyon, the two tugs, and every skiff in their wake.

"This is the Savannah," Joe began. His voice carried clear, even through the static. "All hands, all vessels, copy this transmission."

He looked once at the captain's body, then back to the horizon.

"Captain Elias Raines... has passed. He died on the bridge, at course heading zero-zero-eight, northbound toward the signal we've been tracking for seventy-eight hours. He held his post until his last breath. We'll honor that by doing the same. The fleet remains under command."

He stopped for half a heartbeat, then said the words he knew Raines would have wanted. "We move forward. We move quiet. We move kind."

The comm stayed open for a moment longer. He didn't say his name — not yet. There would be time for that later. He closed the channel and stood still until the static faded into the sound of the sea.

Morales wiped her face with the back of her sleeve. "He'd be proud," she said softly.

Joe's voice was rough. "He'd tell me I took too long to say it."

Manny let out a humorless laugh. "He'd tell you to stop talking and steer."

That earned a faint smile from Valery. "Then we'd better steer."

Joe turned toward the viewport, where dawn had turned the water into a field of molten silver. He touched the console. "Helm, maintain heading zero-zero-eight, speed five knots. Keep formation tight. We've still got a signal to find."

Morales rose to her feet, pulling the blanket from the back of Raines's chair. She laid it gently over him, smoothing the edges, her hands shaking. "Rest easy, Captain," she whispered. "You brought us this far."

Charlie's voice broke the stillness — small, trembling, coming from the doorway. "He's with the sky now."

Everyone turned. The boy stood there barefoot, hair mussed, clutching Lenny against his chest. The little dog whimpered once, then gave a soft bark, as if agreeing.

Amy was just behind him, pale but steady, a hand on the doorframe for balance. "He was a good man," she said. "One of the last."

Joe walked to her, voice low. "He's the reason we're still moving."

She nodded. "Then keep moving, Joey. Don't let the world stop just because it's quieter."

He turned back to the viewport, to the sea that had taken everything and still offered the faintest hum of promise. "We won't stop," he said.

Morales checked the horizon monitor. "Signal strength increasing — five nautical miles and closing."

Valery adjusted the gain. "Reading a structure at bearing zero-one-two. Still faint, but steady."

Joe took one last look at the captain's covered form. Then he set his hand on the helm. "All ships," he said into the comm, his voice calm and sure, "this is Alvarez. Maintain course.

He paused, then added quietly — only for the bridge, maybe only for the dead. "Let's make him proud."

Outside, the fleet answered not with words but with motion — engines rumbling, wakes aligning, hulls slicing north through the flooded heart of what used to be a country.

The Savannah led the way, her flag at half-mast, her people awake and grieving, and her new captain staring into a dawn that suddenly felt heavier and brighter all at once.

The beacon kept calling. Pip-pip. Seven seconds.

This time, it didn't sound like goodbye.

It sounded like destiny.

CHAPTER 6
COMMAND TRANSFER

Fleet Log – SAVANNAH

Day 467 Post-Fire

Position: 34° N, 91° W (approx.)

Course: 008° toward signal origin / formation steady

Weather: Low swell, overcast sky, visibility moderate

Status: Captain Raines committed to sea / Command transferred to Alvarez

Operational Note: Turbine #2 in green / Fleet holding 5 knots north-by-east

They called the crew to the main deck just after sunrise. The Savannah held steady on bare steerage way, bow into a slow swell, engines idling low so the ship wouldn't drift across the line of the fleet. Halcyon kept position a cable astern. The two tugs, Marlin and Kite, sat off the quarters like small, patient guards. Helo One and Two were chained down and silent. No one wanted rotor wash today.

The flag was folded once, then twice, then laid open again across the canvas-wrapped stretcher. It wasn't fancy—sailcloth cut from a service tent, stitched last night by two deckhands who had good hands and better hearts. The flag on top was one Raines had kept in a drawer under the chart table. It was sun-faded at the edges from years of use, but the stars still showed bright. Someone had wiped the salt stains from the stripes.

Morales ran the detail. She didn't give a speech. She set positions the way she always did—clear and simple. "Side party, two to a line. Rails manned from fore to aft. No talking. Phones off. Cameras away. We're here to do a thing right."

Guido handled the rigging. He and Manny had put up a short derrick at the starboard break and slung a quiet block so the stretcher could lower clean without banging the side. Manny triple-checked the cleat. He didn't trust knots on a day like this. He didn't trust his hands, either, but they held steady.

Joe came out last from the bridge. He wasn't in a captain's coat. There wasn't one. He wore his work jacket zipped to the throat and the same black ball cap he always wore when weather was iffy. His face showed the kind of night that doesn't end just because the clock changes. Behind him, Amy walked slowly, color back in her cheeks but still thin. Donnie was there with a blanket over his shoulders. Valery kept close, not hovering, just there if anyone needed her.

Charlie stood between Amy and Donnie, hands jammed in the pockets of a too-big deck hoodie. Lenny stayed glued to his shin until Amy patted her thigh. The dog took that as orders and sat pressed to her leg, eyes on the stretcher.

Morales lifted her chin to Joe. "All hands present who can be spared."

He nodded. "Thank you, XO." He looked down the deck. Faces from every job—the welders, the mess crew, the med techs, the skiff teams. Even the helo crews with grease on their sleeves. Perez and Dave weren't there; they were in the plant holding power and vacuum so the ship wouldn't hiccup. Joe had told them to stay put. Raines would have done the same.

Guido and Manny lifted the stretcher to the rail. The flag didn't slip. The canvas didn't sag. Manny had padded the under-strakes so it wouldn't slap the hull on the way down.

Morales set her watch on the rail and spoke low enough that only the first rows heard it and passed it back. "No eulogies yet. We stand. We listen. Then Joe speaks. Then we commit. That's the sequence."

Wind came light off the bow, taking the smell of diesel and the sour tang of last night's coffee with it. The water had that flat, pewter look you get when the weather hasn't decided what to do. A line of gulls—which shouldn't be here and were here anyway—rode the updraft in a loose V and then broke apart. One circled the masthead once and drifted off.

Joe stepped to the rail and put one hand on the stretcher's frame. He didn't look at anyone for a few seconds. He looked at the flag, the canvas, the rope, the ocean. You could hear small things: a boot scuff, a cough someone tried to swallow, the slow click of a halyard block.

"Okay," he said finally. "Let's do this right."

Morales raised her hand. The side party squared up. Heads came off—caps and helmets tucked under arms. Even Guido took his cap off, and Guido hated sun on his scalp. Manny straightened until his shoulders clicked.

Amy reached for Donnie's hand and found it. Charlie shifted closer to her hip. Lenny didn't make a sound. He watched, ears half-cocked like he understood the rules.

Joe didn't speak yet. He looked toward Halcyon. On her bridge wing, someone lifted a hand and held it steady. The tugs dipped their bows as the swell rolled under them. Helo Two's crew chief stood at attention on the hangar roof with a headset around his neck and his jaw set tight.

Morales gave a small nod to Valery. Valery clicked the PA to deck-only and set the gain low. No music. Just the circuit open so voices would carry without shouting.

Manny checked the lowering line one last time. "Ready on the fall," he said, voice level.

"Ready," Guido answered.

Joe rested his palm on the flag. He didn't close his eyes. He said nothing yet. The ship did the only respectful thing a ship can do: she kept still. Even the engines seemed to settle.

Morales spoke for the log, her voice the same one she used for bearings and numbers. "Savannah, day four hundred sixty-seven since the fires. Latitude unknown, heading north by east. Captain Raines committed to the deep with honors by his crew."

She let out a breath, barely audible. "Stand by for words."

No one moved. The fleet held its line. The stretcher waited at the rail, flag bright against gray water. The morning didn't warm, but it didn't get colder, either. It just held there, like the world was willing to give them the time they needed to say goodbye.

The deck was silent except for the slow wash of the sea and the soft tap of the flag as it caught the wind. Joe stood at the rail, his reflection faint in the water below. He took off his cap and held it against his chest. His eyes looked hollow from the sleepless nights, but there was a steadiness in him now — something that had settled since Raines's last breath.

He cleared his throat, but the first few seconds passed without sound. When he finally spoke, it wasn't loud. It didn't have to be. The crew leaned in, drawn to the quiet.

"Raines taught us something I don't think any of us will forget," Joe said. "He didn't tell people what they wanted to hear. He told us what would keep us alive. When food was low, when the turbines coughed, when storms tried to drown us, he stood right here and said, 'Steady her.' And we did. Because he said it."

His hand tightened around the cap brim. "He used to walk the decks at night, just to listen to the hull talk. Said the ship tells you what she needs if you stop trying to outthink her. I didn't believe that at first. Thought it was just one of his sea tales."

He smiled faintly, eyes still on the water. "Turns out he was right. The sea talks, the ship listens, and the people in between… we're the echo that keeps her moving."

The wind shifted, carrying a hint of rain. Somewhere aft, a rope creaked.

"When the world burned," Joe continued, "he didn't give up. He could've taken one of the early boats, found a quiet shore, lived out what time he had left. But he didn't. He stayed. He said, 'If we keep moving, we'll find something worth finding.'"

He looked back at Mom, at Donnie, at Charlie holding Lenny tight against his chest. Then to Morales, Valery, Manny, Guido — each one of them part of the living proof that Raines's belief hadn't been wasted.

"He told me once that command isn't about being the loudest

voice. It's about knowing when to shut up and let the sea teach you something." Joe's voice went softer now, the kind of tone that forced people to listen harder. "I think we've all learned a lot from the sea. Mostly that it doesn't care. But he did. He cared about every damn one of us."

The flag fluttered against the stretcher. The red and white stripes rippled like breath.

"I don't have fancy words," Joe said, his throat tight. "I don't have the right ones, anyway. But I'll say this — we go on. Because he said we would. And that's the promise we're keeping."

For a heartbeat, no one moved. Then Morales lifted her chin and said clearly, "Aye, Captain."

Others echoed it, voices quiet but solid. "Aye, Captain." "Aye."

Even Charlie whispered it, barely louder than a breath. Lenny gave a small bark that startled the quiet into a smile.

Joe looked out across the fleet — steel hulls, patched decks, weather-beaten faces. The ships rode easy in the morning swell, lines straight, formation perfect. They weren't just following a course. They were following a memory.

He nodded to Manny. "Ready the line."

Manny stepped forward and loosened the cleat. The rope slid smooth through his gloved hands, the stretcher beginning to lower. Guido guided it out and down, slow enough that the flag never dipped.

Joe spoke one last time, voice barely carrying above the water's murmur. "Fair winds, Captain. You brought us this far. We'll handle the rest."

The stretcher kissed the surface, the flag spreading out one last time like a breath leaving the lungs of the sea.

Morales gave the final order, her tone firm but breaking at the edges. "Commit."

The line slipped. The canvas and flag sank together, the water closing over them with a soft, final sigh.

No one clapped. No one spoke. They just stood — a ship full of survivors saluting one of their own, listening to the sea take him home.

And when the last ripple faded, the Savannah seemed to breathe again.

The silence after the sea swallowed Raines felt heavier than any storm. Nobody wanted to be the first to move. Morales stood rigid at attention, the wind tugging strands of hair from her braid. Valery wiped her eyes quickly and pretended she hadn't. Guido leaned on the rail like the weight in his shoulders had doubled.

Then Morales finally spoke, her voice steady but rough. "All hands —return to stations."

The crew obeyed, slow and quiet. Boots scuffed the deck. The soft shuffle of movement carried through the ship like the sound of rain that hadn't started yet. One by one, they filed back toward the hatches until only the command crew remained — Morales, Valery, Manny, Joe, and a few others who'd been closest to Raines.

Morales turned toward Joe. For the first time since the firestorm, she looked uncertain. "He should've had more time," she said quietly.

"He knew," Joe answered. "That's why he kept the log straight until the last page."

Morales nodded. "He left instructions. Signed it two nights before he..." She didn't finish the sentence. She just handed Joe a folded page sealed in plastic. "You should read this."

Joe didn't open it. He already knew what it was. The crew had all known Raines's handwriting — those bold, square letters that always ended in a small curve, like they didn't want to stop leading. "He left me the fleet."

Morales's jaw tightened, but she nodded once. "He did."

"You outrank me," Joe said.

"Not anymore." She crossed her arms, eyes level with his. "He made it clear. Seniority doesn't matter when the crew already looks to you. I'm XO, and I'll stay that way. You're the captain now. Congratulations and condolences."

He gave a humorless laugh. "Doesn't feel like either."

Morales stepped closer. "You think anyone else could hold this together?" She gestured toward the ships outside the glass. "They follow you already. You think that's coincidence? You keep the reactors humming, the med bays supplied, the water clean. You've earned it."

Joe rubbed the back of his neck, eyes on the floor. "I just fix things."

"Exactly," she said. "The fleet's broken in a thousand ways. You're the one who fixes things."

Valery had been quiet, arms crossed against the chill. Now she stepped forward. "She's right," she said softly. "You already command without realizing it. Raines saw that before anyone else did."

Manny stood by the aft rail, watching them. His hands were still wrapped around the rope he'd used to lower the stretcher, knuckles white, callouses split. "He told me something once," he said. "Said the ocean doesn't want heroes, it wants steady hands. Guess that's you, boss."

Joe met his eyes. "Don't call me that."

Manny half smiled. "You're gonna have to get used to it."

The corner of Joe's mouth twitched, almost a smile but not quite. "We'll see."

Morales took a slow breath and squared her shoulders. "All right. Orders?"

Joe looked toward the horizon where the sun had burned a thin line through the gray. "We stay on course toward that signal. We'll hold formation until we know what's ahead."

"Fuel's still good for another week at this speed," Morales said. "But turbine two's showing stress creep again. I'll have Manny take a look."

"I'm already on it," Manny said. "If she groans, I'll know before she screams."

Valery was studying the sea beyond the bow, arms tight around herself. "The signal's stronger this morning. I can clean it up once we stabilize the grid."

"Do it," Joe said. "I want to know who's talking to us."

He finally opened the sealed page. The handwriting was neat but fading, pressed deep into the paper like every word cost effort. If you're reading this, Alvarez, then I'm gone. Keep her steady. They'll follow you because you listen before you speak. That's what a captain does. Remember—land is hope, not salvation. Don't trade one for the other.

Joe folded it back up, slid it into his pocket, and whispered under his breath, "I'll keep her steady."

Morales turned to go, then stopped at the hatch. "Captain?"

He looked up.

"You don't have to be Raines," she said. "You just have to be you."

He nodded once, not trusting himself to speak.

When she was gone, he stood alone at the rail, staring down at the water that had taken the only man he'd ever seen as unshakable. The sea looked calm, but it was the kind of calm that waited to test whoever tried to cross it.

Joe gripped the railing tighter. "All right," he muttered to himself. "Let's see what you've got left for me."

Behind him, Manny gave a small nod. "Aye, Captain."

And this time, Joe didn't argue.

By afternoon, the decks were nearly empty again. The fleet had resumed formation—Savannah leading, Halcyon gliding silent behind her, tugs keeping the flank, and the helicopters locked down for weather. The sky had gone a washed-out white, the kind that made the horizon disappear, and the water beneath it looked more like melted metal than sea.

Morales and Valery stood in the bridge, shoulders touching as they watched the radar flicker. They didn't talk much; they didn't have to. The hum of the ship, the soft rhythm of the pings, and the faint scratch of pencil against paper were language enough.

Joe came in quietly, rubbing the fatigue out of his eyes. He hadn't slept since Raines's burial. He didn't look like he was trying to. He moved straight to the navigation console, checking coordinates, the course line, the steady five-knot pace northward.

Morales spoke without looking up. "Turbine two's back in the green. Manny says she's running smoother than she did yesterday."

"That's something," Joe said. "We can't afford another breakdown. Not now."

Valery pointed at the sonar feed. "The signal's holding steady, five degrees to port of our original course. I boosted the gain, filtered out reflection noise. It's definitely a transmission, not geological or mechanical."

Joe leaned over her shoulder, studying the readout. "Still the same seven-second gap?"

"Still there," she said. "And the double pulse at the start."

He frowned. "That double pulse—it's communication, not coincidence."

"Agreed," Morales said. "But from what? Could be another fleet. Could be one tower someone powered up. Could be survivors who don't know how to shut it off."

Joe crossed his arms. "Or someone who wants to be found."

The three of them stood silent for a few moments, each lost in the thought of what that could mean. After everything they'd seen, hope was as dangerous as any weapon.

Below deck, the clang of a wrench echoed up the ladder. Manny was working again, somewhere under the turbine housing. His voice drifted through the comm line, half muttering, half singing some old tune that might've once been a prayer.

Valery smiled faintly. "At least he's in good spirits."

"That's how he works," Joe said. "If he's humming, it means the ship's still listening to him."

Morales scribbled another note in her log, then set her pencil down. "You know, I've served under six captains. Some were born for the job, some broke under it. Raines was the only one who made it look easy."

Joe gave a tired chuckle. "That's because he never let us see when it wasn't."

Morales looked over, her tone softer now. "And for the record, I wasn't Navy before all this. I was RP—radiation protection—at Turkey Point, working with you, Dave, and Perez. When the world went sideways and we got aboard Savannah, Raines read my quals, watched me run casualty drills, and put me in the chair. Field commission. No stripes, no ceremony—just responsibility. He said, 'After the fire, competence is rank.' So if anyone wonders why a civilian's wearing the XO hat, that's why."

She hesitated, then added, "He promoted me because he saw what this ship needed. And the same goes for you."

Joe didn't answer right away. He looked through the forward glass instead, where the sea stretched in endless gray folds. "Quiet's all we've got left some days."

Valery stepped back from the console, folding her arms. "Then let's make sure it's the kind that means calm, not the kind that comes before something worse."

Joe finally smiled. "I'll take that deal."

The ship shuddered once as a low swell passed beneath the hull, a reminder that even in stillness, the world was moving. Outside, a few gulls tracked their wake, skimming the surface like white ghosts.

Morales closed her log. "If this signal's what we think it is, we'll have landfall within a week—maybe less. You ready for that?"

Joe hesitated. "Ready? No. But willing."

Valery looked at him. "That's close enough."

A sharp rap sounded on the bulkhead—Manny's voice came over the speaker, louder now. "Bridge, this is engineering. Pressure's stable, feed lines locked. If you're feeling vibration, that's just her stretching her legs. Don't panic."

Morales keyed the mic. "Copy that, Chief. Nice work."

"Always," he said, and the line clicked off.

For a moment, it almost felt like normal life—a captain, an XO, a crew, a working ship, a mission. Then the sea rolled again, and the memory of the burial washed back over them like undertow.

Valery straightened her jacket and turned for the door. "I'll relieve Guido on sonar watch. He looks ready to fall asleep standing up."

"Thanks," Joe said.

Morales lingered a moment longer. "You're doing fine, Joe. Better than you think."

He didn't answer, but when she left, he exhaled like he'd been holding his breath since dawn.

The bridge went quiet again, the ship humming beneath his boots. Joe turned to the glass, looking out at the horizon where the sky met the drowned world. Somewhere beyond it was that signal, that voice that refused to stop calling.

He whispered it to himself like a promise. "We're coming.

CHAPTER 7
AFTER THE CAPTAIN

Fleet Log – Savannah
Day 467 Post-Fire
Position: 34° N, 91° W (approx.)
Course: 008° north-by-east toward signal origin
Weather: Low swell, overcast
Status: Command – Capt. J. Alvarez

The ship held a steady hum underfoot, turbines balanced and even. Morning came slow, a dull gray light stretching over flat water. Nobody spoke much on the bridge; the only sound was the rhythmic ping coming through the comms—two short bursts, seven seconds of silence, then two again. It had become the heartbeat of the fleet.

Joe stood by the forward glass, mug in one hand, scanning the horizon as if the signal might materialize there. Morales was at the chart table, pencil moving with practiced precision, marking their new line of travel. Valery sat at her station, headset on, adjusting dials millimeter by millimeter until static faded to clean tone.

She said quietly, "Still the same, Captain. Double pulse, seven-second gap. Strength up by one tenth since 0400. That means we're closing."

Morales didn't look up. "Anchored transmitter then—on land, not drift."

"Could be military," Valery added. "The pattern matches old Air Force emergency code. They used that double-ping for fixed-base homing beacons."

Joe nodded once. "Hold five knots. Keep formation tight. I want the tugs and Halcyon within comm range at all times."

Morales keyed the internal channel. "Engineering, bridge—maintain current speed and turbine load."

Donnie answered first, voice calm and deep through the speaker. "Copy, bridge. Turbines steady, temps normal. She's running smooth, Joe."

Joe allowed a breath of a smile. "Thanks, Chief!"

Manny's voice followed, lighter, joking. "Engineering to bridge—no sparks, no smoke, no miracles needed. For now."

Then Guido's laugh came over the same line. "Speak for yourself. I'm the miracle."

Morales shook her head, half-smiling despite herself. "Copy, Engineering. Keep it together down there."

Joe leaned over Valery's shoulder, watching the waveform scroll across her monitor. "You're sure it's not an echo?"

She shook her head. "Too clean. Echoes smear on return. This is precise—same amplitude every cycle."

"Designed," Morales said from the chart table.

"Exactly," Joe replied. "Someone built that to last."

Valery kept working. "If the maps I pulled are right, the bearing lines up with what used to be Arnold Air Force Base. Elevation's high enough to stay above water."

Morales plotted it. "That puts us three to four days out at this speed."

Joe rubbed his jaw, eyes on the display. "Good. Maybe we'll find more than a signal—maybe supplies, records, something useful."

Valery leaned back slightly. "First solid lead since we left the ridge. Feels different."

Joe nodded. "It is different. Someone meant this to be found."

Morales finished the entry in the ship's log. Fleet maintaining five-knot course toward inland beacon. Source likely military. All systems green.

The bridge lights dimmed automatically as clouds thickened outside. Joe set his coffee aside and folded his arms, watching gray water roll ahead of them. Behind him, the crew worked without chatter, every motion deliberate. The fleet moved as one, small lights cutting through the fog, chasing a sound that refused to die.

He exhaled slowly. "All right," he said. "Let's see what's waiting for us out there."

The pings came again—steady, measured, patient—as if something in the dark north was listening back.

By noon, the sea was smooth as glass, the wind dead calm. Savannah's deck crews worked in silence, tending lines, wiping salt from rails, anything to keep busy. Up on the bridge, Morales had the fleet log open on the chart table. She'd been staring at the "Acting Captain" line for ten minutes.

She turned the tablet toward Joe. "We need to make it official."

He didn't look up right away. "I already wrote the order last night."

"You wrote it," she said. "But the log still calls you acting. Crew knows better, but regulations don't. Let's make it clean."

Valery sat at comms, headset off one ear. "She's right, Joe. You're already doing it. The only difference now is ink."

Joe gave a slow nod. "All right. Let's do it."

Morales began typing. Her voice stayed steady as she read: "Command transfer from Captain E. Raines to Captain J. Alvarez. Effective day four hundred sixty-seven post-Fire. Endorsed by Executive Officer C. Morales." She looked up. "That's it. Sign below."

He hesitated a second, then took the stylus and signed under Raines's name. The screen blinked, timestamp confirmed. Just like that, it was done.

Valery turned in her chair, gave a small smile. "Congratulations, Captain. It suits you."

Joe huffed softly. "We'll see."

Morales tapped the side of the tablet. "It's official whether you like it or not."

The intercom crackled. "Engineering to bridge," Donnie's voice said. "Just got the update from Morales. Proud of you, captain. Feels right."

Joe's voice softened. "Thanks, Dad."

Manny cut in right after, teasing. "Guess that means I've got a new boss to ignore."

Guido followed, laughing. "Not a chance. The man kept us alive through the Fire. I'll take orders from him any day."

Morales smirked. "That's more endorsement than most captains get."

Valery adjusted her headset, half-listening to the chatter, half-watching Joe. "They already follow you. You don't have to prove anything."

He looked at her, the corner of his mouth twitching. "That's not the point. Raines earned their trust by never pretending to know everything. I just need to keep them believing this means something."

"It does," she said quietly. "You're the reason it still does."

Morales broke the moment with a new report. "Halcyon holding formation, tugs flanking steady. Fleet spread minimal. Engineering output constant."

"Good," Joe said. "Keep all ships in sync. We stay at five knots until we reach the beacon."

"Copy that," Morales said. "Entry logged."

The bridge settled back into rhythm. Joe moved toward the glass again, eyes tracing the faint reflection of his own face against the horizon. Captain. The word still didn't feel real.

Valery came to stand beside him. "You'd hate the ceremony anyway," she said, soft enough that only he heard. "Too much talking."

He almost smiled. "You know me too well."

She nodded. "Three years of practice."

He looked at her then—long enough for understanding, not long enough for anyone else to notice.

Outside, the sea rolled quietly, a gray sheet stretching north. Somewhere beyond it, that beacon kept calling.

By midafternoon, the sea had flattened to a dull mirror. The wind dropped off completely, and the air felt heavy, the kind that warned of weather coming. On the bridge, the only sound was the repeating ping. Two short bursts, seven seconds, then two again. The pattern was so steady it almost sounded alive.

Valery sat forward in her chair, eyes locked on the comm screen. "Still tracking the same line," she said. "Bearing zero-zero-eight. Signal clarity up again—barely noticeable, but it's stronger."

Morales drew a line on the map, matching her direction. "Range?"

"Still too far to read coordinates," Valery said. "But I've filtered the static. Frequency sits dead center in old U.S. Air Force emergency bandwidth."

Joe leaned on the rail beside her. "No chance it's an echo?"

She shook her head. "No. It's intentional. And it's still powered by something stable. That's not drift or reflection."

Morales looked up. "Automated then. If the Air Force left any beacons running, it was for survivor recovery."

Joe nodded slowly. "Or supply recovery."

He pressed the intercom button. "Engineering, bridge. You copy?"

Donnie's voice came through immediately. "Loud and clear."

"You hearing the same frequency in your system bleed?"

"Yeah," Donnie said. "We've got a faint carrier in the background noise. Whoever's transmitting is pushing steady voltage. It's not solar. Fuel cell maybe. You don't get that clean a waveform after a year and a half unless the power source is designed for long haul."

Manny's voice cut in next. "If it's military, they used hydrogen stacks. I helped service one before the Fire—thing could run unattended for years."

Guido whistled low. "So we're talking a beacon that could've outlived everyone who built it."

Donnie replied, "Could be more than a beacon. Might be a cache. They used to pair them with deep storage bunkers—fuel, MREs, ammo, water purifiers. If that's what we're closing on, it could change everything."

Joe looked at Morales. "Mark that in the log. Possible supply installation."

"Aye, Captain," she said, typing it cleanly into the record.

Valery turned in her chair. "I'm pulling up terrain data from before the floods. The signal's on high ground if I'm right—Tennessee range. Arnold Air Force Base used to sit at about a thousand feet elevation. That's above the flood line."

Morales nodded. "Four days at this speed. Maybe five if we hit current drift."

Joe glanced toward the window, where the horizon blurred into haze. "Keep us steady. I want every ship in tight formation."

"Already relayed," Morales said.

Donnie's voice came back one last time before he cut the line. "You get eyes on that thing, captain, you call it. I'll have the team ready to offload if it's worth something."

"Copy that," Joe said. "We'll find it."

The intercom clicked off, and for a moment, only the steady ping filled the bridge again. The crew moved around quietly, every one of them pretending they weren't listening to that sound—the proof that somewhere out there, something built by their own hands was still working.

Below deck, the air was thick with heat and the smell of oil. Turbine #2 ran low and steady, the rhythmic thrum echoing through the steel deck plates. Donnie stood over the main housing, sleeves rolled up, checking the coolant pressure gauge. The glow from the monitors painted his face orange.

"Pressure's holding at sixty-eight," he said. "That's the sweet spot. Any higher, and she'll start to complain."

Manny crouched nearby with a wrench in hand. "We could replace that regulator next downtime. It's showing wear."

"Yeah," Donnie said. "But not till we're through this run. Last thing I want is to strip the threads and stall us mid-ocean."

Guido climbed down the ladder from the upper catwalk, grease on his hands. "Fuel feed's clean. Filter was half-clogged again, but I flushed it. If I ever find out who keeps skipping rotations, I'm locking them in the bilge."

"That'd be you, last month," Manny said, smirking.

"Details," Guido said, tossing him a rag. "Anyway, she's good now."

The intercom crackled overhead. Morales's voice came through, calm and clear. "Engineering, this is the bridge. Captain Alvarez requests turbine status."

Donnie pressed the talk switch. "Bridge, this is Engineering. Both turbines stable, load balanced. We're running smooth."

"Copy that, Chief Engineer," Morales said deliberately. "Entry confirmed in the ship's log."

Donnie paused a second before answering. "Appreciate that."

Guido raised an eyebrow. "Chief Engineer, huh? Guess it's official."

Manny grinned. "About time. Should've happened when he saved himself and Amy on Mercy Two."

Donnie shook his head. "Titles don't keep turbines spinning. Work does."

"Still," Guido said, leaning against the rail, "feels right hearing it."

Donnie gave a small smile. "All right, back to business. You two finish that intake inspection before shift change. I want every line checked twice."

Manny wiped his hands on a rag. "What do you think that signal really is, Donnie? Valery says it's Air Force."

"Could be," Donnie said. "They built automated depots all over the inland states—food, fuel, survival gear. If that's what we're heading toward, it could set us up for months. But it could also be something abandoned and dead."

"Or not abandoned," Guido said quietly.

They all fell silent for a few seconds, the hum of the turbine filling the space. The thought didn't need to be explained. Not every survivor they met had been friendly.

Donnie adjusted a valve and watched the gauge settle. "Whatever's waiting, we'll handle it. That's what we do."

"Same plan as always," Manny said. "We fix what's broken and pray it doesn't explode."

"Exactly," Donnie said, smiling faintly. "Now check that feed line again before it decides to prove you wrong."

Above them, the ping from the beacon echoed faintly through the bulkheads—two quick beats, seven seconds of silence. Even here, deep in the ship, they could feel its rhythm.

Guido looked up at the ceiling. "If that's a beacon for survivors, we're about to be the first ones to answer it."

Donnie nodded slowly. "Then let's make sure this ship's ready when we do."

He flipped the switch to cycle the secondary pumps, and the low hum deepened through the deck plates—steady, reliable, alive.

Night settled slow over the fleet. The horizon disappeared into gray mist, and the ocean took on a dull metallic glow under thin cloud cover. The ships moved in perfect line—Savannah leading, Halcyon behind her, the tugs off each quarter keeping pace. Five knots steady. No talking on the bridge now, just the faint tap of rain starting against the glass.

Joe stood at the forward window, arms folded, eyes fixed on nothing in particular. The ping still came through the comms, soft but steady. Two quick tones, seven seconds of silence. It filled the background like breathing.

Valery sat at her station, headset half-off. "Signal strength holding," she said quietly. "No fluctuation for the last two hours."

Morales jotted the note into the log. "Confirmed. Navigation locked. Fleet spacing consistent."

Joe nodded. "Good. We'll hold this heading until 0400. Rotate bridge watch every two hours."

Morales stood, stretching her shoulders. "Copy that. I'll take first rotation."

Valery muted her channel, leaned back, and looked at Joe. "You've been on your feet since dawn."

"I'll rest when the shift changes," he said.

She studied him for a moment. "You said that yesterday."

Joe turned just enough to meet her eyes. "We're closer now. I'll sleep when we hit land."

Valery gave a small nod. "Then I'll keep you awake until then."

Morales smirked faintly from the chart table but didn't comment. She knew better than to interrupt that kind of quiet understanding.

Down below, Engineering was still alive with noise. Donnie,

Manny, and Guido ran their final checks before shift change. The turbines hummed deep and low, a heartbeat beneath the hull.

"Pressure stable," Donnie said, checking the gauge. "Hold her at this load till morning. Let her rest a bit."

Manny wiped his hands on a rag. "We'll take turns keeping watch down here. If that beacon's real, we'll want full power when we reach it."

Guido nodded. "We've come too far to limp in tired."

Donnie gave a small grin. "That's the spirit. Now get something to eat before your stomach starts chewing on its own wiring."

Back on the main deck, Amy moved between the compartments carrying a tray of mugs—water and what passed for coffee now. She handed one to a young deckhand and another to Morales on her way up. "You need to keep your strength up," she said softly.

"Thanks, " Morales said.

Amy looked toward the bridge, saw Joe at the glass, and smiled faintly. "He doesn't stop, does he?"

Morales shook her head. "No. He learned that from you and Donnie both."

Amy nodded, the pride clear in her eyes. "He was like that even when he was little. Always had to fix what was broken."

When she was gone, Joe stayed where he was, watching the empty horizon. The fleet lights stretched behind him like a string of stars. He keyed the intercom one last time.

"Bridge to fleet—all ships hold course and speed. We'll reach that beacon together."

He clicked the mic off, leaned against the rail, and listened. The pings came again, faint but clear—two short bursts, seven seconds of silence. Somewhere out there, something was waiting. For the first time in weeks, it didn't sound like a warning. It sounded like hope.

CHAPTER 8
TIGHT FORMATION

Fleet Log – Savannah

Day 468 Post-Fire

Position: 35° N, 92° W (approx.)

Course: 008° north-by-east / following beacon signal

Weather: Hazy sky, calm seas

Status: Fleet in fuel-conservation drill / all systems green

The beacon's signal still sat far to the north-northeast, a fixed point they were crawling toward instead of charging. They were still days from its source, stuck in the drowned middle, so today was about saving fuel rather than buying speed.

The morning air was heavy and damp, the kind that smelled like rust and old rain. Savannah cut a clean wake through flat water, her reflection sharp in the gray. Halcyon and the two tugs followed close behind, each ship running quiet and deliberate. The fleet looked small but disciplined—like a single thought stretched over miles of ocean.

Morales stood at the bridge console, her voice even as always. "Drill begins at oh-nine-hundred. Goal is twenty percent fuel reduction while keeping formation. If anybody drifts, we waste diesel and I yell."

Joe didn't look up from the chart. "And if yelling doesn't work?"

"Then I promote Guido to morale officer," she said.

Valery smiled faintly from the comms desk. "That's cruel, even for you."

"Desperate times," Morales said. She keyed the intercom. "Bridge to fleet—commencing fuel-conservation drill. Maintain four-point-five knots. Fifty-meter spacing. Confirm."

Replies came one by one. "Halcyon copies." "Kite copies." "Marlin copies." The rhythm of discipline settled in. The ships stretched out but stayed tight, a little constellation moving through washed-out light.

Below deck, Engineering was alive with the deep hum of turbines. Donnie leaned over the console, checking RPMs. "Hold her steady, Manny. Sixty-three percent load. Don't let her drift."

Manny adjusted a dial. "Got it. She's running smoother than a church choir."

Guido snorted. "Yeah, the kind that drinks during rehearsal."

"Wouldn't be a bad idea," Manny said. "Rations are thinner than Morales's patience."

"She can still hear you," Morales said over the line. "And she's logging every word."

"Worth it," Guido muttered.

Charlie crouched beside his father, tightening a coolant line with both hands. His wrench was nearly as long as his arm. "Like this, Dad?"

Donnie nodded. "Perfect. Not too tight. Feel the resistance, then stop. That's how you know it's right."

Charlie grinned. "Like when Mom says stop eating from the pot before dinner?"

"Exactly that," Donnie said. "One of life's great lessons."

Manny leaned over the control panel. "Kid's a natural. He can have my job in ten years."

"Five," Guido said. "He already listens better than you."

Morales came through again. "Engineering, bridge. Fuel flow report."

Donnie pressed the button. "Holding steady. Both turbines balanced. Charlie's on coolant. He hasn't broken anything yet."

"Copy," Morales said. "Logging progress and child labor."

Charlie laughed, proud of himself. "Did she really write that?"

"She did," Donnie said. "You're officially part of the crew now. Sorry, no paychecks."

Back on the bridge, Valery scanned her monitors. "Fleet efficiency up eighteen percent. Fuel flow synchronized."

Joe nodded. "Good work. Keep formation. I want the fleet acting like one ship."

Morales updated the log, her pencil scratching quick and clean. "We'll hold this until eighteen hundred, then rotate watch."

Joe stared out through the bridge glass. The sea looked harmless, which usually meant trouble was warming up somewhere.

"You feel that?" Morales asked quietly.

"Yeah," Joe said. "It's too quiet. Something's due."

Valery half-smiled without looking up. "Don't jinx us, Joe."

He didn't answer. The ping from the northern beacon ticked on in the background—steady, far, patient.

The mess smelled like boiled beans and resignation. Trays clinked, spoons scraped, and the low murmur of crew voices never quite reached "cheerful."

Guido walked in with a clipboard tucked under his arm and the kind of face that said nobody was going to like what he had to say. The room was half full—off-duty crew finishing breakfast that tasted like cardboard and regret. He cleared his throat. "All right, people, you might wanna put your spoons down before I ruin your morning."

Morales looked up from her mug of fake coffee. "That bad?"

Guido flipped a page. "Worse. We're down to fifty-seven sealed ration packs fleet-wide. That's one per person for five days if nobody blinks too fast."

Manny, sitting across the table, groaned. "You count morale rations?"

"Morale rations are gone," Guido said. "Somebody decided to celebrate last week's thunderstorm with extra stew."

"That was me," Manny admitted. "Thought we might all die in the lightning. Figured full stomachs would make a better memory."

Morales leaned back. "Noted. If you do it again, I'll make you eat the ration labels next time."

Guido kept reading. "We're also low on coffee powder, tea, sugar, salt, and apparently, optimism. The only thing we've got plenty of is dried beans, and I'll be damned if I eat another bite of that without an apology from whoever invented beans."

Valery slipped into a seat, still wearing her headset around her neck. "You could always grind them up and pretend it's coffee."

"Don't joke," Guido said. "I tried that once. My tongue still twitches when it rains."

Laughter rippled through the room, tired but genuine. Morales sipped what passed for coffee and made a face. "If we survive long enough to rebuild civilization, I'm starting with real coffee and working backward from there."

Guido checked another line on his sheet. "Water's fine. Filters holding. Condensers working. As long as nobody showers for the next hundred years, we're golden."

"Good," Morales said. "I was worried about morale dropping."

"Already dropped," Guido said. "Bounced twice."

Joe stepped in from the corridor, catching the last part. "We'll make do," he said. "Once we reach the beacon site, we'll resupply. It's military—there'll be stockpiles."

Guido nodded, but his eyes stayed wary. "Hope their definition of 'stockpile' is better than ours."

Before Joe could answer, Valery pressed two fingers to her earpiece, head tilting slightly. "Bridge is calling me."

Morales frowned. "We're calling you and you're standing right here?"

Valery's expression sharpened. "Not the bridge. Sonar."

She was on her feet before anyone could ask, already jogging for the hatch.

By the time Joe and Morales reached the bridge, Valery was bent over the sonar console, fingers dancing over the controls. The low, steady ping had changed—it came with an extra texture now, a softer cloud under the sharp returns.

"What've you got?" Joe asked.

She pointed at the screen. "Look here. Depth about sixty feet, bearing zero-four-two, running parallel to us. Dense returns, moving slow. Too tight and too uniform to be debris."

Morales leaned closer. "School of fish?"

"School of something," Valery said. "Big concentration. And they're hugging a line."

"A line of what?" Joe asked.

She zoomed out. The darker band beneath the school sharpened into angles. "Looks like elevated roadway. Could be an old highway or overpass. Whatever it is, they're using it like a wall."

Morales hissed out a breath. "Fish using a drowned interstate as a feeding lane."

Guido's voice crackled over the comm. "Bridge, this is Guido. Sonia in galley says sonar's doing something weird and Valery ran out mid-bean. Should I be worried or just hungry?"

"Both," Morales said. She hit the broadcast switch. "Guido, get to deck three forward. If this is what we think it is, we might be able to net it."

There was a pause, then: "Copy that. On my way. If this is a false alarm, I'm eating the net."

Joe watched the ghostly cluster drift along the screen. "We're in a fuel drill," he said. "We start trawling, we burn more."

"We burn more now," Morales countered, "or starve slow later. Five days of sealed rations isn't a plan, it's a countdown."

Valery glanced back at him. "They're staying close. If we swing wide and drop net along the outside edge, we can make one pass. We don't have to chase, just let them run parallel."

Joe exhaled through his nose. Fuel charts flashed behind his eyes; so did the faces in the mess. "All right. One pass. No broken gear, no heroics."

He keyed the intercom. "Bridge to all ships—Savannah will briefly adjust course for capture run. Halcyon and tugs hold formation at current heading and speed. Do not follow. Stand by."

Morales switched channels. "Guido, Manny, Donnie—prep the forward trawl. Short run, sixty-foot column. We've got a school of something big under our left hand. I want it in the freezer and not on sonar."

"Copy that," Donnie replied. "We'll wake the net up."

On deck, the air had a bite to it—a raw, wet chill that cut through shirts. Crew hustled around the forward trawl winch while Guido checked the net lines for the fifth time.

Manny yanked on a shackle. "If this tears, Morales will use the holes to bury us."

"Then don't tear it," Guido said. "We only get one surprise party today."

Charlie hovered near the rail, eyes wide. "You really think it's fish?"

"Could be old shopping carts," Guido said. "But if it swims, we're gonna meet it."

Donnie finished the last connection and checked tension. "Net's ready. Short drop, slow pull. If we snag on anything, we cut loose in under three seconds. No arguing."

Morales's voice came over the deck speaker. "All stations, this is the bridge. Net run beginning in thirty seconds. Stay off the edge and keep clear of the cables. Joe's taking us in along an elevated structure. Eyes open."

Savannah's bow eased into a shallow arc. From above, the change was barely visible; from the deck, they could feel it—a sideways slide that sent faint ripples down the hull.

Valery's voice followed. "Target band dead ahead. Depth holding. Density increasing. Looks like big bodies—could be stripers or catfish. Whatever they are, there's a lot of them."

"Good," Guido said. "I'm tired of beans judging me."

"Drop," Donnie ordered.

The winch whined, cables paying out. The net plunged into the gray water with a heavy splash, the floats bobbing at the surface while the weighted bottom line sank. Savannah's forward momentum did the rest, drawing the net out like a curtain beneath her bow.

On sonar, the school brushed the edge of the screen, then slid fully into the mouth of the net.

"Contact," Valery called. "They're in. Holding depth. No structure interference yet."

Yet.

On deck, the cable creaked as the weight grew. The ship's speed dropped a fraction, hull vibration changing in a way only engineers and worriers could feel.

"Got 'em," Manny said under his breath. "Come on, come on…"

A warning ping sounded from Valery's console. "Be advised, faint hard returns beneath the lane. Could be cars or signage under the school. Nothing tall yet."

"Copy," Morales answered. "If we get a snag…"

"Three seconds and we cut," Donnie said, hand resting on the emergency release lever. "I like this ship more than I like fish."

The tension climbed. The net line thrummed like a plucked string. Then the cable lurched as if something huge had grabbed hold.

"Easy," Guido breathed. "Just the load settling."

The hull shivered again, a low scrape beneath.

Valery's voice sharpened. "We've got a low obstacle at ten o'clock—looks like an overhead sign or gantry on the old highway. You're right on the edge of it."

Joe altered course a hair. "Adjusting. Keep me a meter clear."

"Clear," she said a moment later. "You're good. The school's turning with us."

Manny grinned despite the strain in his arms. "Smart fish. Dumb day to be smart."

"Haul," Donnie ordered. "Bring 'em up."

The winch screamed as the cable took on full weight. The net rose slowly, dragging the water into a churning boil around the bow. Silver bodies flashed just under the surface—first a few, then dozens, then a roiling mass of scales and tails.

Charlie whooped, unable to hold it in. "We got 'em!"

The net broke the surface, bulging with life. Fish flapped and thrashed, water pouring off them in shining sheets. Some were as long as Charlie's arm; some longer. Striped bass, fat catfish, a couple of weird cross-breeds the ocean had invented on its own time.

"Holy—" Guido cut himself off and laughed. "Look at that. We just mugged a buffet."

"Get them over the rail," Donnie barked. "Fast. Before half of them remember which way down is."

Crew swarmed the net, hooking and pulling, shouting to each other as the load swung inboard. Fish hit the deck in a slippery, flopping flood. Mendel from galley yelled something about knives and cutting boards. Someone slipped, cursed, and kept working.

Morales watched from the bridge, one hand over her mouth, eyes bright. "Valery, log the catch estimate."

Valery scanned the returns trailing off the screen. "Hard to say. A few hundred kilos at least. Enough protein to stretch rations a lot longer than five days."

Joe let out a breath he hadn't known he was holding. "Then today, the drill paid for itself."

On deck, water and fish blood ran to the drains. Charlie held up one enormous bass with both hands, grinning like he'd caught the sun. Manny ruffled his hair with a grease-stained palm.

"Tonight," Guido announced to anyone listening, "we dine like kings. Very tired, smelly kings."

The infirmary smelled faintly of antiseptic and metal, a mix that never quite left no matter how often Amy wiped the counters. The tang of fresh fish drifted in from the open corridor—sharp, oily, alive.

"Well," she said, half to herself as she stacked bandage packs, "if nothing else, the ship smells like a real port again."

Charlie sat on the floor sorting packets into piles—gauze, tape, bandages. His hands still carried faint scales from the deck. He held up a small vial and squinted. "Mom, this one says iodine. That's for cleaning, right?"

"Right," Amy said. "Not for drinking, no matter what Manny says."

Charlie laughed. "He told me it'd make my hair grow faster."

"Yeah, and when he was your age, he tried to make coffee out of engine coolant," Amy said. "You see how that worked out."

Donnie leaned in from the hallway, hands still greasy and faintly fish-scented. "Smells better in here than down there," he said. "You two saving lives or just making a mess?"

"Both," Amy said. "He's organizing the kit for the next supply check."

Charlie puffed out his chest. "Morales said I'm on training record now."

"You are," Donnie said. "And if you keep this up, I'll put you in charge of cleaning turbine filters next week."

Charlie made a face. "That's not a reward."

"It's a promotion," Donnie said. "Congratulations."

Amy zipped the med bag and shook her head. "He's eleven, Donnie. You could let him be a kid once in a while."

"He is," Donnie said, tapping his son's shoulder. "He's just one that doesn't explode things when I'm not looking."

Amy handed Charlie a roll of tape. "Put that in the side pocket, sweetheart."

He stuffed it in, proud. "There. We've got three bottles of peroxide, two sutures, and a mountain of gauze."

Amy leaned against the counter. "Good. We'll need it. The way this crew handles tools, I should open a full clinic."

Donnie laughed softly. "You already do." He looked at her for a long moment, that quiet kind of look that spoke of everything they'd survived. "You doing all right?"

"I'm fine," Amy said. "As fine as anyone still floating in a soup can."

He smiled, wiped a smear of grease from his wrist, and kissed her temple. "You're tougher than the ship."

She nudged him. "You're just saying that because you need clean shirts washed."

"Guilty," he said, heading back for the hatch. "Engineering's got its own heartbeat right now. Keep the med bay ready just in case."

Amy called after him, "In case what?"

"In case Murphy shows up," he said over his shoulder. "He always does."

Charlie looked up from his work. "Who's Murphy?"

Amy smiled. "He's the guy who makes your dad's day interesting."

Charlie grinned, proud to understand, and zipped the bag shut with a crisp sound that seemed louder than it should've been. The world might be broken, but for now they had fish on ice, medicine sorted, and a ship still moving. For a day, that felt like winning.

By late afternoon, the fog thickened until the horizon vanished completely. Savannah's hull groaned as she pushed through still water, her reflection lost in gray. The bridge lights were dimmed to red for visibility. Morales stood over Valery's shoulder, eyes fixed on the radar screen.

"You seeing that?" Morales asked.

Valery adjusted the gain and leaned closer. "It's not weather. The echoes are too sharp."

Joe came up behind them. "Show me."

Valery pointed to the cluster of marks on the display. "Bearing zero-one-five. Range about six klicks. Looks vertical—tall and irregular. I'm not calling it land yet, but it's not open water."

Morales frowned. "Could be a ship graveyard. Maybe old towers."

"Could be anything," Joe said. "How's the beacon?"

"Still north-by-east," Valery said. "These returns are just west of that line. Could be connected, could be debris."

Morales looked at Joe. "We need a visual check before we run into something solid."

He nodded. "Launch drone recon?"

"Negative," Valery said. "Signal interference too heavy. I'm not risking another drone until I can see a clear return path."

Morales tapped her pencil against the table. "Then we slow her down."

Joe hesitated, weighing it. "Drop to three knots. Keep formation tight. Tell the tugs to stay in visual range only."

Morales relayed the order across the fleet. The responses came back crisp—no chatter, just confirmation.

Valery zoomed in again. "It's rising out of the water. Whatever it is, it's tall—five hundred feet maybe. The shapes aren't even."

"Could be the tops of skyscrapers," Morales said quietly.

Joe nodded once. "That's what it sounds like. Could be the drowned city grid. If that's true, we're closer to the old Tennessee ridge."

Valery glanced up. "So, congratulations—we've officially reached civilization's tombstone."

Guido's voice came over the intercom. "Bridge, this is Engineering. You seeing the spikes on sonar?"

Joe pressed the talk key. "We've got eyes on them here. Could be buildings. Keep turbines steady and alert your teams. If there's debris ahead, I want fast shutdown options."

"Copy that," Donnie's voice followed. "We'll be ready. And Joe—tell Morales not to scratch my ship."

Morales rolled her eyes. "He says that every time."

"Yeah," Joe said. "But this time he means it."

The radar screen pulsed again, sharper now. Valery leaned closer. "Returns increasing. I think we're closing on the outer edge."

Morales picked up the intercom. "Bridge to all ships—reduce speed to two knots. Maintain visual. Prepare to maneuver around obstacles."

Outside, the fog glowed faintly from the ships' lights. Every so often, a dark shape loomed in the mist—jagged and silent, like the skeleton of something enormous. The crew at the windows whispered without realizing it.

Valery lowered her headset. "Those aren't ships. They're buildings."

Joe stared at the shadows ahead, rising straight out of the water like monuments. "Tall returns confirmed," he said quietly. "Mark position."

Morales logged it, her handwriting small and deliberate. "Marked. And Captain…"

"Yeah?"

She looked at him. "If we hit one of those, we're not patching it with duct tape."

Joe didn't smile. "Then let's not hit anything."

The fog thickened until the ships vanished from sight. Savannah felt like she was sailing through smoke. The radar showed more vertical returns now—towers, antennas, jagged metal sticking out of the sea like the bones of another world. The crew spoke in low voices, as if sound itself might wake something beneath them.

Joe stood at the front of the bridge, one hand braced on the rail. The glow from the radar painted his face pale green. Morales hovered nearby, her eyes flicking between the map and the scope. Valery adjusted the gain again, muttering under her breath. "Returns dense at zero-one-five and zero-one-six. Narrow channel in between, maybe a mile wide."

Joe studied it. "That's our path. We'll slip through slow."

Morales hesitated. "You want to announce a light blackout?"

"Yeah," Joe said. "No sense glowing like a target." He keyed the intercom. "Bridge to fleet—kill all running lights. Repeat, all ships dark. Maintain visual on radar only. Confirm."

The answers came back steady but tense. "Halcyon copies." "Kite copies." "Marlin copies." The fleet went dark one vessel at a time until Savannah was the last source of light. Then Valery flipped the final switch, and the bridge sank into dim red glow. Outside, the world disappeared.

Morales broke the silence first. "Remind me again why we always do this at night?"

Joe's voice stayed calm. "Because if something goes wrong, at least we won't see it coming."

Valery smirked without looking away from the screen. "Comforting."

Guido's voice came through the intercom from Engineering. "We're in full blackout mode down here, Captain. It's like working inside a toaster. If I trip over Manny again, we're filing for hazard pay."

"Denied," Morales said automatically.

"Worth a try," Guido replied.

Donnie's voice followed. "All systems steady, Joe. Turbines throttled low. We've got manual override if debris shows up."

"Copy. Keep her quiet," Joe said.

On the deck, even the sound changed. The engines barely whispered. The water against the hull was louder than the turbines. The fog pressed against the windows like wet concrete. Every crew member felt it—the sense that they were gliding through something ancient and wrong.

Valery broke the stillness again. "Radar echo at starboard edge. Thirty meters off."

Morales leaned in. "Structure?"

"Steel," Valery said. "Vertical. Looks like a tower frame. Might've been a bridge."

Joe moved closer to the window though he couldn't see a thing beyond it. "Mark it. Keep us clear."

Morales logged it, her handwriting careful even in the red light. "Marked."

Charlie's voice came over the internal channel from Engineering. "Dad says we're under power minimum. He says she's humming like she's sneaking."

Joe smiled faintly. "That's exactly what we're doing."

Valery glanced up from the console. "Signal beacon's still straight ahead. Stronger every hour."

Joe exhaled. "Then we're close. Keep her steady."

The bridge fell silent again except for the steady heartbeat of the radar. The ship moved forward through black water, unseen towers rising on either side. Somewhere above the fog, the stars were still there—but nobody looked up.

Joe leaned on the rail, voice low. "If the world's still got ghosts, this is where they live."

Morales nodded. "Then let's pass quiet and not wake them."

The beacon's faint pull still pointed somewhere beyond this graveyard, but for now all they could do was feel their way through the bones.

And with that, the fleet slipped deeper into the drowned city, silent as breath, every light extinguished.

CHAPTER 9
GHOST SKYLINE

Fleet Log – Savannah
Day 469 Post-Fire
Position: 36° N, 93° W (approx.)
Course: 009° north-northeast / over submerged structures
Weather: Dense fog, low light
Status: Fleet holding formation / reconnaissance and salvage prep

At sunrise, the fog peeled back just enough for the skyline to appear—black spires jutting from the water like the ribs of a dead world. The sea was glass-flat, pale gold in the first light, and the towers reached out of it at angles that made the crew's stomachs tighten. Some leaned, some stood perfectly straight, and some broke off just below the surface where the waves whispered against them. The tops of skyscrapers marked what had once been the coast. Now, they were the only thing left of it.

Morales stood at the bridge glass with her hands clasped behind her back. "Mark bearings for all visible structures," she said quietly. "We'll log them as obstacles and landmarks."

Valery ran her scanner along the horizon. "Picking up reflected echoes from steel and concrete. They're everywhere, Captain. It's like we're sailing over a city made of ghosts."

Joe didn't look away from the window. "That's exactly what we're doing."

Below them, the sonar pulsed, each return painting a map of what used to be civilization. Streets, intersections, parking decks—all gone, buried under five hundred feet of water. The bridge crew fell silent, every person aware of the weight of what they were seeing.

Morales finally broke it. "Hard to believe people used to live here."

"Harder to believe we still do," Joe said.

On the forward deck, the salvage teams watched from their stations. Guido leaned on the rail with a scope, scanning the spires as they glided past. "Looks stable enough from here," he said into his mic. "But I wouldn't lean on any of it."

Donnie's voice came back through the comm. "We'll anchor between structures three and four. The current's light. Manny, prep a skiff team. Keep your tethers tight—we're not losing anyone to curiosity."

"Copy that," Manny said. He checked his harness, then looked at Guido. "You ever get the feeling the ocean's watching us?"

Guido smirked. "She's probably wondering what kind of idiots built this high just to drown lower."

From the bridge, Valery turned, her expression distant. "Radio static's thick. The signal beacon's still north, but there's a lot of reflection here. It's like the city's trying to talk."

Joe nodded slowly. "Or remember."

The fog drifted again, and one of the tallest towers came into view—a jagged silhouette of glass and steel. Birds nested on the ledges near its peak, the only signs of life in the drowned skyline. A faded logo halfway down the building read FIRST NATIONAL, the letters ghost-pale against black metal.

Morales logged the coordinates. "First National Tower. Visible height, approximately two hundred feet. That used to be a landmark."

Joe folded his arms, eyes still locked on the ruins. "Still is," he said quietly.

The fleet passed between the towers like pilgrims through grave markers. No one spoke for a long time. Even the water seemed to hush itself. Somewhere far below, sunlight flickered off broken windows, turning the flooded city into a mirror that refused to forget.

The skiff bay smelled like diesel and brine. Cables rattled overhead while the crane arms swung into position. Donnie checked his list twice, then slammed it against the bulkhead for the crew to see. "All right," he said. "We're taking one skiff, four hands—light load only. Salvage priority is copper wire, solar panels, and anything marked power storage. No heroics, no souvenirs."

Manny zipped up his survival jacket and grinned. "So, no grabbing the flat-screen off the twenty-seventh floor?"

"Not unless it floats," Donnie said. "You break formation, and Morales will feed me your bones."

Guido was checking the winch lines. "She'd probably alphabetize them, too."

"Stop helping," Donnie said. "You're both on rope detail. Keep chatter low once we're out there. I don't trust the echo in this place."

A quiet voice came from behind them. "Need an extra set of hands?"

They turned. Ashley stood in the hatchway, gloves tucked into her belt, hair tied back under a cap. She looked thinner than before Mercy Two went down, but her eyes were steady.

Donnie straightened. "You sure you're ready, kiddo?"

She nodded. "I'm not sitting around anymore. Morales cleared me for light duty. Said I could start in Engineering."

Manny hesitated, pretending to check a gear strap to hide his surprise. "We're working with open lines and unpredictable current," he said carefully. "You comfortable with that?"

Ashley shrugged. "If I waited to feel comfortable, I'd still be sitting in that bunk. Besides, I know my way around electrical panels. If there's power storage down there, I can help identify it."

Donnie looked her over for a second, then nodded. "All right. Stay close to Manny and Guido. If they say back off, you back off. No arguments."

Guido handed her a safety harness. "You break it, you fix it," he said.

"Fair trade," she said, buckling in.

Manny tried not to stare, but the small copper bolt on her wrist caught his eye—the one he'd given her months ago. She still wore it like a charm.

"Skiff two ready," Donnie said into the mic. "We'll launch at your word, Captain."

Joe's voice came through, calm but clipped. "Approved. Keep comms open and tether lines visible. We'll track your route on sonar."

The bay door opened with a hydraulic groan. Cold wind swept in, carrying salt and silence. The drowned city loomed just beyond the mist—metal towers rising like broken fingers from the deep.

Manny took the controls, Guido unhooked the moorings, and Ashley climbed aboard. The skiff dropped into the water with a slap.

Donnie leaned over the rail. "You know the rule—if it creaks, leaks, or glows, don't touch it till I say."

"Copy that," Manny called back.

Ashley smiled faintly. "I'll take the glowing ones."

Guido shook his head. "Perfect. We're all gonna die stylish."

Donnie laughed, low but genuine. "That's the spirit." He watched the skiff push out into the fog until it vanished between the skyscrapers, then checked his watch. "All right," he muttered. "Let's see if this city's got anything left worth stealing."

The skiff drifted between two half-submerged towers, the water glassy and dark. Rust streaked the steel beams where sunlight once hit them decades ago. Birds scattered from the ledges as the salvage crew approached. Manny throttled back the engine, steering carefully through floating debris. "There," he said, pointing to a building half under the surface. "Roof's mostly intact. Solar array still mounted."

Guido checked the scanner. "Confirmed. Looks like twenty panels. If half still conduct, we've hit jackpot."

Ashley leaned forward, scanning the structure. "We can pull them with the short rig. Keep the tethers clear, or they'll shear off when we lift."

Manny grinned. "Someone's been reading the manuals."

"Somebody has to," she said.

They eased alongside the building, its walls streaked with algae and corrosion. The salvage hooks clanked against the frame. Guido climbed to the bow to secure the harness, muttering, "If this thing falls, I want it on record that I told you both it was a bad idea."

"Noted," Manny said, checking tension on the main line.

Ashley clipped herself to the safety rail and started removing bolts from the solar frames. Her hands were steady, her eyes sharp. Every move precise. She looked more confident than anyone expected—especially Manny.

"Hey," he said. "Don't rush it."

She smirked. "Relax. I've been taking things apart my whole life."

"Yeah, but I'd like you to keep putting them back together."

Before she could answer, a low creak came from below. The metal beam under their skiff vibrated like something alive. Guido froze. "Uh... you hear that?"

"Yeah," Manny said. "Everybody off the roof. Now."

Ashley dropped her wrench and lunged for the tether. Guido pulled the release pin, and the winch snapped the skiff backward just as the tower groaned again—a sound deep and slow, like the ocean taking a breath. Then the structure gave way.

Steel twisted. Glass shattered. A whole corner of the building collapsed into the water, sending up a wave that slammed the skiff sideways. Manny grabbed the throttle, fighting to keep them upright. Guido cursed and clung to the rail. Ashley lost her footing and hit the deck hard, but her harness caught before she went overboard.

"Hold on!" Manny shouted, spinning the controls. The skiff bucked once, then steadied as the wave rolled past. Spray rained down over them.

"Everyone okay?" he called out.

"Still breathing," Guido said. "My pride's dead, though."

Ashley sat up, coughing, her hands shaking slightly. "I'm fine. Just bruised."

Manny crouched beside her, checking the strap that saved her. "That line held. You did everything right."

She met his eyes for a moment, then looked away, embarrassed. "Lucky guess."

"Lucky charm," he said quietly, glancing at the bolt tied to her wrist.

She noticed but didn't reply.

Guido kicked a piece of debris overboard. "Well, half the building's gone. But look on the bright side—we don't have to climb for the lower panels anymore."

"Yeah," Manny said. "Gravity's our new engineer."

Ashley cracked a smile despite the shake in her hands. "Remind me to never volunteer for rooftops again."

Manny smirked. "You're hired permanently."

"Damn it," Guido muttered. "That's how it starts."

They all laughed—short, shaky, but real. The city around them moaned as if in reply, the sound rolling through the steel skeletons like a whisper from the deep.

The skiff came back riding low in the water, loaded with salvaged copper coils and broken solar frames. When they docked, half the crew was soaked, and all of them were grinning like they'd gotten away with something. Morales met them at the deck rail with her log tablet. "Report."

Guido dropped his gloves on the deck. "One building fell, nobody died, and we found enough copper to make the ship glow if we ever get desperate."

"Good," Morales said. "Next time, try not to redecorate the ocean floor."

Manny climbed out, stretching his back. "Hey, at least we didn't sink the skiff. That's progress."

"Low bar, but I'll take it," Morales said. She glanced at Ashley. "You all right?"

Ashley nodded. "Bumps and bruises. Nothing serious."

Morales studied her a moment, then tapped her stylus. "Noted. Glad to have you back on duty, civilian or not."

Ashley smiled faintly. "Feels good to work again."

Donnie came up behind her, wiping grease from his hands. "You did fine," he said. "You kept calm, didn't freeze up. That's what matters."

"She also nearly flew," Guido said. "Ten out of ten for form."

Ashley rolled her eyes. "You scream louder than I did."

Manny coughed to hide a laugh. Guido pointed a wrench at him. "Don't start. You were white as a ghost, too."

"Yeah," Manny said, "but at least I looked good doing it."

Morales gave them the kind of look that ended most conversations. "If you're done comparing near-death styles, help Valery map the safe routes through this sector. I'm designating it the Ghost Belt."

Valery's voice came from the bridge over the comm. "Already working on it, XO. I'm patching visual returns from drone two. Sending coordinates to your pad now."

Morales nodded. "Copy."

Ashley followed Guido up to the mapping console. Valery's data streamed across the display—hazy outlines of towers and debris, the fleet's path winding like a thin thread between danger.

"Looks like a maze," Ashley said.

Guido leaned closer. "Yeah. One wrong move, and we get to be modern art."

She smirked. "I'll stay on the right side of history, thanks."

Manny joined them, leaning on the rail. "You did good today," he said to her quietly.

Ashley looked up at him. "You mean besides almost breaking my neck?"

"Yeah," he said. "Besides that."

She smiled, small but real. "Thanks."

Valery's voice cut in again. "Captain, mapping complete. We've got a corridor about one mile wide through the densest area. Recommend we travel during daylight only. Some of these towers are leaning hard."

Joe answered, his tone calm. "Copy that. We'll call this section secure. Morales, keep eyes on sonar until we clear it."

"Already on it," Morales said.

The bridge went back to work, the hum of the ship returning to its steady rhythm. On deck, Ashley helped Guido coil salvage lines. Manny tossed her a rag, and she caught it without missing a beat.

Guido watched them and muttered, "Great. Another engineer who flirts with tools."

"Beats arguing with you," Ashley said.

He pointed the wrench again. "You're officially one of us now."

"Guess that's good," she said.

"It's terminal," Guido replied.

She laughed quietly and went back to work while, far above, the sun burned a pale hole in the fog. The Ghost Belt stretched ahead—dead city, live crew, moving forward.

Night came soft and heavy. The fleet floated over black water so still it looked solid. Most of the crew had turned in, engines throttled low to a gentle hum. The air smelled faintly of salt, rust, and machine oil—home for people who no longer had one.

Joe sat alone in the bridge, the log tablet glowing faintly under his hand. Morales had left him with fresh notes, Valery's maps were spread out beside the console, and the radar spun lazy circles over empty space. Every few seconds, a ping echoed through the hull—steady, patient, alive.

He started typing, his voice quiet as he recorded. "Fleet Log – Savannah. Day 469 post-Fire. Course nine degrees north-northeast. The Ghost Belt mapped and cleared for passage. Salvage operation successful. One structure collapse, minor injuries only. Crew performance: steady and calm."

He paused, thinking. "Not much left of this city, but it still knows how to talk. You can hear it if you listen—the metal flexing under the water, the old towers breathing when the waves pass. It's like the past won't shut up, and maybe it shouldn't."

He rubbed a hand over his jaw and looked through the bridge glass. The skyline was just shadows now—black shapes rising from deeper black. The lights from Savannah's windows reflected faintly off broken glass far below, shimmering like stars trapped under the surface.

Morales came in quietly, a mug in her hand. "Still writing?"

"Trying to," Joe said. "Feels like the city's watching."

"Let it," she said. "It's got better manners than half our crew."

Joe almost smiled. "How's Engineering?"

"Donnie's doing rounds. Guido's trying to convince Manny that duct tape counts as structural reinforcement. Ashley's helping clean gear. She fits in."

"She's tougher than most," Joe said. "Took her a while to come back, but she's doing it."

Morales nodded. "Raines would've liked her."

Joe leaned back in the chair. "Raines liked anyone who didn't quit. She's one of those."

Morales sipped her drink. "We all are." She set the mug down and looked out at the dark. "When we reach that beacon, what then?"

"Land," Joe said. "Mountains, trees, dry ground. A place that doesn't move when we sleep."

"You sound like you almost believe it."

"I do," he said. "Because if I stop, they stop."

She nodded once, quiet understanding passing between them. Then she left him alone again with the hum of the ship.

Joe closed the log and sat for a long time in the red glow of the instruments. Outside, the drowned towers swayed just below the surface, their reflections breaking and reforming with each small ripple. He could almost hear something under the water—a low groan that might've been the sea or memory itself.

He switched off the mic but spoke anyway, just for himself. "Fleet's still moving. Crew's still breathing. The city beneath us still whispers. And for tonight, that's enough. Somewhere past this ghost skyline, the charts say the seafloor will rise again, and the beacon keeps tugging us that way."

He set the tablet aside and leaned back as the Savannah drifted onward through the ghost skyline, silent but alive.

CHAPTER 10
CLEARER VOICES

Day 470 Post-Fire

Position: 37° N, 93° W (approx.)

Course: 009° north-northeast toward signal source

Weather: Low haze, calm seas

Status: Beacon repeating / formation steady

The radio room stayed quiet except for the soft hiss of static. Valery sat with her headset angled just right, eyes fixed on the frequency monitor. Every few seconds, the pulse came—two short bursts, a pause, then another seven seconds of silence before it repeated. The rhythm hadn't changed since the first time they picked it up. She logged the timing, marked the coordinates, and played it again through the ship's speakers.

Joe stood near the console, arms crossed. "Still seven seconds?"

"Exactly," Valery said. "Two fast pulses followed by silence. It's consistent. Has to be automated. No human could keep that timing perfect this long."

Morales leaned over the chart table, her pencil scratching. "Any shift in signal strength?"

"Up by point-two decibels in the last hour," Valery said. "That means we're closing in. If I had to guess, maybe another hundred miles out."

Joe nodded slowly. "Same direction?"

"Still due north-northeast," she said. "Beacon's not drifting."

Morales looked up. "Could be a repeater tower or a military distress beacon. The pattern's clean enough for high-end equipment."

Guido's voice came through the comm line from Engineering. "Bridge, this is Engineering. We're steady at five knots. Turbine temps in the green. No complaints from the old girl yet."

Joe pressed the switch. "Copy, Guido. Keep her quiet. We're tracking a military-grade signal up ahead. Could be worth the trouble."

"Understood, Captain," Guido said. "If it's the military, tell them to save us a coffee pot."

"Noted," Joe said, smiling faintly.

Morales moved closer to the radar screen. "If this is what we think it is, it could be one of the pre-Fire survival nodes. Some bases had automated emergency systems. They were supposed to broadcast for up to two years."

Valery adjusted her headset, filtering the static again. "This one's still going. I can almost hear something under the tone—like faint modulation."

Joe stepped closer. "Voice?"

"Maybe," she said. "Could be compression noise or a recorded loop. I'll clean it up."

Morales looked over. "If it's a node, the last coordinates before shutdown would be archived in the beacon's data. Could lead to resources—fuel, food, medical."

Joe's tone stayed level. "We'll track it until we're sure. Keep the fleet at five knots. If this thing's been running for over a year, it's not going anywhere fast."

Down below, Donnie's voice came through from Engineering. "Bridge, Chief here. Confirm we're following that same beacon?"

"Confirmed," Joe said.

"Copy that. I'll keep the turbines humming. If there's fuel at the end of this trail, I'm ready to meet it."

"Understood, Chief. Keep your teams sharp."

Valery turned from her console, pulling one earpiece off. "Signal repeating clean again. Two short, seven seconds, constant. I'll boost the recording through filters next pass. Maybe we'll hear what it's saying."

Joe nodded. "Do it."

The sound came again—two quick tones and silence. The crew listened without speaking. It was simple, mechanical, patient.

Morales looked up. "Feels like someone left a light on and never came back."

Joe didn't answer. He just stared through the forward glass where the sea met the gray sky and said quietly, "Then let's see what's waiting under it."

Valery had been at the console for two hours straight. The signal hissed and clicked through the speakers, a flat pattern of sound she was slowly peeling apart. Morales stood nearby with her arms folded, watching the waveforms roll across the screen. Joe leaned against the bulkhead, patient but alert. "Anything new?" he asked.

Valery adjusted the filter dial another notch. "I've isolated the carrier tone. It's military bandwidth. Old, but solid. Whoever built this system knew what they were doing."

Morales checked her notes. "How long has it been broadcasting?"

"If it started right after the first fallout storms," Valery said, "then at least a year and a quarter nonstop. Power's still strong. That means it's either nuclear-backed or running off long-cycle solar storage."

Guido's voice came through the intercom. "Bridge, this is Engineering. We can hear your radio hiss down here. Any chance it's playing something good?"

"Stay on turbines," Morales said. "You'll get your concert when we hit land."

"Copy that," Guido replied.

Valery kept working. "All right, I'm dropping the static by forty percent. Listen to this." She hit playback. The noise thinned, and a voice broke through—tinny, recorded, calm.

• • •

"This is Commander Holt, United States Air Force, Station Arnold, Tennessee. Automated emergency node active. If you can hear this, you have survived. Protocol Charlie-Nine engaged. Supplies include food, water, fuel, and medical. This broadcast will continue until power loss or manual shutdown."

Then a short beep, and the message repeated.

Morales looked up sharply. "Arnold Air Force Base," she said. "That's inland, Tennessee. About two hundred miles northeast from here. High elevation—should be dry ground."

Joe nodded slowly. "You're sure about the source?"

Valery turned her screen so he could see the coordinates. "Positive. The beacon's transmitting from grid thirty-five-point-three north, eighty-six-point-zero west. Matches the base location exactly."

Morales tapped the map overlay. "That ridge system should still be above water. The topography holds at six hundred feet average elevation. If this thing's still running, the base itself might be sealed and stable."

Donnie's voice came through the line from below deck. "You're telling me there's fuel up there?"

"According to Commander Holt, yes," Morales said.

"Then I'm in," Donnie replied. "We're one busted seal away from rowing the rest of the way north."

Joe looked back at Valery. "Replay it again, clean."

She nodded and pressed the key. The voice repeated exactly as before—no change, no emotion, just duty frozen in time.

"If you can hear this, you have survived. Protocol Charlie-Nine engaged. Supplies include food, water, fuel, and medical..."

When the loop ended, no one spoke for a moment.

Morales exhaled. "So it's not a live person. Just a recording still doing its job."

"Still doing it right," Joe said. "That's more than most people left behind managed."

Valery looked over her shoulder. "You want me to trace the path inland?"

"Yeah," he said. "Plot it. If that base is still standing, we're going there."

Morales gave a short nod. "Then it's settled. Arnold Air Force Base is our next destination."

Joe turned back toward the window. The water stretched north, calm and endless. "All right," he said. "Let's go claim what they left us."

Engineering was loud again. The hum of the turbines rolled through the deck plates like a heartbeat, steady but strained. Donnie crouched near the main feed line with a flashlight, watching a fine mist spray from a hairline crack. "All right," he said. "That's the leak. Small, but she'll get worse if we ignore it. Guido, grab the sealant kit. Manny, kill pressure on the secondary valve."

"On it, Chief," Manny said, moving to the control panel. He spun the valve until the hiss died away. "Pressure dropping. Feed isolated."

Ashley arrived a second later, tool bag slung over her shoulder. "Morales said you might need a hand."

Donnie looked up. "Morales finally let you out of paperwork jail?"

She nodded. "Light duty, yeah. I can reseal the feed once you prep it."

"Good. Keep your gloves dry, it's slippery down here."

Guido knelt beside the leak, squinting through the beam of Donnie's light. "We'll need to cut a patch and re-weld. I can't reach the seam from this angle."

Ashley set her tools down and slid closer. "I can."

Guido smirked. "She's smaller, Chief. Might as well let her earn hazard pay."

"None of us get paid," Donnie said. "That's the trick."

Ashley ignored the teasing. She pulled off one glove, traced the crack with her finger, and nodded. "It's not too deep. We can clean it and reinforce it with compound before we weld. Should hold."

Manny handed her the solvent cloth. "Here. Careful, it burns through skin."

She gave him a look. "Thanks for the confidence boost."

Guido grinned. "He means well. He just doesn't know how to talk to women who can fix things."

"Still trying to learn," Manny said.

Ashley smirked and went to work. She wiped the crack clean, dried it, and applied the sealant paste in one steady motion. Her hands didn't shake. Donnie watched closely. "Good job," he said. "You've done this before."

"On smaller engines," she said. "Never on a ship."

"Same principle. Just louder."

Charlie poked his head down from the catwalk above. "Chief, want me to log the repair?"

"Yeah," Donnie said. "Write it up under your name. You're getting close to earning your first mechanic stripe."

Charlie grinned and disappeared back up the ladder.

Manny prepped the welder, the blue light flickering across their faces. "All right, patch in place. Ready?"

Ashley gave a thumbs-up. He started the weld, steady arcs sealing the metal closed. The smell of burnt compound filled the air. When the light faded, the crack was gone.

Guido checked the pressure gauge. "Line's holding. No leaks."

Donnie nodded, satisfied. "Good work, all of you. Bring pressure back to fifty percent, Manny. Let's see if she hums."

The turbine wound up slow, the hum deepening until it leveled out to its usual rhythm. No vibration, no spray.

"Perfect," Donnie said. "You'd never know she almost blew."

Guido wiped his forehead. "Another crisis prevented by caffeine and luck."

Ashley smiled faintly. "Mostly luck."

Manny looked at her. "You're good at this."

She shrugged. "Guess I needed something to fix."

He smiled. "You came to the right place."

Donnie caught the look but said nothing, just turned back to his gauges. "All right, break's over. Let's keep her running. I'd like to make it to that base before we invent any new leaks."

Morales leaned over the chart table, the glow from the monitors painting her face in soft red light. Valery sat beside her, overlaying sonar scans on top of old digital topography. The screen flickered between what the land used to look like and what it looked like now—

valleys turned into deep channels, mountain ridges into islands. "All right," Morales said. "Let's start connecting these depth readings. The signal's originating near the old base coordinates. That means we're looking for ground six hundred feet above sea level or higher."

Valery traced a finger along the map. "Here. The terrain rises gradually. It's not a cliff—it's a climb. There's a ridge system under the waterline that leads north. It's narrow, but navigable."

Morales marked the route with a stylus. "That'll be our corridor. From here to here."

Joe stepped into the bridge, drying his hands with a rag. "Status?"

"Mapping the approach," Morales said. "We've got something that looks like an old interstate submerged along this rise. It'll guide us through the Tennessee Shallows."

Valery zoomed in on the image. "It's eerie. You can see the old road signs on sonar reflections. Looks like the world's longest grave marker."

Morales ignored the comment. "Depth decreases steadily as we move north. If our readings hold, we'll see dry ground within forty-eight hours."

Joe nodded slowly. "What about obstacles?"

"Plenty," Morales said. "Bridges, debris, collapsed buildings, probably aircraft wreckage. The usual."

Valery entered the calculations into the navigation computer. "I'll call it The Ascent Path in the log. Sounds better than 'Route Through Rubble.'"

Morales half smiled. "You've got a gift for naming things."

"Years of watching the world fall apart," Valery said.

Joe leaned on the table, studying the map. "All right. We keep five knots until we hit the ridge rise. Then we cut to half speed. I don't want to lose a single hull to bad navigation."

Morales nodded. "Understood."

Below deck, Donnie's voice came through the intercom. "Bridge, Chief here. Turbines stable. We can hold five knots easy. The patch we laid earlier's holding fine."

"Copy that, Chief," Morales said. "Stay on that pressure gauge. I don't want surprises when we hit the shallows."

"You and me both," Donnie said.

Joe watched the sonar return scroll across the main screen. The terrain was changing, rising beneath them. "It's strange seeing the ocean get shallower."

Valery tilted her head. "It's like the world's giving us back a little piece at a time."

Morales marked another point on the chart. "Don't get poetic now. Save that for after we find land."

Joe smirked faintly. "She's right. We're not there yet."

Valery glanced up from her console. "Still, I like the sound of it—land."

Morales saved the chart and straightened. "Course locked. The Ascent Path is plotted."

Joe looked through the bridge glass at the horizon, the faint hint of dawn light cutting through the haze. "Then let's climb."

By dusk the fleet ran quiet again. The water around them shimmered bronze under a flat sky, and the only sound was the low drone of the turbines. Valery sat at her console with the beacon loop running, scrubbing every byte of its signal. Morales and Joe stood beside her, listening as the mechanical voice repeated the same calm words they already knew. Then Valery caught something new — a faint click between loops. She frowned and adjusted the spectrum filter.

"There's data buried in here," she said. "Low-frequency text packet. Probably part of the original protocol."

Joe leaned closer. "Can you decode it?"

"Working on it." The screen filled with lines of distorted symbols before resolving into text. Valery read aloud, her voice steady.

"This is Commander Nathaniel Holt, United States Air Force, Station Arnold, Tennessee," he began. His voice filled the silent booth. Steady, grounded, unshakable. "Emergency Protocol Charlie-Nine is now active. If you are receiving this broadcast, you have survived the initial Rain of Fire."

"Emergency supply sectors one through five contain food, water, medical equipment, and fuel reserves. These caches are available to any survivors who reach this location. This is not a rescue beacon. It is a survivor beacon."

"Use what you need." "Take what keeps you alive. These supplies exist because we planned for this moment. You are not alone. You are not abandoned. This message will continue until power loss or structural collapse."

The message ended there, looping back to static.

No one spoke. Morales finally said, "It's not a distress call. It's a handoff."

Joe nodded slowly. "He knew nobody was coming. So he left the lights on for whoever might."

Valery saved the transcript and logged it into the ship's archive. "Location matches exactly. Coordinates unchanged. If the base stayed sealed, there could be fuel tanks, generators, maybe even solar backups still charged."

Morales straightened. "Then that's our priority. We'll set course tonight and approach at daylight."

Guido's voice came through the comm, cheerful despite the static. "Bridge, this is Engineering. Heard the recording down here. Guess we've got ourselves an invitation."

"Copy that, Guido," Joe said. "Tell the Chief to keep the turbines trimmed. We'll need full efficiency for the climb."

"Already on it," Guido replied.

Ashley's voice followed, softer. "Engineering standing by. Feels good to have a destination."

Joe looked toward the window. "Yeah," he said quietly. "It does."

Morales updated the log while Valery leaned back in her chair, exhaustion written across her face. "Message decoded, archived, and verified," Morales said. "Beacon confirmed as automated. Source: Arnold Air Force Base, Tennessee. Purpose: survivor support."

Joe opened his own recorder and began to dictate. "Fleet Log — Savannah, Day 470 post-Fire. Transmission confirmed automated. Commander Holt's final message received and recorded. He left behind food, water, and fuel for whoever reached him. We're heading there now. The fleet remains stable. Crew morale high." He paused before finishing. "If the Commander could see us coming, I think he'd call this mission complete."

Morales closed the last entry of the day. "We start ascent at first light."

Valery shut off the beacon loop. The silence that followed felt clean.

Outside, the ocean darkened around the fleet as they turned north toward the unseen ridges. The ships moved slow, steady, and sure — following the voice of a man long gone, carrying hope that refused to die.

CHAPTER 11
LAST PUSH TO LAND

Joe stood on the starboard deck before sunrise, one hand braced on the railing, watching the faint outline of land in the distance. It wasn't much to look at through the haze—just a dark ridge against gray water—but the entire Savannah buzzed behind him like someone had plugged hope straight into the hull. Voices carried through the air. People were awake early, excited, pacing, whispering. After over a year floating between storms and ruined coastlines, land meant something real. It meant tomorrow could be different from yesterday.

Valery walked up beside him with the drone case under her arm. "You ready?" she asked. Her voice was tired but steady. She hadn't had a full night's rest in days, but she looked sharper now that she had something new to aim at.

"Yeah," Joe said. "Let's find out what's waiting for us."

Morales joined them next, adjusting the strap on her rifle. "Guido's got the ops room warmed up," she said. "Manny's finishing the link to the drone feed. They're waiting on you."

Donnie arrived with coffee mugs in hand. He handed one to Joe, one to Morales, and gave Valery a nod. "Mom's with Charlie," he said. "Keeping him busy. Lenny's sitting right on the toolbox like he's supervising."

Joe smiled. "Also sounds like him."

They headed below to the cramped ops room where the ship's main screens had been repurposed for drone work. Inside, Manny and Guido were bent over the console, running diagnostics. A map of the flooded valley and the ridge glowed on the wall, lines and markers scribbled over it in grease pencil.

Manny straightened when Joe walked in. "Drone's ready," he said. "Ten-mile range, full battery, link's clean."

Joe nodded. "Do it."

Valery powered up the drone, the rotors humming to life on the launch pad outside the bulkhead door. Manny routed the video feed to the main screen. The camera view jolted, then steadied as the drone lifted into the fog, the Savannah shrinking below it.

"Take her forward," Joe said.

Valery eased the joystick. The screen showed the ridge first—sloping ground, dry at the top, trees clustered along the sides. No obvious washouts. No collapsed structures. Just high ground that hadn't drowned.

Past the ridge, the land dipped into a valley that was half flooded, with pockets of higher ground scattered like islands. Trees stood thin but alive. In the far distance, broken roads curled around the valley border like faded scars.

Morales leaned closer. "Pan along the front face of the ridge."

Valery adjusted the camera. The "shoreline" was no beach—just a broken band of rock ledges and gravel shelves where the drowned valley met the water. Old concrete chunks and twisted rebar showed where something man-made had once been.

"That rock shelf runs a good stretch," Manny said. "Depth drops slow. We can anchor a floating pier along there."

Joe watched the contours, thinking in grids and loads. "Any obvious slides?" he asked.

"None yet," Valery said. She flipped to thermal overlay. The view shifted into ghostly colors. "No hot spots. No machinery. No recent fires."

They pushed the drone inland. The terrain shifted from ridge slope to scattered clearings. The drone hovered over a shallow stream cutting down from the ridge and winding toward the flooded basin. It was narrow, but the water ran clear instead of stagnant.

"That's a good sign," Manny said. "We can work with that. Filters will handle the rest."

Guido tapped the map on the table. "Tree line on the ridge gives natural windbreak. Enough open ground up top to lay out tents and huts without hacking through a forest."

"Any signs of people?" Morales asked.

Valery shook her head. "Nothing. No camps. No vehicles. No trails I can see."

The overhead speaker crackled. Perez's voice came through from the reactor deck. "Ops, this is reactor. Savannah's core is steady at low output. Dave and I have her in safe mode."

Joe hit the intercom switch. "Copy, reactor. We're looking at the ridge now."

"Understood," Dave's voice followed. "Whatever you decide up there, we'll keep her humming until you say otherwise."

He turned back to the main screen. The drone circled the ridge one more time, tracing the rocky edge, the slope, the flat top.

Everyone in the room watched it in silence.

Joe straightened. "This is it," he said. "We build here."

Valery powered the drone into a holding pattern. "Then that's home."

Guido let out a low breath. "About time."

Morales nodded once. "Next step is boots on that rock."

Joe looked at each of them—Manny, Guido, Valery, Morales, Donnie—and felt the weight of the moment settle in. After over a year fighting just to stay afloat, they finally had somewhere to land.

"Gear up," he said. "We're taking a skiff in."

The skiff rocked as Manny eased it away from Savannah's hull, the rope ladder still swinging above them. Joe sat at the bow with Morales beside him. Petty Officer Craig, one of the remaining Navy men, took the second set of oars opposite Manny. Valery sat mid-boat with the drone controller in her lap, headset on, and Guido rode at the stern with a handheld radio.

The air was cool and damp. The ridge ahead grew sharper with every pull of the oars—a dark spine rising out of the flooded basin, the rock face jagged where old shoreline had collapsed.

"Water's shallow along that ledge," Manny murmured. "I'm seeing submerged rubble, not open drop. Good anchorage if we don't get stupid."

"Let's not get stupid," Joe said. "We need these skiffs working for a long time."

Morales swept the ridge with her eyes. "No movement on the rocks. No birds flushing from the trees."

Valery kept her gaze on the drone feed. The small craft hovered ahead of them, its view gliding along the ridge line like a scout. "Nothing above you," she said. "No fresh paths. No heat signatures."

The skiff bumped gently against a slanted slab of rock and fractured concrete. Manny and Craig hopped out first, boots splashing in the shallow water, and hauled the bow higher until it rested against the rough edge.

Joe stepped onto solid ground for the first time in months. It wasn't sand; it was shattered stone and leftover man-made debris—broken pilings, rusted steel, gravel, and chunks of what might once have been a shoreline road. His boots slid for a heartbeat, then held.

"Feels real," Guido said, climbing up behind him, breathing harder than he wanted anyone to notice. "Not like those drowned ports we kept pretending weren't sinking."

Valery guided the drone lower, letting it hover just above the rock shelf. "Shelf runs long enough for a full pier," she said. "Depth stays friendly for floating sections as long as we don't go wandering into the deep."

Joe looked up the slope. Grass and scrub clung stubbornly to broken soil. Trees began farther up—enough cover to break the wind, not so dense they'd have to chop their way through.

"Path first," Morales said. "No one else comes ashore until we know where it goes."

Joe nodded. "Manny, you and Morales take the left. Check for loose rock, sinkholes, anything that'll twist an ankle. Craig and Guido, you take right—clear debris, flag anything unstable. I'll take center. Valery—"

"Overwatch," she said. "If I see anything I don't like, I'll say it before you step in it."

They climbed in staggered pairs. The ground was damp but firm. Rocks shifted in places but nothing collapsed. No boot prints. No ashes. Just old storm scars and stubborn green returning in patches.

A small flock of birds burst from scrub when Morales got too close. Joe watched them arc into the sky, and something inside him loosened —life was here.

At the top, the ridge leveled into a wide spread of ground dotted with brush and scattered boulders. From here, they could see back to the ships and across the half-drowned valley inland. The elevation was good—high enough to stay dry even in another heavy rain. The wind blew steady, but not merciless.

"This works," Morales said. She turned in a slow circle. "Tents here and along that curve." She pointed to a gentle arc that stayed clear of the steep edge. "Kitchens there. Shared space close to the slope for access."

Manny dug his fingers into the soil. "Decent layer. Won't flood unless the universe decides to mess with us again. Latrine trenches will hold."

Craig scanned the tree line. "Good sight lines. Easy to run patrols without blind corners."

Guido shaded his eyes. "Enough lumber out there to build something respectable, but not so much we spend weeks clearing it."

Joe turned slowly, mapping the space—sleep lines, kitchens, med, walkways, watch posts. "We build the floating pier first," he said. "Anchor sections along that rock shelf. Once it's stable, we start moving families ashore in batches. Until then, only work crews come up."

Valery's voice crackled through the radio. "Drone confirms the ridge is isolated. No trails leading in. Nothing man-made beyond what you're standing on."

Joe nodded. "Then this ridge is phase one."

Morales adjusted her gear. "You want markers set now?"

"Yeah," Joe said. "Let's lay out the bones before the muscles show up."

They spent the next hour pacing out rows, tying cloth markers to stakes hammered into stubborn soil—future tent corners, future kitchens, future med spaces. Simple lines on open ground, but to Joe they already felt like the start of something real. Something theirs.

When they'd seen enough, Joe nodded. "That's it. We've got what we need."

"Back to the skiff?" Manny asked.

"Yeah," Joe said. "We've got two hundred fifty-seven people waiting to hear if this place can hold them."

They headed back down the slope, double-checking footing as they went. Valery brought the drone into a landing pattern, the whine fading as it touched down on the pad at the boat's bow.

Joe paused one last time before stepping into the skiff, taking in the ridge—the slope, the tree line, the wide top.

"Yeah," he murmured. "We can work with this."

The skiff ride back to Savannah felt heavier than the trip out, though nobody said it. Even Manny—usually joking—kept his eyes on the oars, matching Craig's rhythm with steady pulls. Valery replayed drone footage on her handheld screen, scrubbing through angles, marking shelf depth and potential anchor points. Morales flipped through her notebook, already sketching layouts. Guido sat uncharacteristically quiet, staring at the ridge fading behind them.

Joe didn't say much. He was thinking too far ahead.

The moment they reached the ladder, hands reached down to pull them aboard. As they climbed onto the deck, two hundred fifty-seven people waited in a tight cluster near the rails—faces worn, hopeful, starving for good news. The noise dimmed the second Joe stepped forward.

"Listen up," he called out. His voice carried across the deck like a promise. "We set foot on the ridge. It's solid. High ground. Dry. Enough space for a camp. We're building our first settlement here."

The reaction wasn't loud—no cheering, no shouting. Just a deep ripple of sound. A woman in the front covered her face with her hands.

Someone else let out a breath that sounded like a sob. A few kids grabbed each other, wide-eyed.

"We're not rushing this," Joe continued. "We need a floating pier built first. That'll take days—maybe a week. Once it's secure, we start moving families ashore in groups. You'll all have time to pack and prep. Nobody's going over that ridge until it's safe."

People nodded. They understood the difference between hope and haste.

Mom pushed her way through the crowd, wiping flour on her apron. She wrapped Joe in a tight hug without a word. He leaned into it for a second before she pulled back and looked him over.

"You're filthy," she said.

"So are you," Joe answered.

She swatted his arm. "Good. Means you're working."

Donnie arrived next, Charlie pressed against his side with Lenny prancing in front of him. Donnie's eyes went straight to Joe.

"You found us something solid," Donnie said.

Joe nodded. "Solid enough to build on. We'll need a full pier and a ton of labor to get everyone up that slope."

"Good," Donnie said, rolling his shoulders like he'd been waiting for something heavy to lift. "I was starting to lose my mind staying on this ship."

Charlie lifted his little hammer. "I'm ready to build huts!"

Joe crouched. "Oh yeah? What's your job?"

"Stakes," Charlie said proudly. "Dad says I'm really good at stakes."

Lenny barked once, as if confirming it.

"That's because you are," Joe said. "You and Lenny stay with dad when the time comes. No wandering."

Charlie nodded seriously. Lenny nodded too, or maybe he just wiggled.

Morales stepped up beside Joe, notebook still open. "We need to assign pier teams now," she said. "People are restless. We keep structure or this whole deck gets chaotic."

Joe nodded and turned to the small cluster of volunteers hovering nearby—deckhands, mechanics, carpenters, all practically vibrating with purpose. "Here's how we start," Joe said. "Manny runs float construction—he knows the structural limits. Donnie handles framing. Guido handles rope and hardware. Craig helps coordinate with the tugs. Morales and I will run security and overall flow."

Valery lifted her tablet. "I'll keep the drone in the air every daylight hour the battery allows. Nobody comes within half a mile of that ridge without us knowing."

Joe nodded. "Good. This is day one. We don't screw up day one."

He turned, raising his voice to the whole deck. "Pier crews—meet at the foredeck. Everyone else—eat, rest, stay out of the way of the cranes unless you're assigned. And remember: we do this right or we don't do it at all."

People began moving, not frantically but with purpose. After so long drifting, structure felt like oxygen.

Donnie stayed behind a moment. "You still need a name for the place," he said.

Joe looked toward the ridge—a green-brown rise at the edge of a ruined world.

"Ridge Camp," he said. "At least until something better sticks."

Mom took his arm and squeezed. "Ridge Camp," she said softly. "Sounds like a beginning."

The reactor deck felt like stepping into the belly of a sleeping giant—warm metal, steady vibration, the low hum of machinery doing exactly what it was born to do. Joe ducked under a pipe and stepped onto the grated platform where Dave and Perez stood over the main control panel. The four remaining techs moved in quiet, careful rhythm behind them.

Dave glanced up. "You touch dirt today?"

Joe smirked. "Rock. But yeah."

Perez wiped his hands on a rag. "So what's the plan for the core?"

Joe braced his hands on the rail and kept his voice steady. "We keep the reactor on low output. Enough to run the ship's systems, comms, pumps, and whatever lines we tie to shore while we build Ridge

Camp. No heavy loads. No experiments. Once we've got solar up and stable, we start planning a long-term shutdown."

Dave nodded. "Good. She'll behave if we treat her like a cranky grandmother."

Perez added, "Four of us stay here full time. No rotations. We babysit the core until you tell us otherwise."

Joe met both their eyes. "You've kept us alive this long. I trust you. But if anything smells wrong, you tell me before it becomes a problem."

"You'll know before it sneezes," Perez said.

"That's what scares me," Joe muttered.

Dave cracked a grin. "Go build your village. We'll keep the heart beating."

Joe headed back up the ladder, the hum of the reactor lingering in his bones.

When he reached the lower deck again, the last of the sunlight filtered through the open bay door. The smell of brine and steel drifted in with the evening air. For a moment he leaned on the bulkhead and closed his eyes.

He didn't see the ridge.

He saw blue water and the Valentina.

The memory hit so clean it almost startled him—the old 50-foot boat rocking under his boots, salt spray stinging his face, Valery shouting over the wind as she tried to keep her playlist from skipping.

"Joe!" she'd called out, laughing as her hair stuck to her cheeks. "If this engine coughs again, it's your turn to sing to it!"

"You don't want that," he'd yelled back. "The ocean will spit us out."

"Then sing anyway!" she'd insisted. "Everything runs better when you talk to it!"

Isaac had been eating chips straight from the bag, Manny tinkering with a broken radio, the sun going down in streaks of gold across the water. Joe had sat on the edge of the deck, arm draped across the rail, watching Valery steer their creaky, stubborn boat toward nowhere special.

She'd glanced back at him then—wide grin, face sunburned, eyes bright.

"We're alive," she'd said. "That's the whole point."

He hadn't realized until later how right she was.

Joe opened his eyes again. The ocean around Savannah wasn't that ocean. And the world they were building now wasn't that world. But the people? The stubbornness? The grit?

Same damn crew.

He headed for the flight deck.

The helicopter waited there—patched and ugly, but reliable. Morales checked the rotors. Manny double-checked the gauges. Valery stood with her headset around her neck, tablet glowing in her hands.

"You're all fueled and balanced," she said. "Route is ready."

Joe nodded. "Tomorrow we take a closer look at the ridge from the air. Not just the path—we need alternate routes, potential hazards, drainage lines. The full picture."

Morales strapped in. "Then we should move at dawn."

"Dawn works," Joe said.

Behind the safety line, Mom, Donnie, Charlie, and Lenny waited. Charlie waved the little hammer he never dropped. Lenny barked once, impatient.

Joe tapped two fingers to his helmet. Charlie mirrored it, serious as a soldier.

Joe strapped in beside the open door. Manny sat behind him with a crate of emergency gear. Valery stepped back, one hand shading her eyes.

The blades spun, chopping the air.

Savannah rocked gently under them.

Morales lifted the chopper, and the deck fell away.

The ridge rose beneath them—broad, solid, waiting.

A place to begin.

A place to become something more.

The helicopter skimmed low along the ridge, giving Joe a clean look at everything they'd only seen from the ground. In the early light, the rock shelf glimmered where the water met the broken shore—solid, jagged, uneven, but workable. The ridge top stretched wider than he remembered, a natural spine with enough room to build rows of tents, huts, and all the bones of a real settlement.

Morales shouted over the engine, "North side looks stable. Southern edge drops into the basin. We'll mark both."

Joe nodded, gripping the strap beside the open door. "We'll need access routes, watch posts, and a clean line for hauling supplies."

Manny leaned forward between the seats. "We also need to track drainage. That creek cuts straight from mid-ridge down into the basin. Good for filtering water, bad if we build too close to the low spots."

"Then we stay high," Joe said. "Dry land is the whole point."

They looped twice more before Morales banked toward Savannah. As the ship came into view—quiet, steady, waiting—Joe felt the shift inside him. A year ago, they had nothing but moving water and luck. Now they had dirt. Now they had direction.

The helicopter thudded onto the deck. The rotors spun down, wind flattening stray papers and lifting dust along the deck.

Morales shut off power. Manny unclipped. Joe stepped out into the warm light and the waiting crew.

Guido jogged up first. "How's it look? Still there? Didn't sink while we were gone?"

"Still there," Joe said. "We got work to do."

Valery met him second. "I mapped your flight path. I marked all elevation shifts and erosion points. And the creek. And the four best tent rows. And the windbreak. And—"

Joe raised a hand. "Good. We need all of it."

Mom stepped in next. She didn't hug him this time—she just put a hand on his cheek, searching his eyes. "You ready to put 257 people on that ridge?"

Joe answered honestly. "Yeah. For the first time… I think we can."

Donnie appeared behind her, sleeves rolled up, sawdust in his hair even though he hadn't even touched a tree yet. "You say the word, and we start building."

"Not yet," Joe said. "We build the pier first."

Morales blew her whistle, drawing the teams close. "Pier crew, gather in. We're laying out the first section today."

The deck snapped into motion. Tools clanged. Voices rose. Manny and Guido hauled barrels and pontoons to the forward staging line. Petty Officer Craig called out spacing measurements. Volunteers lined up to move lumber and rope.

Joe stepped onto a crate and spoke over them all.

"Listen up. We do this right. No shortcuts. No guessing. The pier is our lifeline until Ridge Camp stands on its own. When this is built, we start landing people in waves. Until then, nobody goes ashore except assigned teams."

Heads nodded. People listened. No panic. No arguments. They trusted him—and that weight settled deep in his chest.

"Tomorrow," he said, "we start our first real day on land."

Work started hard.

Barrels were strapped under deck frames. Bolts hammered. Lines tied. Every section was built twice as tough as needed because failure wasn't an option—not with everything depending on it.

By midday, the first pontoon assembly was ready.

"Lower," Manny called.

The crane swung the massive section out over the water. Ropes tightened. Feet braced. Guiding hands eased it down until it hit the surface with a heavy splash and bobbed in place.

Guido whooped. "It floats! Not pretty, but it floats!"

"Good," Joe said. "Pretty comes later."

They pushed it toward the ridge using two skiffs. Morales and Craig handled the anchor lines, hammering bolts into the rock shelf until their arms shook.

"First section secured!" Morales shouted.

Cheers erupted across the water.

Even Savannah's deck crew applauded.

By sundown, section two was nearly built. By day two, four sections. By day five, the pier stretched halfway to the ridge. On day seven, the full walkway stood—floating, sturdy, imperfect, but theirs.

Joe walked the length of it at sunrise with Manny, Morales, Valery, Guido, Donnie, and Charlie trotting behind with Lenny at his heels. The sections swayed gently under their feet.

At the rock shelf, Joe stopped. Everyone gathered behind him.

"This is the line," he said. "From ship to land. From surviving to rebuilding."

Mom's voice drifted down the pier behind them. "Breakfast is up here waiting. Don't make me feed the gulls instead."

Laughter rolled across the group.

Joe looked at his people—his crew, his family, his hope.

"Let's go build home."

They stepped forward, boots on rock, hearts steady.

Ridge Camp had begun.

CHAPTER 12
FIRST FOOTING

The ridge looked different after a week of work. What once felt like a question mark now looked like a construction zone carved into the shoulder of the hills. The rocky slope that first scraped the skiffs now held rope lines, carved paths, and flat sections widened by pickaxes and sweat. Tents dotted the higher ground like beige brushstrokes. Smoke curled from the first cooking frames. People moved with purpose instead of nerves. Ridge Camp was starting to look like a place that might last.

Joe stood halfway up the ridge path, watching the morning fog lift in pale sheets. Morales climbed beside him, boots crunching gravel. "Skiff teams are ready," she said. "Load lists finalized. Pier's holding. Today we start moving the big equipment."

"Good," Joe said. "We get the bulk stuff ashore now, we won't spend the next month fighting tides."

Morales smirked. "There are no tides here."

"Feels like there are," he said. "Everything moves whether we want it to or not."

Below them, the floating pier stretched out from the rock shelf into the deeper water where the ships sat anchored. Tugboats were positioned at both ends, ready to steady each load. The water wasn't a beach—just a narrow, angled ledge of stone and hard-packed mud where the ridge met the lake. Not friendly terrain, but workable.

Donnie stood near the waterline with a team of riggers, tightening the pulley system they'd bolted into the ridge face. "Keep that tension even!" he called. "If it slips, we lose the whole load!"

Charlie stayed beside him with his little clipboard, marking each item as if the survival of the camp depended on his handwriting. Lenny sat at his feet, tail sweeping the dirt.

Up the ridge, Ashley and Manny configured the cable anchors that would haul supply crates straight up the slope instead of dragging them by hand. Manny tested one of the grip clamps. "If this holds," he said, "we'll move twice as many loads in half the time."

Ashley tugged the line sharply. "Feels solid. Just don't get cocky."

"Me? Cocky?" he said. "Never."

"Uh-huh," she said.

Valery's voice crackled from Joe's radio. "Comms active. Drone uplink ready. I'll monitor movement on both sides of the ridge."

"Copy," Joe said.

Guido ran up beside them, breathing hard. "Fuel drums secured. Water filters staged. Where do you want the generator frames?"

"Top plateau," Joe said. "Flat ground behind the tent rows. Keep the heavy stuff on the ridge, not the slope."

Guido nodded and jogged back down.

Joe turned to Morales. "It's coming together."

"Yeah," she said. "Feels like a real camp now."

People were everywhere—hauling crates, digging drainage trenches, carrying tools, stacking lumber. Voices carried up the ridge, steady and confident. No one asked if they'd survive. They acted like they already knew the answer.

Joe watched the first skiff approach the pier, tugboats holding it steady as the crane on Savannah lowered a bundled set of tool racks. A week ago, people shook when they stepped off the ship. Today, they were racing the clock to build a home.

This was the real beginning.

The first crate of the day rose from the skiff like a stubborn beast, swaying on the pulley line as Donnie and Manny kept tension even from opposite ends. The crate swung against the rock face, thudding once before Manny steadied it with a grunt.

"Ease it! Not drag it!" Donnie barked.

"I am easing it!" Manny shot back. "The ridge hates me. That's what this is."

Ashley leaned into the line beside him, boots braced on the uneven ground. "Everything hates you. Pull."

Together they guided the heavy supply crate upward until it reached the carved platform where Guido and two volunteers waited. Guido clipped the rope off, hands moving fast. "Got it! Send the next one!"

"Next one's heavier," Manny warned.

"Then send Morales up here to lift it with her bare hands," Guido said.

Morales walked by just in time to hear him. "If I have to lift anything with my bare hands, I'm throwing you off the ridge."

Guido saluted sharply. "Yes, ma'am."

She kept walking, hiding a grin.

Joe reached the top platform and took in the view. The tents sprawled across the ridge now—thirty on the main rise, ten more scattered along safer slopes. Smoke from the kitchen tent curled upward in thin blue ribbons. The med tent stood near the highest point where the wind kept the air clear. Solar panel frames lined the flatter section, ready for installation. Everything was still rough—raw wood, mud, ropes, sweat—but it was organized.

"Looks good," Joe said.

Morales nodded. "People work faster when they know it's home."

On the lower path, Mom helped organize the incoming medical crates with two corpsmen. "Set the trauma kits in the shade," she said. "We'll unpack those first. And don't leave the antiseptic in the sun unless you want it to explode."

Charlie came running up the ridge with Lenny at his heels. "Dad! They need more stakes!"

Donnie wiped his forehead. "Bring me the smaller ones, not the long ones. And slow down before you break your ankle."

"Yes, sir!" Charlie sprinted off again, Lenny barking as if supervising.

Valery's voice returned over the radio. "Heads up—thermal shows two deer about a mile north. Good sign for hunting later."

"That's food," Morales said.

"That's hope," Joe answered.

Manny's crew began assembling the solar panels near the tree line where morning light hit strongest. Ashley knelt beside the first frame, tightening bolts. "If one of these cracks during install," she warned, "I'm blaming Manny."

Manny held his hands up. "Anything that breaks is definitely Guido's fault."

Guido froze mid-lift. "Excuse me? I'm carrying the weight of civilization right now!"

"You're carrying a solar panel," Ashley said. "Don't drop civilization."

Joe shook his head, amused. "Keep going. Once the panels are in, we test the first line of lights tonight."

The work didn't stop. It didn't even slow. Every crew moved as if the world depended on it—which, in truth, it did. For the first time in a long time, their survival didn't hinge on running. It hinged on building.

And they were ready for that.

By midmorning, Ridge Camp was a living engine. The rope haul line up the rock face moved constantly—gear up, instructions down, people shouting to keep rhythm. The ground was still uneven in places, muddy in others, but paths began to form where countless boots had passed.

Morales kept the pace organized. "Crate lane on the left! Personnel lane on the right! Nobody walks in the middle unless you want to get crushed!"

"Yes, ma'am!" came a scattered chorus of voices.

Joe worked beside Manny and two sailors, hauling a long crate that held water filters. The crate slid over the rocks, rough but steady.

"On three," Manny said. "One, two, three—lift!"

They heaved it onto a flat slab where Donnie and his carpentry team were staging supplies.

"Filters ready," Joe said.

Donnie nodded. "Good. The west side needs the water line set today. I've got Charlie digging the trench."

Joe raised a brow. "You let him dig a trench?"

Donnie smirked. "The kid's built like a shovel. Might as well put it to use."

Charlie wasn't far—covered in dirt, swinging a wooden-handled trenching tool half his size. Lenny helped by barking at the dirt each time Charlie moved a stone.

"Perfect," Joe muttered with a tired smile.

Farther upslope, Ashley stood on a crate tightening wiring along the first solar frame. Manny held the panel steady, muttering about his boot sinking into mud again.

"You're standing wrong," Ashley said.

"I'm standing like a human being," Manny shot back. "The mud is standing wrong."

"Lift your heel."

Manny lifted it—and the boot suctioned free with a wet slurp.

Ashley grinned. "See? You're welcome."

Manny pointed a finger. "If I fall and die, you're giving my eulogy."

"Gladly," she said.

Guido walked past them shaking his head. "Stop flirting and finish the panel."

"We are not flirting," both said at once.

From the ridge crest, Valery called down, "Power grid anchor line's ready to run!"

Joe climbed up to her. A spool of heavy cable lay at her feet, the end clipped neatly to a metal grounding rod.

"That'll reach the first panel bank," she said. "And if the charge holds through sundown, we'll have test lights."

"Good," Joe said. "How's comms?"

She adjusted her headset. "Crystal. No interference. No outside chatter."

Near the tree line, Morales oversaw the perimeter stakes being driven in. Every ten meters a bright orange flag fluttered. She marked each location on her map with deliberate care.

"We'll set motion trip lines tomorrow," she told Joe. "Nothing fancy. Just bells and bright cloth. Basic, but effective."

"Good. We'll rotate patrols once families move in fully."

Down in the central clearing, the kitchen team lit the first stove. Steam rose as they boiled water for lunch. People stopped just to smell it—real food, not ship rations heated by engine coils.

Mom approached Joe with a tin cup. "Drink," she ordered. "Hydration is survival."

He took it gratefully. "Thanks, Mom."

"You look better," she said. "Less gray around the edges."

"Flattering," Joe muttered.

She laughed softly. "Keep going. They're watching you, you know."

He knew. Every pair of eyes in camp carried hope and fear in equal measure.

He lifted the cup toward her in thanks, then headed back to work.

Because the day wasn't slowing.

The settlement wasn't done.

And the future wasn't waiting.

By late afternoon, Ridge Camp was officially taking shape. What had started as tarps and crates looked—finally—like a place meant for living, not surviving. The tents formed neat rows along the upper ridge, anchored deep into soil that held steady even after the storms of the past. Ropes stretched tightly. Stakes held firm. Windbreak lines formed natural corridors between the tent rows.

Children ran between them like it was a playground instead of a survival camp.

Morales supervised the installation of the sanitation lines along the southern slope. She checked the trench depth with a measuring stick. "Good. Keep that angle. Rain'll run downhill instead of through camp."

The sanitation crew nodded and kept digging.

Near the center of camp, the medical tent stood fully assembled. Amy worked inside with two corpsmen, unpacking and sorting equipment. "Put the sterilization kits on the left," she said. "And stack the blankets by the cots. We'll have people coming in tired and scraped up all week."

The corpsman nodded. "Yes, ma'am."

"And somebody bring me the water drum," she added. "We're not reusing anything unwashed."

Charlie, coated head to toe in dust, dragged the drum to her with Lenny trotting alongside him like a tiny security guard.

"Thank you, sweetheart," Amy said.

Charlie puffed his chest and ran back to Donnie's crew.

Donnie's carpentry team was already building early frame pieces for the first permanent huts. Nothing fancy—just simple triangular wall braces, stacked lumber, and the start of roof trusses.

Manny hammered a beam into place and stepped back. "When's the last time any of us built something that wasn't falling apart?"

Guido answered, "Mississippi. And that shed collapsed on me."

Joe walked up the ridge toward them. "How's the lumber holding?"

Donnie wiped his forehead. "Better than the junk we salvaged last year. Good quality. Dry. Solid. We can build small huts within two weeks if we keep this pace."

Joe nodded. "Good. Start with the big families. Give them priority."

"Already got a list," Donnie said. "Charlie wrote it out. Neater handwriting than mine."

Guido pointed a thumb toward Charlie. "Kid's got leadership potential."

Joe smiled faintly. "Maybe he gets it from his old man."

Donnie grunted. "Let's hope he gets your stubbornness instead."

Up on the ridge crest, Valery stood with her tablet, tracking the first power charge readings from the solar array. "We're getting low voltage," she called down. "Enough for lanterns and comms. Not enough for long-term use yet."

Joe climbed up beside her. "Panels need another day of full sun."

"Yeah," she said. "But the grid's holding steady. Tomorrow we can finish the second row of panels."

Morales joined them, brushing dust off her boots. "Perimeter stakes are done. Tomorrow we set trip lines. I want patrols split along northeast and southwest arcs by dawn."

"Good," Joe said.

Valery looked out over the valley. "No movement. No smoke. No heat signatures. It's quiet."

"Too quiet?" Morales asked.

Valery shrugged. "Quiet for now. But we stay ready."

Joe scanned the ridge line, the newly raised tents, the people working hard without complaint, the families settling in. The weight on his shoulders didn't disappear—but it changed. Became focused. Became purpose instead of dread.

"We're doing it," he said quietly.

Morales nodded. "Yeah. We are."

Then, from the forest below, someone called out: "Dinner's ready!"

The camp erupted—real laughter, tired but bright. For the first time since the Firestorm, the sound felt like it belonged.

As the last of the light slipped behind the trees, Ridge Camp settled into a kind of tired, satisfied hum. Lanterns glowed along the tent rows—soft, steady light powered by the small charge the solar panels had managed to pull. It wasn't much, but after two years of darkness at sea, it felt like luxury.

Joe stood at the ridge edge again, boots planted firmly in dirt he still couldn't quite believe was theirs. The valley below lay quiet—no movement, no threats, just the faint shimmer of water and the skeletal remains of the old world.

Morales walked up beside him, her report clipboard tucked under one arm. "Perimeter's locked for the night. No one wandering past the markers. Patrol teams know their shifts."

"Good," Joe said.

"You should rest," she added.

"Later."

She didn't argue—just nodded once and headed back toward camp.

Valery approached next, wiping her hands on a rag after testing the comm relay. "Solar's giving enough for the overnight systems," she said. "Not full power, but enough to keep us connected."

"Anything on the lower bands?" Joe asked.

"Nothing," she said. "Not even static worth worrying about. Looks like the valley wants us to have one quiet night."

Joe didn't fully believe it, but he let the quiet be what it was.

Down below, the cooking crew ladled hot stew into tin bowls. Workers lined up with weary smiles. Kids ran between tents with bread chunks and cracked jokes. Amy sat on a crate outside the med tent, handing out cups of water to tired hands. Donnie joined her with Charlie and Lenny, both of them looking exhausted but proud.

Charlie plopped down beside his mother with dramatic flair. "Dad says I'm the best helper on the ridge."

"Of course you are," Amy said, kissing his hair.

Lenny curled against Charlie's side, giving a sleepy little huff.

On the far side of camp, Manny and Guido still worked on stabilizing the tool shed frame. Ashley held the lantern for them, teasing the two men as they argued over whether the beam needed shifting "one inch" or "exactly one-eighth of an inch." Donnie would settle it in the morning.

People were tired—bone tired. But they weren't defeated. They weren't drifting anymore. They were building.

Joe walked through the center of camp, nodding to crew members, checking tent lines, tightening small things that didn't need tightening just for the comfort of working with his hands. Every few steps someone stopped him:

"Chief, need extra nails?"

"Chief, panel three is charging slower—what should we do?"

"Chief, kids want to help carry water tomorrow—is that okay?"

"Chief, do we really get breakfast on land tomorrow?"

He answered each one calmly, patiently. People needed a leader they could touch, talk to, look in the eye.

By the time he reached the edge of the camp again, the stars were out—real stars, not reflections on black water. The forest whispered in the breeze. Someone had started a quiet song near the fire, and others joined softly. It wasn't celebration.

It was relief.

Valery joined Joe one more time, hands in her pockets, eyes on the half-built village.

"Feels different," she said.

"It is different," Joe replied.

"You ever think we'd see this again?" she asked.

"Every day," he said. "Even when I didn't believe it."

She gave a tired smile. "Tomorrow will be worse."

"Oh yeah," he said. "Tomorrow we actually start building a town."

"Good," she said. "People need something to belong to again."

He nodded. "And they will."

As the camp dimmed into sleep, Ridge Camp breathed for the first time—a deep, collective exhale.

Joe stood watch a little longer, letting the quiet settle around him.

Tonight, they had land. They had shelter. And they had each other.

For now, that was enough.

CHAPTER 13
RIDGE CAMP DAY ONE

By morning, the ridge looked alive. The ships sat steady in the narrow bay while small skiffs ferried supplies across the shallow water to the rocky shoreline below. The sound of engines, shouting, and laughter mixed with the cry of gulls overhead. It was organized chaos—257 people working together to unload everything that had kept them alive at sea. Steel grated, crates thudded over gravel, and someone started a rough tune on a harmonica that made the long line of workers laugh as they passed boxes hand-to-hand.

Morales stood near the shoreline directing traffic, clipboard in hand. "Medical crates go to the west section, food near those pines. Tools and power lines stay by the skiffs until we finish sorting."

On Savannah's deck, Donnie managed the crane controls, his voice crackling over the radio. "Keep those ropes tight. If that drum rolls, I'm swimming after it."

Standing knee-deep in the shallows, Guido shouted back, "You swim about as well as that drum does!"

Charlie ran between them, chalk marker in hand. "Dad, this one's fuel, right?"

"Correct," Donnie said. "Label it red. And stay out of Guido's way—he's allergic to work."

Guido winked. "Only break out in competence when you're looking, Chief."

Up the slope, Amy handed out bottles of water and helped the med crew set up their supply tent. She paused beside a family from Louisiana—Willa, her parents, two younger brothers, and their cousin—who were marking out the shape of their hut using sticks and string.

"You sure you want to build under that pine?" Amy asked.

Willa's dad nodded. "Closest thing to shade we've seen in two years."

Willa's mother added, "Feels like home already."

Morales gave them a soft smile before turning back to her board. "Storage perimeter needs a place. Joe, where do you want it?"

Joe stood beside Valery halfway up the ridge, studying the slope. "High ground. Above the flood line. Close enough for access, far enough from tents for safety. We start there."

Valery nodded. "If we grid it properly, we can run solar cables clean once Donnie's panels are charged."

Manny and Guido started hauling heavy boxes of wiring up the ridge. "You heard the boss," Guido said. "Let's make ourselves an empire of extension cords."

Manny grunted as he lifted one crate. "Careful. That's lithium, not Legos."

Morales barked, "Watch the language! We've got kids!"

The laughter that followed cut through the morning fog like sunlight.

Charlie climbed onto a boulder and looked over the emerging layout. "It's starting to look like a village."

Donnie smiled without looking up from a tape measure. "That's because it is."

By midday, the sun was warm but the pace didn't slow. Skiffs kept moving, ropes kept flying, and crates stacked up under makeshift tarps. A rhythm formed—work, lift, pass, repeat.

Morales radioed the bridge. "Captain, shore camp is stable."

Joe answered from the waterline. "Copy that. Keep them moving. We're almost through."

He turned to Valery. "Once the major offload is done, we take the helo north."

"Arnold Air Force Base," she said. "Let's find out what's left standing."

Joe nodded. "If we're going to rebuild, we need more than hope. We need a supply chain."

Engines roared again, and the survivors kept moving—257 souls building the first pieces of a new life.

The helicopter blades kicked up a storm of grit and dust as Helo One lifted from the clearing above camp. Work halted for a moment as people shaded their eyes and watched the machine rise, banking toward the ridge and then northeast toward Arnold Air Force Base. It was the first outbound helo flight from Ridge Camp—every pair of eyes followed it until it vanished beyond the tree line.

Morales stood near the radio post, headset pressed tight. "Telemetry stable. Bearing north-northeast. Thirty klicks to target." Her voice carried calm authority, grounding the crew even as excitement buzzed around her.

Joe sat in the co-pilot seat, headset snug, eyes fixed on the horizon. "Valery, how's she handling?"

Valery adjusted the collective. "Like she's been asleep a year and just woke up cranky. But she'll fly."

Manny and Guido sat behind them, surrounded by strapped-down gear—fuel cans, med kit, rifles, and open space for whatever they recovered. Guido looked out the window at the green-brown sprawl below. "Hard to believe this was Tennessee farmland."

"Looks more like a swamp bottom," Manny said. "But the ridge saved us."

Below, drowned forests spread like twisted fingers poking through standing water. Broken highways cut jagged scars through half-submerged terrain. The world they once knew felt distant—tilted, warped, and drowned, but still somehow recognizable.

Joe studied the taped map on his knee. "Arnold AFB should be ten miles ahead if the old satellite data is still worth a damn."

Valery nodded toward the glass. "There. The ridge line. That's the perimeter fencing—what's left of it."

As they descended, the base took shape through the haze: a cracked runway stitched with weeds, hangars slumped but standing, metal glinting from scattered supply pallets. No smoke. No tracks. No movement.

Valery eased the helo down beside the old control tower. "We're down," she said. "Welcome to whatever's left of the Air Force."

Guido unbuckled. "Looks like civilization died and never bothered to clean up."

Joe pushed open the hatch and stepped onto the dry, ripped concrete. "Then we figure out what the Commander left behind."

Inside the first hangar, dust coated everything like a burial shroud. But the pallets—rows upon rows—were intact.

Manny whistled. "MREs. Thousands."

Guido cracked open another crate. "Fuel cells. New and sealed."

Valery brushed dust off stacked water drums. "All untouched. Nobody's been here."

Joe spotted a yellowed sheet pinned beneath a wrench on an abandoned desk. He unfolded it carefully.

"For survivors. Use it well. Keep moving north. — Commander R. Holt, USAF."

The words hung in the dead air like a ghost's whisper.

"He knew this would happen," Manny said softly.

Joe folded the note and tucked it into his vest. "He knew someone would survive."

They split through the compound. Valery cataloged crates of ammunition and first aid supplies. Manny located sealed seed vaults—beans, corn, barley, wheat. Guido found a generator bay wired to portable solar arrays. "Donnie's gonna lose his mind," he muttered.

They loaded as much as Helo One could carry—fuel, medicine, MREs, and a crate of seeds.

Joe looked back at the vast stacks left behind. "We'll send crews tomorrow. This could feed Ridge Camp for a year."

As they lifted off, sunlight glinted off the pallets like a silent promise.

Valery looked down. "Feels strange," she said. "Finding hope in ruins."

Joe replied quietly, "That's where it's always been."

By the time Helo One appeared as a dark speck over the ridge, the entire camp gathered near the landing zone. Someone shouted, "There they are!" and a wave of relief swept across Ridge Camp. People dropped tools, ran from tent frames, and climbed boulders for a better view. After years of drifting and days of heavy work, hope came thundering back on rotor blades.

The helicopter descended in a swirl of dust and grit. Morales stepped forward, arm raised. "Clear the landing zone! Nobody rush the pad until rotors are at full stop!"

The helo settled hard, skids digging into the rocky dirt. As the rotors slowed, silence shifted into anticipation, then into cheering the moment Joe stepped out.

His jacket was streaked with grease and dust. Valery followed with a tired smile. Manny hopped down next, waving to the crowd. Guido jumped last, holding a crate over his head like it was treasure.

"Somebody call the kitchen!" Guido yelled. "Dinner's finally exciting!"

Morales met Joe halfway, eyebrows raised. "You find anything?"

Joe set the small crate at his feet. "Better. Enough to feed us for a year."

Gasps and murmurs rippled through the crowd.

Valery lifted her voice. "Arnold Air Force Base is intact. MRE pallets—thousands. Water drums. Fuel. Medicine. Seeds. All untouched. All dry."

The camp erupted—cheers, clapping, laughter, people hugging strangers and crying. After everything they had survived, after storms, starvation scares, and long nights staring into dark water—this felt like a miracle.

Morales raised her hand. "Quiet for a moment!" As the noise settled, Joe pulled the yellowed paper from his jacket.

"We also found this," he said.

People leaned forward as he read.

"For survivors. Use it well. Keep moving north. — Commander N. Holt, United States Air Force."

No one spoke. Even the wind seemed to pause.

Finally Donnie stepped out from the crowd. "Then we don't waste his sacrifice."

Morales nodded sharply. "We've got work. Joe, get your crew unloading and inventorying. Dad—secure those fuel drums near the ridge boulders and keep them covered. Manny, Guido, form a chain line for the med crates."

They didn't need to be told twice. Volunteers poured forward. Manny and Guido handed out crates like bartenders slinging drinks. Ashley joined the med line, helping Amy stack boxes under a tarp, wiping dust from her face.

"Feels like Christmas," Amy said.

Guido barked a laugh. "A very ugly, very dusty Christmas!"

Valery climbed back into the helicopter, handing down smaller sealed packs. "Medicines first! Then food! Then fuel!"

The sun dipped lower as crates piled up. Children ran with messages between teams. The logistics crew scribbled notes, crossed out old numbers, added new ones. People worked with a joy that hadn't appeared in years.

At the top of the ridge, lanterns flicked on one by one, faint glows marking the shape of a real town.

Joe watched the last crate come off the helo. Morales stood beside him, arms crossed.

"We'll go back tomorrow?" she asked.

"Yes," Joe said. "And the next day. And the next. We're not leaving anything behind."

She nodded toward the paper in his vest. "Commander Holt would be proud."

Joe looked out over Ridge Camp—tents rising, fires crackling, families laughing for the first time in months.

"We honor him by building something that lasts."

He tucked the old note away. For the first time since the Rain of Fire, he believed it.

By sunrise, Ridge Camp looked less like a landing zone and more like a settlement finding its rhythm. Smoke from breakfast fires drifted through the trees, and the sound of tools echoing off the ridge carried a kind of music—hammers, saws, voices shouting directions, laughter breaking through exhaustion.

Donnie stood at a crude wooden table he'd built in under an hour, a map spread across it. Morales, Valery, Joe, Manny, and Guido circled around as he tapped a pencil against the sketched layout.

"Water lines run from the gravity tanks here," Donnie said. "Filters and pumps go beside the creek. It's clean runoff and elevated enough for gravity to help."

Morales nodded. "I'll assign maintenance shifts by noon."

Valery pointed to the top ridge line on the map. "The radios from Arnold go here. Best elevation, no obstruction. Once the mast is up, we'll reach at least fifty miles."

"Good," Joe said. "Set it up. If survivors are out there, we want them to find us."

Guido arrived with a box of radio parts, muttering. "You know what I love about building civilization from scratch? Not a damn thing."

Manny thumped the heavy antenna base beside him. "Keep complaining. It distracts from how slow you lift."

Guido shot back, "My lifting speed is classified."

"You mean embarrassing," Manny said.

Ashley walked by carrying rope, rolled her eyes, and kept going. "If you two are done acting twelve, we've got work."

Down below, families staked out their hut areas. The Louisiana family—Willa, her parents, her two brothers, and their cousin—had already framed their walls using bamboo poles and driftwood stakes. Willa spotted Charlie hauling rope and waved.

Amy worked with the med crew, organizing supplies into sorted crates. She directed younger volunteers as they set up cots inside the med tent. "Don't lose track of the first-aid kits," she warned. "You'll regret it the minute someone slices a thumb open."

"Yes, ma'am," the girl answered, securing a box under the tarp.

Donnie's voice crackled over the camp channel. "Water filters online! We've got flow!"

Cheers erupted across the ridge. It echoed off the tree line like the place itself approved.

Morales grinned. "One system down. Twenty to go."

Near the new radio mast, Valery connected wires while Guido tightened bolts. Manny crouched by the transmitter box, flipping switches until the unit hummed to life. The radio sputtered, then filled with crisp static—beautiful, alive, full of possibility.

"Signal's up," Valery said. "Clear band. Strong output."

Joe stepped forward and took the mic. "This is Ridge Camp calling any survivors on frequency nine-one-three. We have food, water, shelter, and medical aid. You are not alone. Repeat—you are not alone."

They waited through the hiss of open air.

No voice answered.

Morales exhaled softly. "We'll keep listening."

"Every hour on the hour," Joe said. "Round the clock."

The rest of the day surged forward with purpose. The cooking tent rose beside the fire pit. A line of garden soil was cleared and marked for the seed crates from Arnold. Children ran between work crews delivering messages and collecting scraps for makeshift toys. The air smelled of sawdust, smoke, pine, and something new—life moving forward.

Ridge Camp wasn't just surviving.

It was beginning.

By nightfall, Ridge Camp glowed like a small frontier town waking from a long sleep. Lanterns lined the walkways between tents and early hut frames. Smoke from cooking fires curled into the dark sky. The steady hum of the new generator blended with the crackle of firewood and the murmur of voices settling into something like peace.

People sat together in clusters—sharing food, trading stories, arguing about tools, laughing louder than they had in months. The smell of soup drifted through camp, mixed with pine sap and warm dirt. It felt real. Earned. Alive.

Manny tightened the final bolt on the generator housing and wiped his forehead with a rag. "That's it," he said. "Chief better kiss this thing goodnight. I practically hand-stitched it together."

Ashley handed him a water bottle. "Funny. You said that five times today."

He grinned. "Yeah, but now I mean it."

She sat on the workbench beside him, watching the lantern glow spread across camp. "It's weird seeing lights that aren't from the ship or emergency lamps."

"We built this," Manny said. "All of us."

Ashley twirled the little bolt charm on her bracelet—her nervous habit. "Didn't think I'd ever fit in anywhere again. Not after Mercy Two."

Manny didn't push. He let the quiet sit between them, warm and heavy.

"You do fit," he said finally. "You've got the skills. And the guts. Nobody here cares about the past—only whether you can help keep the place running."

She smiled softly. "Guess I'm lucky for that."

He tapped her elbow gently. "You're here because you earned it."

Ashley breathed in the ridge air—pine, smoke, dirt, something almost sweet. "My dad used to fix things for fun. Radios, tractor engines… anything with gears. I think he'd like what we're building."

Manny studied her hands, steady even after a full day's work. "I know he'd be proud of you."

"You think so?"

"I know so," he said.

They both looked toward the center of camp. Joe stood alone on the ridge rise, staring into the dark forest. His silhouette was still, but his posture held weight—responsibility, fear, stubborn hope. Morales walked the perimeter with two guards, checking the new stake lights. Valery fine-tuned the receiver beneath the radio mast, her headset glowing green in the dark like a beacon.

Down near the water rig, Donnie and Guido were still arguing over hose fittings.

"They're backward!" Guido insisted.

"They're labeled!" Donnie shot back.

Charlie sat beside them holding Lenny, who watched the whole argument like a referee waiting to bark a ruling.

Ashley chuckled softly. "They'll never change."

"Good," Manny said. "Those two arguing means camp's running right."

The radio crackled faintly in the distance—just static, but everybody paused for a heartbeat, listening. Hope made people freeze like that now. Waiting for any scrap of a voice.

No reply came.

Light conversation slowly picked up again.

Ashley looked back at Manny. "You think anyone's out there?"

"I hope so," he said. "But even if there isn't... we've got enough right here to start over."

She nodded. "That's enough for me."

They sat quietly, watching the lanterns flicker as Ridge Camp settled for the night. A child laughed somewhere near the fire pits. Someone played a soft tune on a homemade flute. Wind pushed gently through the trees above, stirring the leaves into a whisper.

The world was still broken.

But it felt like it wanted to heal.

CHAPTER 14
THE GUARD

Halcyon drifted at anchor offshore, a dark shape pressed into the morning fog. The submarine had been mostly stripped—engines silent, decks cold, hatches sealed except for the main access. A three-person maintenance crew rotated shifts to check systems and pull anything still worth the sweat.

This run was supposed to be simple.

It wasn't.

The moment Guido stepped through the hatch, he felt it. The air was still—too still. The faint background hum of the auxiliary batteries was gone. Something hot and burned rode in the metal smell.

"Lights," he said.

He flipped his headlamp on. The beam cut through dust, catching floating motes in the stale air.

"You hear that?" one of the men behind him asked.

"No," Guido said. "That's the problem."

They worked their way aft. The corridors were quiet except for the tick and creak of cooling metal. The brig door stood half open, one hinge bent, lock plate scarred.

Guido's stomach dropped. "Ah, hell."

The bunk inside was empty. A uniform lay folded on the mattress, neat as a funeral offering. The locker next to it had been pried open.

One rifle missing. Rations gone. A portable radio gone. The log board on the wall was blank for two days.

"He's gone," one of the crew whispered.

"Yeah," Guido said. "And he took lunch."

He swept his lamp across the compartment again. The second bunk was stripped bare. The guard's locker beside it hung open, rack light dead. A crate under the bench had a torn inventory tag; the serial number matched one of the "missing at sea" entries from months back. The duty sheet on a clipboard showed smeared initials over two nights.

"Private Jensen was last on watch," Guido said, voice flat. "He's not on today's roster. Harrow didn't walk alone."

He flipped the supply ledger pages; a couple of lines were scraped and re-entered in the same stiff block letters they'd flagged during the earlier thefts.

"Same handwriting games as the ration losses offshore," he muttered. "This wasn't a panic run. It's a plan."

They searched the sub top to bottom, but the story didn't change. Captain Harrow—the ex-officer turned guard they'd locked away—had vanished sometime in the last forty-eight hours. The aft hatch showed fresh scuffs on the ladder, one missing life raft, one empty fuel can, survival kit compartment open.

Guido keyed his handheld. "Morales, this is Halcyon team. We've got a problem."

Her voice came back steady but sharp. "Define problem."

"Your favorite ex-officer took a walk. Harrow's gone."

Silence. Then Morales: "You sure?"

"Brig's open. Gun, rations, and a comm set are missing. Jensen, too." Guido glanced at the stripped guard bunk. "Ledger's doctored. Same pattern as the thefts at sea."

"Copy," Morales said. "Return immediately. No chatter on open bands."

"Roger." Guido lowered the radio. "Pack it up. We're not sight-seeing today."

They locked down the sub as far as they could and launched their skiff, the silence of the water pressing in around them.

The fog thinned as they neared shore. Ridge Camp was already stirring—smoke from breakfast fires rising, voices threading through the trees, the distant clang of metal on metal.

Morales stood at the rock shelf when the skiff nudged in, boots planted, jaw set. Joe was beside her, arms folded, eyes on the incoming boat.

Guido jumped off and handed her the report sheet. "He's gone, Commander. Took food, a rifle, and a comm set. Jensen's missing, too."

Morales read fast, her expression hardening with each line.

Joe looked past her toward the faint gray smudge of Halcyon offshore. "How long?" he asked.

"Two days, maybe three," Guido said. "Guard logs stop after that. We found a crate tag that matches the old missing serials. Ledger lines scratched and re-written. Same pattern as the thefts."

Joe exhaled through his teeth. "We'll sweep the coast and check for tracks."

Amy climbed up the rock shelf, wiping her hands on a towel, eyes on Joe's face. "How bad is it?" she asked quietly.

"Bad enough," he said. "But we caught it now."

She searched his eyes, then gave a short nod. "Then we deal with it," she said, squeezing his arm once before stepping back beside Donnie.

Morales folded the report and tucked it into her jacket. "This stays tight for now," she said. "I don't want panic."

Guido hesitated. "What about Ashley?"

"Not yet," Joe said. "She's finally breathing easy. We don't take that away until we have to."

Morales's gaze stayed on the water. "We find him before he finds us."

The briefing tent was quiet except for the distant hum of the generator and the occasional clink of tools from camp. Joe stood at the center table, staring down at the report Guido had written. The map of the shoreline lay open beside it, marked in grease pencil.

Morales leaned against a support pole, arms crossed. "We should've kept him sedated," she said flatly.

"We didn't have the meds or the right," Joe replied. "He was still a man, not an animal."

"A man with a record," she said. "And now he's out there with a rifle and someone who thinks he's worth following."

Joe rubbed the back of his neck. "We don't know how far he got. Maybe he drowned. Maybe he's still close."

Guido ducked into the tent, wiping his hands on a rag. "If he's smart, he's following the shoreline. Trees start about three klicks east—enough cover to move unseen. He'll avoid open ground."

"He's smart," Morales said. "And desperate."

"We've got two tugs fueled," Guido added. "I can run a quiet sweep around the ridge before sunset. If he's nearby, we'll see tracks or smoke."

"Make it quiet," Joe said. "No engines past the inlet. Use paddles once you hit shallow water. Last thing we do is advertise we're looking."

"Understood."

Morales walked around the table and tapped the map. "He can't live off the land for long," she said. "He's not a hunter. The rifle's for leverage, not meat. He'll look for someone to lean on."

"Raiders?" Guido asked.

"If any are still out there," Morales said.

"That's what I'm worried about," Joe said.

He fell silent for a moment, thinking, then added, "We don't tell Ashley yet."

Guido frowned. "You sure? She deserves the truth."

"And she'll get it," Joe said. "But not the same day we find out he's armed and loose. She's finally sleeping. We buy her a little more time."

Morales took a long breath through her nose, then nodded once. "All right. I'll double the perimeter watch tonight. No one leaves camp after dark without clearance. Guido, I want trip alarms on both northern approaches. Cans, wire, whatever you've got."

"I can handle that," he said. "I'll even make it look pretty."

Donnie stepped in from outside, voice low but steady. "We should bring all the fuel drums in under guard. If he's still thinking like an officer, he'll target our power first."

"Do it," Joe said. "Move the drums up near the rock face and fence the generator."

Morales already had her radio in hand. "I'll have Valery monitor the bands for anything weird. If he uses that comm unit, we'll hear it eventually."

Joe stared out through the tent flap at the ridge—rows of tents, half-built huts, smoke drifting. "We'll find him," he said quietly. "Or he'll find us."

Morales met his eyes. "Let's make sure it's on our terms."

Outside, the camp hummed with the sound of work and small talk. Nobody beyond that tent had any idea what was coming.

By evening, the ridge turned dim and watchful. The wind came off the flooded basin in cool bursts, tugging at canvas walls, rattling Guido's new tripwire cans with a soft, nervous clink. Campfires burned low. The night watch took their posts.

Inside the comm tent, Valery sat at the radio console with a mug of black coffee cooling by her elbow. The radio hissed and popped—white-noise breathing that filled the space.

She turned a dial a fraction. Static thickened, then thinned. At exactly 22:00, something slipped through.

Three quick bursts of tone. Pause. Then three more.

Valery stilled. "That's not weather," she murmured.

Morales stepped inside, flashlight tucked under one arm. "You've been at it for hours," she said. "Get some rest before you fall over."

"Wait," Valery said, holding up a finger. "Listen."

Morales tilted her head.

The pattern came again—three short bursts, a pause, and a longer one after.

"Morse?" Morales asked.

"Too clean," Valery said. "Too regular. Feels like an encrypted beacon. Same neighborhood of frequency the Air Force used pre-Fire, but the spacing's wrong."

"Timeline?" Morales asked.

"Started just before twenty-two hundred," Valery said. "Repeats every seven minutes. Exactly."

Morales moved closer. "Record it."

"Already am." Valery tapped the recorder. "If it's someone out there, they're disciplined. This isn't somebody leaning on a button by accident."

The tent flap opened again and Joe slipped in, shoulders squared, expression sharper than his tired eyes. "You called it, Val?"

"Yeah." She kept her eyes on the dials. "Structured signal. Repeating sequence. Not ours."

The pattern played again—three short bursts, pause, one long.

Joe froze. "I know that rhythm."

Morales turned. "What do you mean you know it?"

"It's Harrow's old unit tag," Joe said quietly. "He used that sequence during field ops to mark his teams. Raines hated it."

Morales swore under her breath. "He's out there. And he's talking."

"Or answering," Valery said. "Signal strength is low but steady. It's bouncing off the ridge. He's close."

Joe leaned over her shoulder, watching the readout. "Can you triangulate?"

"Not tonight," she said. "I'd need another mast or a mobile set. Right now all I can tell you is it's not coming from across the ocean."

Morales keyed her handheld. "Perimeter posts, report. Any movement outside the lights?"

"Negative," a guard answered. "All quiet."

Morales let the radio drop to her side. "If he's trying to reach someone, it's not us."

"No," Joe said. "That code's meant for people who already know it."

The pattern played again. Three short, one long. This time, something faint rode under the last tone—a ghost of a voice that never quite broke through.

Valery muted the speaker. "Could be bait. Could be a test."

"He's testing us," Joe said. "He wants to know if we're listening."

Morales set her jaw. "Then we pretend we aren't."

Joe nodded. "Keep the frequency open. Log every pulse. Don't respond."

Valery's hand hovered over the mic. "Understood."

Joe stepped back out into the dark. The wind grabbed the tent flap behind him like a hand. Somewhere beyond the ridge, past the trees and broken roads, a code kept beating against the air.

Three short. One long.

A heartbeat you could hear but not see.

By morning, Ridge Camp looked like any other hard-working settlement trying to pretend the world wasn't still on fire.

Smoke from breakfast fires drifted between trees. The sound of hammers and saws rolled across the slope. Kids chased each other between tent lines. Someone started a pot of something that smelled almost like real coffee.

Most people woke up believing the only thing that mattered today was building.

Morales and Joe walked the perimeter at first light, checking Guido's new tripwires. Cans rattled softly where the wind tugged at the lines.

"Crude," Morales said, nudging one with her boot. It jingled in a sharp, bright burst. "But it'll do."

"Simple's harder to break," Joe said. "Batteries die. Wire and tin don't."

They moved past the water lines where Donnie was talking with two engineers about reinforcing the gravity tanks. Charlie trotted at his side with a notebook, scribbling numbers like they were secret codes. Lenny followed, nose full of damp earth and breakfast.

Donnie waved. "Morning. Flow's steady. Filters are clear. No leaks."

"Good," Joe said. "Keep it that way. Nobody wants to go back to ship showers."

Down near a rope-marked patch of ground, Amy helped a young mother lay out blankets for a makeshift daycare—spare medical sheets, a couple of carved toys, one battered stuffed animal that looked like it had survived three apocalypses. Children's laughter drifted up the slope, thin but real.

"They think we're safe," Morales said quietly.

"They need to," Joe answered. "Or they'll never put roots down."

In the engineering yard, Manny and Ashley worked on the solar rig, tightening cables that fed the battery bank. Guido hovered nearby, pretending to supervise while guarding his mug of cold coffee like a relic.

"You keep working like that," he said, "and I'm going to start charging rent."

Ashley shot him a look. "You'd charge for breathing if you could."

"Only premium oxygen," he said. "Comes with a view."

Manny snorted, cinching the last connector. "All right, maestro. Throw it."

Guido flipped the breaker. The panels hummed. The charge light blinked green.

Ashley grinned. "We've got stable feed."

"Miracles still happen," Manny said.

"Don't jinx it," she replied. "You said that about the generator yesterday."

"That was optimism," he said. "This is science."

Guido pointed at him with his cup. "It's hubris. The universe hates that."

"Then you're in trouble," Ashley said, walking off toward the next bank of panels.

Guido watched her go. "She's getting her edge back," he said, softer.

"Yeah," Manny answered. "She deserves to."

Up near the top of the ridge, Valery sat under the radio mast, headset on, clipboard in her lap. The coded channel stayed open but mute, per Joe's orders. Every seven minutes the tiny red indicator flickered as the signal hit—three short, one long—logged by hand, never answered.

Morales climbed the last few steps to the radio post. "Anything new?"

"Same pattern," Valery said. "Same strength. No voice traffic yet. He's patient."

Morales looked down at the camp. "Let them have today," she said. "Whatever he's building out there, it doesn't get to steal this."

Valery nodded, eyes on the gauge. "We'll be ready when he stops whispering and starts talking."

Down below, Ridge Camp moved like it believed in its own future —tents going up, huts framing in, water running, wires humming.

Far beyond the trees, somewhere over broken roads and flooded lowlands, a man with a stolen rifle and an old unit code kept tapping against the air, looking for someone to answer.

No one at Ridge Camp spoke his name out loud that day.

But they all felt the shadow of him moving closer.

CHAPTER 15
THE SEVEN SEATS

The morning air was crisp and cool, the kind that made breath show like smoke. Word had spread fast—Captain Joe Alvarez had called everyone to the main clearing at first light. By the time he and Morales reached the ridge, more than two hundred people were already gathered. Men and women in mixed uniforms and work clothes stood shoulder to shoulder. A few children sat on overturned supply crates near the front, swinging their legs. The smell of pine and wood smoke filled the clearing, and the faint clang of tools from the workshops echoed down the slope.

Someone had thrown together a rough platform out of scrap planks, barely level, but it held a table and seven mismatched chairs. Valery crouched over a radio speaker wired into a battery so everyone could hear. She looked up at Joe and nodded once. "You're live."

Joe stepped forward, hands in his jacket pockets, eyes scanning the crowd. "We've been living by chain of command for almost two years," he began. "That worked when we were running for our lives, but it doesn't fit what we are now. We're not soldiers anymore. We're not a fleet. We're a community."

The murmur that followed was low and steady—agreement mixed with uncertainty. Amy stood near the front beside Donnie, her hand resting lightly on his arm. Morales joined Joe on the platform. "Here's how this works," she said. "We're forming a council. Seven seats. One for each thing that keeps us alive—defense, power, food and water, communications, logistics, health, and welfare. Each seat goes to someone you trust, not someone who volunteers because they like the sound of their own voice."

Guido shouted from the back, "So I'm out?" Laughter rippled through the crowd. Morales didn't smile, but her tone softened. "You can vote, Guido. That's enough responsibility for one day."

Joe waited until the sound died down. "Every hand here counts. There's no rank anymore, no salute, no chain. We do this together or not at all." He gestured to Morales. "Start the nominations."

Morales read names as they came—loud, simple, honest. The process was messy, but it worked. Within an hour, the seats filled themselves. Morales kept Defense and Security; no one argued. Donnie Renfroe took Power and Infrastructure, his experience with turbines and water systems unmatched. Valery was chosen for Communications and Recon, her calm voice already the sound of the camp's airwaves. Amy Renfroe took Family and Civil Welfare; she was the one everyone already went to when they needed comfort or sense. Manny earned Logistics and Resources, his knack for rationing fuel and food saving them more than once. Guido tried to dodge Food and Water, but the kitchen staff shouted him into it.

That left Health and Science. Morales called for names, but silence hung heavy until someone in the crowd said, "Dr. Nkomo." Heads turned toward the tall, dark-skinned woman standing near the med tent. She'd been there since the first months at sea—a Navy field medic who kept quiet and worked harder than anyone. Her sleeves were rolled up, hands still stained from tending cuts and burns. She looked surprised. "I'm not—"

Amy cut her off. "You kept half this camp alive when medicine ran out. You're more qualified than anyone here."

The crowd murmured in agreement. Morales smiled faintly. "All in favor?"

Every hand went up. Dr. Eliza Nkomo blinked once, then gave a small nod. "All right," she said. "If we're building something new, I'll help keep it standing."

Morales looked over the list, reading each name aloud. When she finished, the crowd clapped and whistled.

Joe stepped back to the table. "That's seven seats," he said. "From now on, no orders—only votes. Every major decision goes through this council."

Morales picked up one of the empty chairs and set it in the center. "You forgot one," she said. "The chair."

Joe shook his head. "I'm not taking it."

Amy's voice came gentle from the crowd. "You already have it, Joey. You just need to sit down."

Valery added, "Temporary chair, until the rest of us find our footing."

Joe hesitated, then sighed and sat. "All right. Temporary it is."

Morales turned back to the people. "Then it's official. Ridge Camp has a council."

The cheer that followed rolled through the forest and out over the bay. For the first time since the Rain of Fire, it sounded like hope instead of survival.

The cheering faded into quiet chatter as people drifted toward the edges of the clearing. Morales raised her hands to steady them. "Hold your seats," she said. "We're not done yet." She nodded to Valery, who turned up the small radio speaker so her voice carried through the clearing. "Council members remain," Valery announced. "Everyone else, you're welcome to stay and listen."

Most of the camp stayed. Nobody wanted to miss what came next. The seven new council members took their seats at the long table. The chairs didn't match—one was an overturned crate, another a medical stool—but the symbolism wasn't lost on anyone. They had a government now, rough as it was.

Morales cleared her throat. "All right. First order of business—confirm leadership structure. The chair coordinates meetings, but every vote counts equal."

Donnie leaned forward. "Then let's vote on that, make it official."

One by one, hands went up. Joe's last.

Morales nodded. "Motion passed."

Joe exhaled slowly. "Now that we've got that settled, I need to bring up something serious." The tone in his voice changed. Conversations in the crowd fell away until all that could be heard was the distant crackle of a fire.

Joe rested his palms on the table. "Three days ago, a maintenance crew boarded Halcyon to recover supplies. They found the brig door open and Captain Harrow gone. He took food, a rifle, and a portable radio. He didn't go alone—one of the guards assigned to watch him, Private Jensen, went with him. The brig log shows Jensen's last entry before the book goes blank. They stripped two ration packs and a service rifle from the locker."

The reaction was immediate. Voices rose—shock, disbelief, anger. Ashley froze halfway between the tool shed and the med tent, her face pale. Morales stood, her voice cutting through the noise. "Settle down. Security is already doubled. We have trip alarms on both approaches and patrols posted around the ridge. No one's coming in without us knowing."

But the murmurs continued, thick with fear. Amy stepped forward from the crowd. "Why didn't we hear about this sooner?"

Joe looked straight at her. "Because, Mom, I didn't want panic before we had facts. Last night, Valery picked up coded transmissions —military style. Harrow's using his old unit ID. That means he's out there, close, and he's talking to someone."

Ashley's voice was tight, controlled. "He escaped?"

Joe met her eyes. "Yeah. And if he comes near this camp, he won't like what he finds."

Morales set her hands on the table. "For now, we stay calm. He's one man with one weapon. We're 257 strong. We don't lose our heads over a ghost."

Dr. Nkomo spoke next, her voice calm but steady. "Fear spreads faster than disease. Keep routine. Keep order. Let's not undo everything we've built in one morning."

Her words carried weight. The crowd slowly settled again, eyes shifting back to Joe.

He nodded. "We go on with our day. We build, we post guards, and we listen. That's it."

The meeting broke up slowly, no one wanting to be the first to leave. Conversations buzzed in low voices, some angry, some afraid. Ashley didn't wait. She turned from the clearing and walked down the path toward the creek, her boots crunching over dry needles. The sound of running water drowned out the camp noise behind her. She stopped at the bank and stared at her reflection in the current. The sunlight shimmered over it, distorting her face until she didn't recognize herself.

She didn't hear footsteps until Manny crouched beside her, his shadow falling across the water. He didn't say anything right away, just tossed a small rock into the current and watched it sink. "You left fast," he said finally.

"I've heard that name in my nightmares for a year," she said. "Didn't expect to hear it again in daylight."

Manny nodded, quiet. "I figured as much." He picked up another rock and skipped it once across the surface. "You okay?"

Ashley let out a breath that sounded more like a laugh. "That's a dumb question."

"Yeah," he said. "Guess it was."

She sat down on a flat rock near the water. "I thought I was done being scared. Guess not."

He sat beside her, elbows resting on his knees. "Fear's fine. It keeps you alive. It's what you do after that counts."

She turned to look at him. "You really think he's close?"

Manny shrugged. "Joe wouldn't have told us if he wasn't sure. But don't let it eat you alive. We've got people here who won't let one man tear it down."

"You think I should tell anyone what he did to me? The details?"

"You already told Joe once," Manny said. "That's enough. You don't owe anyone the story again. What you owe yourself is not letting that bastard live rent-free in your head."

Ashley smiled faintly. "You sound like Donnie."

"I'll take that as a compliment," he said. "He's smarter than both of us."

She looked toward the ridge. The wind carried the smell of pine and smoke. "I used to like the sound of helicopters," she said quietly. "Now it just reminds me of leaving everything behind."

"Then let's give it a new meaning," Manny said. "Next time one lifts off, it's not for running—it's for rebuilding."

She studied him for a moment. "You always know what to say?"

He smiled. "No, but I try until it works."

She stood, brushing off her hands. "Thanks, Manny."

He rose with her. "Anytime."

They walked back toward camp together, quiet but side by side. Behind them, the creek kept moving, carrying away the sound of her whisper: "He escaped."

The camp slept light that night. The air smelled of ash and pine, the kind of quiet that only comes after long fear. Joe sat at the center table, a lantern burning low between him and the others. The flame bent with every breath of wind, throwing shadows across their faces. Morales had the roster open beside her, pencil tapping in rhythm. Donnie leaned back with his arms crossed, eyes tired but alert. Valery adjusted the radio set she'd salvaged from Arnold, its faint hum filling the silence. Guido rolled a piece of wire between his fingers, too restless to sit still. Dr. Nkomo folded her hands neatly on the table, waiting.

Joe looked around the circle. "All right," he said. "Let's keep it simple. Patrols rotate in three-hour shifts. No one goes beyond the perimeter after dark."

Morales nodded. "Already posted. East and west lines overlap by fifty yards. Nobody's sneaking through without tripping alarms."

Guido smirked. "Unless they fly."

Valery didn't look up. "Then they'll light up on my scope. I've got the radios running through two frequencies now—one for camp, one for long-range. We picked up faint chatter again an hour ago. Same pattern, same pulse. It's Harrow's frequency."

Joe's jaw tightened. "Distance?"

"Hard to tell," Valery said. "He's masking signal strength, but I'd guess less than fifteen miles."

Donnie swore under his breath. "That close?"

Dr. Nkomo leaned forward. "He'll test us first. Men like that always do. He'll want to see what we'll risk to stay calm."

Joe nodded. "Then we don't give him what he wants. We work. We build. We don't chase ghosts."

Morales said, "People are scared, Joe. They trust you, but they talk. Some think he's got followers."

"He might," Joe said. "Or he might just sound bigger than he is. Either way, if he's broadcasting, someone could be answering. Valery, keep those channels monitored full time. Don't respond unless it's command-approved."

"Copy that," she said.

Donnie leaned forward. "If we get a clear fix on him, what's the plan?"

Joe looked across the table, the light catching the silver in his hair. "Then we deal with it. Quietly."

Morales's eyes met his. "You mean lethal?"

"If it comes to that," he said. "He's a threat. Not just to the camp, but to the kind of peace we're building."

Dr. Nkomo nodded once. "Then make sure whoever faces him does it without hate. Hate's what made him."

Silence settled again, the kind that made the night seem larger than it was. Outside, the wind shifted through the ridge pines, carrying faint sounds of movement—maybe branches, maybe something more. Joe stood, closing the logbook. "All right," he said. "We've got work at dawn. Get some rest."

Morales blew out the lantern. The last thing anyone heard before the dark took the room was Joe's low voice. "Tomorrow we make this camp stronger."

The camp was quiet, the kind of silence that comes after exhaustion. Most of the fires had burned to ash, and only the command tent glowed faintly under the stars. Joe sat inside with Valery, both leaning over the radio console they'd salvaged from Arnold Air Force Base. The low hum of the generator mixed with the faint whine of static.

Valery adjusted the dial one notch. The signal came through again —seven seconds apart, two short pulses, one long. Joe rubbed the back of his neck. "Same pattern?"

"Same as before," Valery said. "But listen to this." She turned another knob, isolating the background. Beneath the first transmission, another faint one echoed in reply—slightly off tempo but identical in rhythm.

Joe frowned. "That's a second source."

"Yep," she said. "Not an echo. Not bounce. Two different transmitters talking to each other."

He leaned closer. "Distance?"

"Hard to pin down, but they're not far. Maybe twenty miles between them. One's probably Harrow. The other could be anyone with a transmitter."

Joe exhaled slowly. "So he's not talking to himself."

Valery looked up at him. "You think he found people?"

He nodded. "Has to be. He was alone when he ran. He must've come across a group of survivors. Could be stragglers from one of the grounded convoys, or deserters. Maybe even a few of his old crew from Halcyon who never made it off."

Valery's expression hardened. "That's not good. He was a control freak. If he found desperate people, they'll do whatever he tells them."

Joe nodded. "Fear's easy to sell when you've got a uniform and a gun." He leaned back, thinking. "How many could he have?"

"From the signal strength, it sounds like he's running off a portable unit. Maybe a dozen people tops. Not a big force, but enough to make noise."

Joe looked down at the table. "He's organizing. Building something. And if he's sending out signals, he's not hiding—he's recruiting."

Valery twisted the gain knob again. "He's calling it 'the Guard.' Listen." She raised the volume just enough for the faint words to form through the static. A man's voice—gravelly, deliberate.

"...Guard moving east. Supplies ready. Keep watch."

Then the transmission cut.

Joe's stomach sank. "The Guard," he said quietly. "He's trying to sound like he's still in command."

Valery nodded. "He's trying to make it sound official—like some lost branch of the fleet."

Joe stood, pacing the narrow tent. "We can't tell the whole camp yet. Not until we know how many are out there. Last thing I need is panic."

He looked out through the tent flap toward the sleeping camp. Lanterns flickered near the center fire, shadows moving as guards changed posts. Beyond them, the dark ridge loomed like a wall against the stars. "He's close," Joe said quietly. "I can feel it."

Valery watched the radio light blink. "He's not hiding anymore. He's trying to draw us out."

Joe looked back at her. "Then we don't take the bait. We build stronger. We find him first."

She gave a small nod, then reached for her mug. "You should get some sleep."

He half smiled. "You've been telling me that for two years."

"Still hasn't worked," she said.

He picked up the folded fleet log from the table and slid it into his jacket. "If he's recruiting, we'll hear more chatter by morning. Wake me if the pattern changes."

She nodded. "I will."

Joe stepped outside into the cold night air. The camp stretched silent beneath the ridge, tents glowing faintly like small stars. He looked east, toward the dark valley below, where a man he once trusted was building something dangerous out of old rules and broken people.

"Not this time," Joe muttered. "You're not taking anyone else." Then he walked back toward his tent, the sound of the radio pulses fading behind him—steady, relentless, and getting closer.

CHAPTER 16
SPLIT SIGNAL

The rain had stopped hours ago, but the forest still dripped with the sound of it. Ridge Camp slept in uneasy quiet—fires burned low, and the sentries on the perimeter walked slow arcs through the mist that clung to their boots. In the communications tent, Valery sat hunched over the radio console, the glow of the dials painting her face in green and amber. She hadn't moved for an hour.

A fresh sheet of log paper sat beside her, already half-filled with time stamps and coded notes. The same pattern had been repeating since dusk: three short pulses, one long. Every seven minutes, almost surgical in precision.

She rubbed her eyes and leaned closer to the speaker. "C'mon, talk to me," she muttered.

The tent's flap shifted as Morales ducked inside, shaking rain from her shoulders. "You've been on that thing since dinner," she said. "Get some sleep before you start hearing ghosts."

Valery gave her a sidelong look. "Ghosts don't use encryption."

Morales stepped closer, listening. The pattern came again—three quick bursts, pause, one long. Clean. Confident. It didn't sound like noise; it sounded like purpose.

"Same signal?" Morales asked.

Valery nodded. "Same rhythm, same amplitude. It's not weather bounce, and it's not random. Someone's keying a transmitter manually."

Morales frowned. "You said before you thought it was Harrow."

"Yeah," Valery said, adjusting the gain. "But tonight, there's something else. Listen."

She turned the dial slightly, isolating the frequency's edge. Beneath the main signal was a faint echo—a mirrored rhythm, offset by barely half a minute.

Morales squinted. "You're getting two of them?"

"Twin frequencies," Valery confirmed. "Same pattern, same tone, but different origin points. The second one's weaker, like a student repeating a teacher."

Morales's expression darkened. "He's training someone."

"Looks like it," Valery said. "They're not in sync yet. One's leading; one's following. It's like... call and response."

Joe entered quietly, drawn by the low murmur of voices. He looked exhausted—uniform jacket half-buttoned, boots streaked with mud—but his eyes were sharp. "Report."

Valery pointed to the receiver. "Two transmitters, same code. Primary signal's east, probably Harrow. The secondary is southwest—closer. Not a reflection, not bounce. It's manual."

Joe leaned in to listen. The first pulse pattern came—steady, precise. The second followed thirty-one seconds later, lighter but unmistakable.

"Hell," he said softly. "He's not alone anymore."

Morales folded her arms. "How close's the second?"

"Hard to say without a triangulation tower," Valery said. "But judging by signal strength—maybe twenty, thirty miles tops. That's inside travel range if they've got transport."

Joe rubbed his jaw. "He's building something. That's why the food thefts started before he escaped."

Valery nodded. "Rations, wiring, portable power units. You could run a small outpost with what he took."

"An outpost," Morales repeated, "or a forward base."

Joe's gaze didn't leave the radio. "He's trying to sound official. Make it look like the fleet's still talking."

Valery looked up at him. "He's not just broadcasting, Joe. He's recruiting. He's saying, there's still order out here."

The next pulse came through—sharper this time, like it was closer. The second followed right after, the delay shorter than before. Valery leaned forward, the hair on her arms lifting.

"See that?" she whispered. "They're syncing faster."

Joe frowned. "He's teaching someone Morse?"

"Not exactly," she said. "It's a variation—an old unit ID rhythm, like the one you used in the field. He's using pattern memory. Even someone half-trained could learn it fast."

The thought landed heavy between them. Morales's jaw tightened. "He's not just talking to survivors, then. He's building a squad."

Joe nodded. "And if he's using our old methods, he knows how we listen."

Morales looked at him sharply. "Then he knows we're hearing this right now."

Joe's voice dropped low. "He wants us to. It's bait."

The pattern started again—three short, one long—but this time the echo arrived almost on top of it. A perfect harmony.

Valery's hand went still on the dial. "He's not testing anymore. He's confirming link."

Joe stood straight, all trace of fatigue gone. "Then the game just changed."

He looked to Morales. "Wake the watch leaders. I want every post double-manned until morning. And no one—no one—uses open bands. We're off the air until further notice."

Morales nodded, grabbing her rifle. "What about the rest of the camp?"

"Not a word," Joe said. "We don't spread fear until we've got facts."

As she left, Valery kept her headset on, eyes fixed on the faint pulse of the radio light.

The two signals beat like twin hearts in the dark—one steady, one catching up.

And somewhere beyond the trees, Captain Harrow was smiling.

The camp moved like clockwork that morning—buckets clanking, hammers ringing, children laughing near the cook fires. From a distance, it looked like peace. Up close, it was anything but. Joe stood in the center clearing, eyes fixed on the ridge line where Morales's team had disappeared an hour earlier. His gut had that old submarine feeling again—the kind before alarms went off, when quiet was too perfect to trust.

Valery jogged toward him from the comm tent, her face pale under the rising light. "He's shifting again," she said, holding the radio log tight in one hand. "Both transmitters moved positions. The second one —southwest—is creeping closer every hour."

Joe took the paper, scanning the time stamps. "Same pulse pattern?"

"Same," she said. "Only tighter now. They're syncing almost perfectly. He's teaching them live."

The sound of boots on dirt made them both turn. Morales and Guido emerged from the trail, streaked with mud and breathing hard. "We've got prints," Morales said. "Two men, fresh. Harrow and his tagalong. They set camp less than a mile east—watching the bay, watching us."

Joe's jaw tightened. "They see the camp?"

"They had to," Morales said. "It's a perfect vantage. They burned rations—our rations. Same wrappers from the missing supply crates on Halcyon."

Guido dropped his pack onto the table and pulled out a scrap of foil, blackened at the edges. "Fleet issue," he said. "Can't fake that tear pattern."

Joe didn't even look at it. "So now we know what he's been eating."

Morales nodded grimly. "And who he's teaching. Jensen's with him. We found both boot sizes."

Joe cursed under his breath. "That guard was loyal to the end. Harrow must've worked him for weeks."

"Worked him or broke him," Morales said. "Either way, he's got a second pair of hands now—and one rifle becomes two."

Joe looked around the camp. Workers carried lumber. Amy and Donnie were helping build a wash area. Ashley sat near the solar rig with Manny, sorting bolts into coffee tins. He watched them for a second too long before turning back to Morales. "We can't let this turn into panic. Keep it quiet."

"Already did," she said. "But you need to set the tone. They'll follow your lead."

Joe's voice hardened. "All right. Double the watch. Every gate, every post. Rotate three-hour shifts, no gaps. I want scouts posted at the creek crossing and near the tree line east. No one walks alone—not even to the latrine."

Valery checked her watch. "You want radio blackout again tonight?"

He shook his head. "Limited comms. You, me, Morales. Everyone else stays on line-of-sight hand signals. If he's close enough to hear, he's close enough to track."

Morales folded her arms. "And if he's already tracking?"

Joe looked her straight in the eye. "Then we make him regret it."

Guido smirked faintly. "That's the Joe I remember."

"Yeah," Joe said quietly. "I remember him too."

He turned toward the open clearing. The sun caught the hull of the Halcyon in the bay—cold and steel-gray. The sight made his stomach twist. He'd spent too many days living inside metal. Now it looked back at him like a grave.

Morales followed his gaze. "You thinking what I'm thinking?"

"That it's bait?"

"Yeah."

He nodded. "He left it floating there for a reason. If I were him, I'd use the wreck as a fallback. Stash supplies, maybe even a relay antenna. He knows we'll check it eventually."

Guido grimaced. "You're saying we leave it alone?"

"Not yet," Joe said. "We watch it from shore. If anyone touches that hull, I want it on record."

Valery checked her notes again. "The second frequency—southwest one—it's changed altitude. Higher. That's not random; that's someone moving up a slope. If they've reached the upper ridge, they'll have line-of-sight over camp."

Morales raised an eyebrow. "You think we pull the civilians in tighter? Keep them near the ridge line, relocate a few huts, concentrate defense around the water station?"

Valery turned her head slowly. "You think hiding from him will save us?"

"Right now?" Morales said. "It'll buy us time."

Valery's eyes flicked back to the map. "We use his rhythm against him."

Morales raised an eyebrow. "Meaning?"

Valery tapped the receiver lightly. "Every sequence he sends is timed. I've been logging intervals—his follower's response is always four seconds late. That means Harrow's giving direct commands. But if we cut a false echo into the return band, we can make his recruit think he's being countermanded."

Morales grinned. "You're talking interference."

"Psychological," Valery said. "Not jamming—ghosting. We send a single altered burst from a decoy transmitter, make it sound like Harrow's orders are being overridden. Confuse them long enough to make a mistake."

"Risky," Morales said. "If he catches the fake, he'll know we're listening."

Valery finally looked up, eyes sharp. "He already knows."

The air was heavy the following morning, laden with the scent of damp earth and flowering pines. Ridge Camp stirred slowly to life, weary faces of survivors emerging from their tents like starving plants reaching for sunlight. Joe stepped outside, scanning the campgrounds while his senses tingled with the weight of the previous night's revelations. Each heartbeat echoed a reminder—Harrow was out there.

A gentle hand came to rest on his shoulder. Amy stepped beside him, her expression a blend of concern and motherly strength. "You didn't sleep last night."

"I'm fine, Mom," he assured her, though fatigue hung heavy beneath his eyes.

"That doesn't mean you can't be honest with me." She tugged at his jacket, straightening its collar as though it could somehow adjust his mind. "You're carrying a lot."

Joe sighed, looking at the clusters of people waking up. They needed hope—even more than he did. "I just need to make sure everyone is safe."

"They'll follow you if you let them," she said gently. "Trust in that."

He nodded, grateful yet overwhelmed. "I know."

Another calm morning enveloped Ridge Camp, yet an unsaid fear loomed just beyond the final tree line. Joe pushed away from the comforting weight of his mother's hand and turned to focus on logistics.

Morales joined them shortly after, her demeanor set with urgency. "First patrols are going out. We're keeping to the routes you identified last night, but there's something I want you to see."

His gaze intensified. "What is it?"

She led him behind the cook tent toward the perimeter, where the trees thinned and the ridge stood stark against the morning sky. In the heart of the clearing, they found Manny and a few scouts assembling supplies.

"Check this out," Manny said, pulling away a tarp that shielded a makeshift gathering of rifles and ammunition. "We've got what we need for a quick response team—extra gear, water rations, enough to move fast."

"Good work," Joe said, though a sense of dread bubbled beneath the gratitude. This was preparation for battle—something they didn't have the luxury of choice over.

But then another patrol returned, streaked with mud and rain. Morales's expression turned grim as the leader stepped forward. "We found more tracks, sir. Two sets, both heading northeast into the ridge."

"More than Harrow," Morales pressed. "If he's out there with followers, we're at risk."

Joe felt the weight of anxiety grow as he considered the implications. Each step Harrow took was carefully planned. "We lock down the perimeter. No one in or out without clearance. This is our camp, and we can't let him break through."

Manny's eyes narrowed, weighing his words. "What if they come looking for food? If Harrow's using us for bait, then..."

"Then we keep watch." Joe turned to Morales. "You've got the scouts—find the trails and tighten security. No increasing risk without concrete intel."

As they spoke, the sun ambled overhead, and the camp buzzed with motion. Amy moved among them, comforting weary families and answering questions. Joe's whole world was shifting, and he wondered if they'd be able to navigate the ebb and flow of Harrow's growing force.

Valery appeared next, sprinting back from the tent where the comm equipment was stowed. "We'll intercept whatever calls Harrow is feeding his recruits, but we need more eyes in the field. If his people are getting organized, we need to know their plans."

"Agreed," Joe said through clenched teeth. "Show us where to listen and respond, and update whatever signals you receive."

"And if they respond?" Morales asked.

"If they do, we'll counteract." Joe's heart raced, but he wouldn't allow fear to dictate their next steps. "This camp is our home. We protect it."

As his resolve hardened, the camp rallied around him. People acknowledged the threats with nods and whispers of solidarity, the tension turning into determination.

And as day turned to dusk, thunder brewed overhead, echoing through the woods, the momentary peace shattered beneath the weight of consequence.

Joe glanced toward the ridge, where the first pulse of Harrow's signal resumed its mechanical heartbeat.

Each wave brought new dread, but together they'd face whatever the storm held.

The camp moved like clockwork that evening. Shadows stretched across the campsite, flickering firelight illuminating determined faces as families huddled near dinner preparations. Joe made the rounds, checking in with each council member. He needed to ensure that while fear lingered, it did not consume them.

"Mom, how are the kids?" he asked, catching her near the

communal cook fires where several children busily sorted through pots of stew.

"They're resilient, Joey," she said, her voice full of warmth. "Even amid this uncertainty, they're leaning into hope. It's what we need."

"That's good to hear." He patted her shoulder affectionately, then turned to check on Donnie. "How's the repair on the turbines, Dad?"

"Almost finished." Donnie wiped his hands on a rag, face set with determination. "We'll have water flowing by morning, I promise. This land needs us to hold strong."

"I believe that." Joe swallowed hard. "We just need to stay sharp and ready."

As he moved through the camp, every creak and crack of branches sent his heart racing. Anxiety became a constant companion. When night fell, silence weighed heavy, broken only by the rustling of nearby tents.

Valery emerged from the radio tent, earbuds tangled in her hair, eyes wide with urgency. "The pulses—" she began, but her voice was cut off by a sharp crack of thunder.

"Hold on!" Joe exclaimed, stepping closer. "What've you got?"

"It's changing again—there's three echoes now, Joe, not two. He's shifted commands multiple times but they're still operational!"

"What do you want me to do?" Morales interrupted, urgency flooding her voice. "He's preparing for something."

Joe glanced outside where rain pattered steadily like percussion, blurring the outlines of tents and trees. "No—Scott, Johnson—go hold the line. If they move, I want eyes on them."

Before they could respond, a new signal cut through the static, not through the radio but beyond the tree line—a shout, then a crack, followed by another shout.

"Get ready!" Joe shouted. "They've hit our perimeter! Move!"

He rushed from the tent, heart pounding as he grabbed his rifle and joined Morales and the team. The tension during the day sparked into chaos as thunder rolled, the noise blending with shouted warnings.

"What's happening?" Valery shouted, pressing closer.

"Recon is down! Stay sharp!" Morales yelled back, scanning the darkness for movement.

Then the first gunfire pierced the night—a series of sharp cracks that echoed from beyond their sight.

"Everyone move back!" Joe shouted, leading his militia toward the perimeter and away from sprinkled fires. "Valery, monitor comms, and don't stop reporting! Morales, we stay close."

With fierce determination, the team surged toward the tree line, heartbeats matching the pulsing rhythm of the signals that had turned from a mechanical heartbeat into a siren call for war.

As they plunged into the trees, the chaos erupted behind them, the storm revealing itself in dark shapes and shadows. Joe's heart pounded in sync with the frantic pace of his thoughts. Every footfall seemed out of tune with the plan he had hoped for—a harmonious defense, a united front. But now they faced the unknown, a tide of danger bearing down on Ridge Camp.

"Valery, keep us updated!" Joe shouted, pushing deeper into the underbrush, the damp earth squelching beneath his boots.

Ahead, Morales led the way, scanning the shadows for signs of movement. The rain-soaked branches became obstacles, but the sound of gunfire pulled them forward, urgency propelling them deeper into the heart of the storm.

Another crack rang out, followed by a cry—a warning followed by chaos. The militia crew pressed on, adrenaline coursing through them as they reached the perimeter. They could hear voices now, punctuated by shouting and the clatter of metal against metal.

"Everyone, hold back," Joe commanded, raising his rifle and gesturing for Morales and Guido to take positions on either side. "We wait for visual confirmation."

"On your call, Joe," Morales whispered, her eyes narrowed, focused.

Through the dense foliage, Joe peered toward the clearing where shadows moved erratically. He could make out figures—two, possibly three—flitting through the darkness, the flashes of weapons illuminating the area. The sight sent a spasm of fear through him, but he forced it down, memory of the previous battles coupled with a fierce determination to protect what they had built binding him.

"Do you see who it is?" Morales asked, voice low.

He shook his head. "Can't tell yet. Let them come closer."

Then the silhouettes emerged from the thick underbrush—two figures, darting toward the edge of the camp. Joe tightened his grip on his rifle, adrenaline sharpening his instincts.

Suddenly, a figure broke through the shadows, stepping into the clearing illuminated by the flickering flames of the campfires behind them. It was a man, tall and imposing, easily fitting the profile of a military officer. Harrow.

"Damn it," Morales hissed.

Joe clenched his teeth. "He's come to taunt us," he growled under his breath, recognizing the gleam in Harrow's eyes that held a sickening mixture of confidence and malice. This was the ghost they had feared—the promise of chaos wrapped in a veneer of authority.

"Ashley, Manny, hold the line!" Joe shouted, pushing back through the underbrush, but as the militia converged, Harrow raised his rifle, pointing it at one of the campsites, calling out into the dark.

"Joe Alvarez!" Harrow's voice echoed through the night, a chilling resonance that sent a shiver down Joe's spine. "You don't have to hide behind your fences! Come out and face me!"

"Don't engage!" Morales ordered, readying her rifle. "We can't afford to draw fire unless we know their numbers."

"Sounds like you're afraid!" Harrow taunted, striding further into view. "You're all just ghosts, shuffling in the dark, pretending to have control."

"Enough!" Joe shouted, stepping forward, his voice booming over the chaotic noise around him. "You've got no right to come here—"

But before he could finish, there was a crack as the tension in the air snapped. Gunfire erupted from the east, a volley of shots that startled everyone into action. The wordless draw of their weapons sounded like a chorus of fury as the militia armed themselves against the impending onslaught.

Guido raised his rifle toward the west side. "Flank's been compromised!" he shouted. "We need to fall back!"

That's when Joe saw it: dark figures pouring through the trees from the east, their silhouettes sharpening in the dim light. Harrow's followers, armed and ready.

"Fall back to the perimeter!" Joe commanded, his heart hammering in his chest as they moved in unison. "Protect the families!"

Valery's voice crackled through his earpiece. "They're coming in waves, Joe! I can give you locations!"

"Do it!" He barked.

Joe rushed past Morales and Guido, positioning himself at the edge of the clearing. He peered through the veil of rain and mist toward the incoming swarm. "Keep your eyes peeled, boys. Aim true!"

The first wave erupted into view—a ragtag collection of troubled survivors, but with each visible face, each rifle, the reality of Harrow's influence became painfully clear. Desperation glimmered in their eyes, and that same fervor echoed within their raised weapons.

Then, in a flash of gunfire, chaos broke loose. Joe raised his rifle, aimed at a target—a towering figure in the back. His finger twitched against the trigger.

"Joe!" Morales hollered, snapping his focus. "Don't shoot until we know who we're facing. Remember, it's not just Harrow!"

A familiar face stepped forward from the turmoil. Shouting, a young man, facial features hidden by the dark like shadows and the flames flickering wildly. "Joe, wait! It's us! We're here to help!"

The surprise crammed Joe's throat. "Jason?" he shouted, recognizing the voice from the dwindling couple of families they'd come to know well. "What are you doing here?"

"We've been running! Harrow found us," Jason shouted over the noise, gesturing frantically at the looming mass behind him. "They threatened to kill anyone who didn't join them! We had no choice!"

"Get back!" Joe commanded, heart pounding as he surveyed the chaos erupting before them. Families, confused and lost, were swept up among the fervor as Harrow's followers pressed on toward the camp's heart. He could see vulnerability in their faces—Harrow's cruel pull that had drawn them in.

"Stop pushing!" Morales shouted, combat-ready as her eyes darted around, assessing each movement. "If you want to survive, you need to get behind us!"

A sudden roar came from Harrow's men, a rallying cry echoing through the trees, instilling fear in the hearts of those who were not prepared to fight.

Joe threw out his hand to steady the chaos. "Listen to me!" he boomed, voice strong over the uproar. "We're not your enemy! We're here for the same reason you are—to survive! But not like this. Deception is what got you into this position, and it's not the way out!"

Harrow stepped forward, swaggering like a vulture among the rise of frantic arms and voices, reveling in the chaos he had orchestrated. "They offer you safety, and you think now is the time to choose? No. It's time to stand and be counted! You can take refuge beneath my command! Will you follow me?"

Joe watched in horror as a handful of those desperate survivors looked up to Harrow, hesitating, but weighing their loyalties against fear.

"No!" Joe shouted, pushing forward. "You don't want this! Fight alongside us, not because of fear, but because you see it's the right choice! This isn't just a conflict; it's your future at stake!"

He raised his rifle, the gesture pulling his militia back to him. In that moment, he caught Valery's eye; she stood in blended shadows, hands gripping her weapon as she calibrated the beams overhead.

"Now!" Joe commanded. "Line up at the perimeter! Stand with Ridge Camp!"

Thunder rumbled overhead, cracking through the clouds as both factions took stock of one another. Debates of loyalty rippled among the confused crowd, and behind him, Valery stepped forward, confidence radiating.

"If you want to move forward," she called out, standing tall amidst the misdirection, "you must do it together! Cast aside fear! We need people who can stand united!"

The tension broke momentarily. Some of Harrow's followers lowered their weapons, confusion blossoming among them. A few shifted backward, questioning Harrow's leadership, while others remained steadfast under his influence—ready and willing to fight.

Harrow's lips curled into a sly smile, reveling in the test of wills. He raised his weapon, gesturing forward. "Join me, or watch your friends fall! This is war!"

At that, Joe felt the stakes rise. "We won't back down!" He shot back, standing his ground. "This is our home!"

In the split second that followed, chaos erupted again, gunfire ripping through the night as Joe shouted for his followers to form ranks. Iron will turned into physical action as Ridge Camp stirred with purpose.

"Now!" Morales shouted, her rifle steady beside Joe's. They braced for this immediate clash—a battle for survival.

The colors of conflict ignited as chaos spun into the night, the echo of gunfire mixing with the anguished cries of battle. Joe surged forward, heart racing in time with the pulse of the camp, refusing to step back as the gales of Harrow's growing army rose against them.

"Stay together!" he yelled, a war cry emerging through the tumult. "We fight for each other! We are our own! Stand strong, Ridge Camp!"

Surrounded by friends, united against the storm, they gripped their weapons—their resolve carved in the firelight of desperation driven by hope. Thunder boomed above, rumbling like a beast awakened as the battle ignited across the ridge.

And somewhere behind the lines, Captain Harrow commanded his followers, his voice rising through the storm as uncertainty twisted into fury—prepared to crush the resolve of anyone who dared stand in his way.

Rain swept through the clearing as the battle raged, mingling with blood and sweat. The heart of Ridge Camp pulsed defiantly against the rising storm beyond its borders—the true measure of unity and stubborn strength not yet fully tested, ready to confront whatever shadows hovered just beyond the darkness.

As the battle unfolded, the rain swept through the clearing, mingling with blood and sweat. The heart of Ridge Camp pulsed, not as an idea, but as boots in the mud and hands on rifles.

"Left side, down!" Morales shouted.

Three figures broke from the tree line, firing wild. Joe dropped to a knee, sighted center mass, and squeezed—one shot, then a second. Bark exploded beside them; one of Harrow's men yelped and fell back into the dark, dragging a leg.

"Cease fire on the families!" Joe barked. "You see a kid, you hold your damn shot!"

Shapes wavered in the shifting light. A woman with no weapon held both hands up and stumbled toward the camp, dragging a boy by the wrist. Behind her, a man in a torn jacket held his own rifle like it was burning his palms.

"Get behind us!" Morales yelled, waving them through. "Move!"

Two militia volunteers lunged forward, grabbing the newcomers and hauling them past the line. A stray round snapped through the air and punched a hole in a water barrel, sending a thin stream spilling into the mud.

"Valery, talk to me," Joe snapped into his radio.

Her voice came back, tight. "Two main groups. Harrow's core is holding east, using the trees for cover. The rest are scattered—half are firing, half are just trying not to get shot."

"Copy. Bring the camp lights up on the ridge," Joe said. "I want him backlit."

There was a beat, then the lamps along the outer poles flicked brighter, cutting through the rain like small, ugly suns. Shadows sharpened. The silhouettes behind Harrow's forward line turned from faceless shapes into people—thin, tired, too many with hands that shook when they raised their weapons.

"Look at them!" Joe shouted, voice cracking across the clearing. "That's who he brings to a fight. Scared families. You really think he's here for you?"

A burst of gunfire chattered from the right. Dirt jumped a foot from Guido's boots.

"Joe, less speeches, more cover," Guido muttered, firing two controlled rounds back into the trees.

On the flank, Ashley and Manny held low behind a fallen log, trading short bursts with a knot of Harrow's men who were trying to crawl closer.

"Don't bunch up!" Manny called, voice hoarse. "Short bursts! You're not mowing a lawn!"

Ashley's cheek was streaked with mud, eyes wide but steady. "They're not aiming," she said. "They're just pointing and praying."

"Yeah, well," Manny said, "pray slower."

Another round cracked. One of the volunteers to Joe's left screamed and went down, clutching his upper arm. Amy was there in seconds, dragging him back with Donnie, both of them low and fast, like they'd done this before. Blood soaked through the man's sleeve, but his eyes were clear.

"I got you," Amy said. "You're not dying in the mud. I refuse."

Harrow stepped forward again, just inside the tree line, barely lit, rifle still in his hands. Rain slicked his hair to his skull; his expression didn't carry any of the wildness his voice did.

"This is what they offer you!" he shouted, swinging the muzzle toward the families huddled behind Joe's line. "Starvation, fear, waiting for someone else to come save you!"

"Better than getting shot in the back while you're sleeping!" Morales yelled back.

"Stand down, Harrow!" Joe roared. "You want to talk, we talk. You want to walk away, you walk. But you come one step closer to my people, we finish this tonight."

Harrow laughed, a short, ugly sound. "These aren't your people, Alvarez. They're inventory. They just don't know it yet."

Behind him, one of his armed recruits—a girl barely out of her teens, soaked, trembling—lowered her weapon an inch. "Sir—" she started.

He didn't even look at her. "Raise it," he snapped. "Or I'll find someone who will."

The girl flinched like he'd struck her. Her eyes went past him, to Joe, to Morales, to the families pressed tight behind their line. She took one shaky breath, then stepped sideways, out of Harrow's shadow.

"I'm done," she said, voice thin but carrying. She slung the rifle over one shoulder and started walking, slow and deliberate, toward Ridge Camp.

Harrow's jaw flexed. "Anyone takes a step with her, you're marked," he said. "You hear me? You'll wear it."

"Keep coming," Joe called. "Nobody shoots you here."

Two more peeled off—Jason's wife, clutching their youngest by the hand, and a man Joe recognized from the supply queue. They moved in a loose line, heads bowed, as if expecting a shot between the shoulder blades any second.

No shot came.

Morales hissed through her teeth. "He's losing his grip."

"Yeah," Joe said. "That's when people like him get sloppy."

For a few heartbeats, the clearing hung on a knife edge. Rain pattered. Someone sobbed, bitten off quick. Valery's updates crackled faint in Joe's ear, numbers and bearings.

Then, from deeper in the trees, a whistle cut the air—three short, one long. Harrow's code.

The firing from his side stuttered, then dropped off. Shadows shifted. A few last desperate rounds went wide and high, more for noise than aim. Figures began to back away into the dark, covering one another in twos and threes.

"He's pulling them out," Morales said. "Coward."

"Strategic coward," Guido muttered. "The worst kind."

Harrow didn't move at first. He stood there, framed by wet trunks and mist, looking straight at Joe. The rain made rivulets on his face that might've been mistaken for something gentler in another life.

"This isn't over," he called. "You can't hold them all. They'll come find me when they're hungry enough."

"Maybe," Joe said. "But next time you come up this ridge, you don't walk away."

Harrow's mouth curled in something that wasn't quite a smile. "Next time," he agreed.

He gave a short hand signal. Two men in his shadow dropped smoke canisters into the mud. Gray plumes surged up, thick and choking. When the wind finally tore them thin, the trees beyond were empty.

Silence rushed in all at once, loud as thunder.

"Hold fire!" Joe yelled. "Nobody chases! We don't run blind into his ground!"

Breathing filled the gap where gunfire had been. The smell of cordite and wet earth settled over everything. Somewhere a child started crying; a second later, someone scooped them up and carried them away from the line.

"Medic to the perimeter," Amy called. "We've got three down, all breathing!"

Manny slumped against the log, hand shaking as he checked Ashley over for holes. "You good?"

She nodded, jaw clenched. "Yeah. I'm… yeah."

He blew out a breath. "Hate that guy," he said softly.

"Get in line," she answered.

Joe walked the length of the makeshift barricade, boots sucking at the mud. He touched shoulders, met eyes, counted heads. The living first, the wounded next, the dead—none tonight, by some small miracle—for last.

Guido shoved his rifle onto a sling and scrubbed a hand over his face. "Call that a win?" he asked.

"No," Joe said. "Call it a warning we survived."

He turned to the cluster of new arrivals—Jason's family, the girl who'd stepped away from Harrow, the man with the torn jacket. They stood just inside the camp edge, unsure if they were guests or prisoners.

"You're safe here," Joe said, voice rough but steady. "If you want to stay, you work, you follow camp rules, you don't answer to him anymore. That's the deal."

The girl swallowed. "He won't forget," she said. "He always… remembers."

Joe glanced toward the dark line of trees where Harrow had gone, then back at her. "So do we," he said. "Difference is what we do with it."

Rain eased to a mist. The fires behind them were guttering low, needing wood and calmer hands.

"Alright," Morales called out, louder now, taking up the slack as the adrenaline ebbed. "Weapons cleared and safetied unless you're on watch. First and second squad, you're rotating on perimeter till dawn. Third, help with the wounded and get families inside the inner ring. Nobody walks alone."

People moved, slowly at first, then with purpose. Rifles were checked and slinged, magazines counted, bandages tied off. The long, stretched note of panic shifted back into the ragged rhythm of survival.

Joe let himself look up at the ridge line. The lamps cast a faint halo over the edge of the camp, just enough light to see that the trees were empty—for now.

Morales stepped in beside him. "He tested our fence," she said. "We're still standing."

"Yeah," Joe said. "He'll hate that."

They stood there a moment longer, listening to the camp stitch itself back together—voices low, the clatter of pots, a baby fussing and then quieted.

"Get some rest, Joe," Morales added. "He'll try again. You'll want a full brain when he does."

He almost laughed. "I'll take half a brain and decent coffee."

"Dream big," she said, and walked off to check the watch posts.

Joe stayed until the last of the smoke blew out of the trees and the ridge was just wet rock and dark pine again. His hands still shook when he unclenched them.

Harrow was out there, somewhere beyond the line of light. Plotting. Remembering.

By the time Joe finally turned back toward camp, the rain had thinned to a cold drizzle and the first hint of dawn was just a paler band on the horizon.

Ridge Camp had held—for one more night.

CHAPTER 17
THE EMBLEM

Joe hadn't slept. Not after the gunfire, not after the wounded were carried in, not after Harrow vanished into the pines like a shadow that learned to breathe. Dawn found him standing at the ridge, soaked to the bone and watching the tide roll in. That's when the kids screamed.

The tide had left something behind.

At first light, two kids hauling driftwood for the breakfast fire spotted it—a plank wedged in the wrack line, bone-white and painted with a slashing red mark that didn't look like any tide had made it.

Willa was the one who screamed, a sharp intake of breath that cut through the early morning stillness like a knife. It wasn't loud, not panic—more like the sound a person makes when a memory grabs them by the back of the neck. Manny heard it from the storage yard and jogged down the slope, toolbox rattling. Ashley was already there, hand hovering over the plank like it might bite.

"Don't touch," Manny said. "Could be laced with something." She nodded once and stepped back. The plank looked ordinary until you saw the paint. It wasn't a symbol so much as a message; a single, diagonal lash, thick at the start, tapering like a whip across pale wood. The bristles had left ragged edges where whoever painted it pressed too hard. Someone had taken their time to make it look fast.

Ashley swallowed hard. "He used to carve it on crates," she said. "Back on the sub. Private punishments. If you didn't pass inspection, he'd mark your locker with a line like that. One stripe for a warning. More if he wanted everyone to see you bend."

Willa's father put an arm around the girl and pulled her back toward the crowd gathering at the waterline. "Back up," he said, voice low. "Give them room."

Manny keyed his radio. "Joe, Morales. You'll want eyes on the shore."

"On our way," Joe replied, his voice snapping through the early morning air.

Smoke from breakfast fires drifted across the rocky ridge as people trickled down, hesitant. The plank lay there like a tooth pulled from something too big to see, the red slash bright against the gray morning. A gull hopped near it, decided better, and took off.

Morales arrived first with two of the volunteer guards. She didn't crouch. She stood over the thing, boots planted, and stared like it owed her an answer.

"Is that paint?" she asked.

"Looks like oil mixed with dye," Manny said. "Thick. It'll hold in salt."

Morales nodded once. "So it was meant to be found."

Joe stepped through the circle and looked hard at Ashley, not the plank. "You know it?"

She kept her eyes on the mark. "He called it the lash. Said discipline was a language. This was his accent."

A murmur rolled through the bystanders, more felt than heard. People remembered different things at once—raids, sirens, nights on watch when the ocean had teeth. The red line pulled them into the same room.

Joe scanned the surf. "Any more?"

One of the guards pointed down shore. "Two smaller pieces in the rocks. Same mark. Looks like they were planted along the tide line."

Morales followed his finger, jaw tight. "Then he was close enough to read our campfires."

Joe didn't rise to the bait. He watched the water the way you watch a door in a bad neighborhood. "Bag it," he said. "Gloves. Don't smudge the paint. I want the grain pattern logged and a photo of the brushwork."

Guido clattered down the steps a moment later with a canvas satchel and a grin he didn't feel. "Beach art show? Nice." Then he saw the red slash, and the grin fell off his face. "Never mind." He pulled on rubber gloves without being asked.

Ashley took a step after him. "He won't stop at driftwood," she said, voice steady. "He'll mark things he wants: fuel drums, food... people."

Willa heard that and flinched, so her mother steered the girl toward the path with a quiet touch. "C'mon, honey. Help me with breakfast. Leave the ugly to the adults."

The crowd breathed again. Work crews peeled off in twos and threes, pretending they hadn't stood there a minute too long. Joe waited until the beach thinned, then crouched at last, eye level with the plank. He didn't touch it. He looked at the edges of the paint where bristles had splayed and the flecks where salt had dried into tiny stars.

"He did this at night," he said. "Tide line says two hours before dawn."

Morales nodded. "Which means he was on this ridge while our guards were making their second rotation."

"Or just outside the lights," Manny said. "Where sound slips."

Morales looked up the slope toward camp. "He wants the civilians to see this first," she said. "He wants to live in their heads before he steps onto their ground."

Joe stood. "Then he doesn't get to." He pointed to the plank. "Tag it, store it, log who found it and when. No rumors. We tell them what it is and nothing more."

Guido slid the evidence bag over the wood. "You want it in the depot?"

"No," Joe said, and his voice went flat as the sea. "In my tent. Until I decide who needs to see it."

Mom came back down the path as he said it, steam curling from a tin mug in her hand. She stopped in front of him and set her palm on his chest like she was checking a heartbeat. "Have you eaten, Joey?"

"Not yet, Mom."

"Then eat while you hunt your ghosts," she said.

She pressed the mug into his hand and kissed his cheek. For a second, the red line on the plank didn't own the morning.

Then Morales's radio crackled. "Perimeter east. We found another mark in the drift. Same paint. Smaller board."

Joe handed the mug back. "I'll eat after."

Mom didn't argue; she squeezed his arm once and stepped aside. Morales looked at the surf again, then at Joe. "We mark this a message?"

"No," he said. "We mark it a confession."

He turned to the guards. "Sweep the shore both ways for a quarter mile. Everything with a red line gets bagged. If you see anything that isn't wood with that paint on it, you don't touch it. You call me."

Ashley stared at the plank one last time. "The Guard remembers," she said, almost to herself. "That's what he'd say. Like memory was a weapon."

Joe met her eyes. "Then we remember louder."

And with that, the day began under a sky that felt one shade darker than it had an hour before.

By midmorning, the red marks had multiplied. Six boards in total were found scattered along the rocky ledge, each painted with the same deliberate stroke—thick to thin, angled slightly left, like the hand that made them wanted the mark to look both human and mechanical. Morales stood over the makeshift table inside the command tent, the planks lined up side by side on tarp. Even through gloves, she didn't like touching them. The paint had a tacky sheen that hadn't dried right, as if it had been mixed with oil or blood.

The tension throughout the camp was palpable. Residents brushed past each other, exchanging worried glances, while a sense of urgency escalated over the morning preparations. Joe paced the command tent, and in a scrawl of ink, he outlined plans.

"Everyone is on high alert." He leaned over the table, scanning the rows of driftwood. "They know enough to be dangerous."

Guido leaned against a support pole, rolling a pencil between his fingers. "You think he's trying to get the camp riled up? Because that seems like a bad idea."

A sudden gust of wind shot through the tent, sending papers fluttering to the ground. Morales shot him a look. "He's probably not looking for a fair fight; he's trying to push us into retreat."

"Retreat?" Joe's voice rose, challenging the very notion. "The moment he thinks we'll back down, he starts believing he's won."

"Then he's painting a bullseye on our heads," Morales pressed, folding her arms tightly across her chest. "We need to strike first before he has a chance to pull together whatever madness he's planning."

Joe stood tall, his gaze intense. "What's the first step?"

"Civilians need to be pulled into the inner layer," Morales said. "If they see him as a threat during the day, it doesn't matter what ghosts the night brings."

"Fine." Joe nodded, rubbing his temples. "We'll center the families around the cook tents, reinforce the outer perimeter and secure lookout points. Make sure they know we're here, and not afraid."

A tap at the tent's entrance broke through their tense exchange. Valery poked her head in, her face pale and drawn. "We've got another signal."

"Another?" Morales shot up, moving closer.

"It's more broadcasting from that secondary frequency—it's moving fast," Valery said, urgency quaking her voice. "He's calling out for recruits."

"Great." Joe ran a hand through his hair. "What's the message say?"

"Listen," she said, activating the receiver. The static hissed before another deep voice broke through.

"Come, join The Guard! Join us and find peace in strength! You are not alone! Obey, and you will be free!"

Joe's stomach twisted as the words resonated throughout the tent, fear gripping him tightly. He turned to Morales and Valery. "He's gathering them like cattle."

"That isn't just intimidating," Valery whispered. "It's a rally."

"We can't let that happen," Morales interjected. "We need to fight against this ideological recruitment. Every moment we waste, he's pulling more people into his influence."

Joe turned sharply, collecting himself. "All right. We need to send a message. Social communication needs to change. We give them our position while pulling the civilians into the center. We need to reassure them."

Morales nodded firmly. "Leave that to me. I'll circulate through the camp and explain everything. Get them ready for the meeting tonight."

Valery's gaze darted toward the receiver, the weight of concern hanging in the air. "What about his followers?"

Joe straightened. "We need to gather volunteers for patrol, a few trustworthy scouts to monitor the trails. Information on both him and his movement is critical."

Murmurs emerged amidst the tension—but this time, they had a plan.

As Valery returned to her station, Joe stepped outside, where the growing morning light pierced through the dense canopies, illuminating the silhouettes of worried community members. He gestured for attention, stepping up atop a crate to speak over the crowd.

"Listen up!" His voice echoed, steady and authoritative. "We have a situation—dockings have compromised our morale. Harrow is moving fast to gather more followers, and we must take action against him."

Gasps echoed throughout, faces pale as the realization of urgency sank in.

"We're reinforcing the perimeter," he continued, hand steadying on the crate. "Anyone who sees anything suspicious, I want you to report it. Every last mark will be accounted for."

He felt the weight of their gazes and offered a reassuring smile. "We've faced worse and come out standing tall. Together, we will create an imprint that Harrow will never forget!"

The crowd stirred, nods of agreement rippling through as energy began to rekindle.

"Now! Let's band together and make sure no one falls prey to fear. We rise as a community!"

Morales joined him, voice swelling with conviction. "Our camp stands united! We will not be ruled by shadows!"

The gathered residents echoed her words, their rallying cries filling the air as they began organizing, their energies aligning toward a common goal—their community's survival against dark forces threatening to reshape the world around them.

That night, tension gripped Ridge Camp like a vice over their hearts. Dark clouds hung heavy, punctuated by flashes of lightning that cracked the sky in jagged lines. Fires flickered cheerlessly, casting odd shadows across the faces of those who remained inside the tent. But as the noise settled down, a blanketed silence encased the camp. People huddled close, whispering rumors of Harrow while seeking comfort among friends.

Joe took a breath, the weight of command pressing down like lead. He scanned the gathered folks before him, familiar faces lit warmly from the glow of campfires scattered across the night.

Standing between them, together they stood as one.

"Tonight, we gather as a community to push back against the horrors we face. Fear is a weapon we won't allow to tear us apart!" he called out, sharing his resolve. "We have all made a choice to fight against it. And together, we will reclaim that which Harrow seeks to destroy."

As his words echoed through the dimness, Valery adjusted the sensors, clicking through frequencies to ensure nothing went missed.

"I've been monitoring his movements," she said, voice reassuring. "We will not let him tear our unity apart."

"What do we do when he comes?" one voice rose from the crowd—a mix of fear and demand.

Joe met their gaze, heartbeat slowing as he calmed their anxious spirits. "We fight to protect our home, our families, and ourselves. When that moment comes, we won't just stand—we will overwhelm."

Throughout the night, plans were formed, rallying spirits igniting the flames of determination once more. The people of Ridge Camp drifted together, sharing stories of loss and the hope that blossomed beneath their journey.

It would be a long night, but amidst the challenges ahead, they found strength in unity, ready to face whatever darkness emerged beyond the trees.

As daybreak approached, the storm clouds lingered in the sky, promising rain anew. Joe stood at the edge of the ridge, overlooking the horizon.

Across the rocky ledge, the faint glow of dawn signaled impending light, but as he turned back towards camp, one thought persisted: they were ready, and they would not back down.

"Together," he repeated under his breath. "We face the storm."

Somewhere in the distance, Harrow continued to weave his web, but Ridge Camp had begun to turn the tide.

CHAPTER 18
FIRST STRIKE

The night cracked open like a shell. The first explosion came from the eastern slope—sharp, close, metal on metal—followed by the roll of something heavy tumbling through the brush. The camp alarms howled, a rising, human wail that turned every head and froze half the hearts listening.

"Fuel depot!" someone shouted.

The smell hit before the smoke—diesel and hot steel. Morales burst from her tent half-dressed, pistol in hand. "Positions! All lights low!"

The ridge came alive in flashes. Civilians stumbled out of shelters, pulling children to the center of camp while armed volunteers fanned toward the fire line. Over the roar of wind and alarms came the distinct clatter of automatic fire—short, controlled bursts, professional. Harrow's men had breached the perimeter. They'd come through the tree line with torches and crowbars, aiming for the tanks first. This stolen discipline of soldiers was now made dangerous again.

Guido's voice crackled through the comm: "East flank! They're rolling barrels downhill—booby traps!"

The words barely cleared the radio before another blast split the dark. One of the fuel drums went up near the generator pit, sending a fountain of sparks that lit the whole ridge in orange. Morales dove behind a sandbag wall, teeth clenched. "Cut the tanks before it spreads!"

Through the smoke, shadowed figures moved—lean, fast, carrying scavenged rifles and knives. The red lash mark was daubed across their coats, wet and glinting like fresh blood. "The Guard remembers!" one of them screamed before a rifle cracked from the watchtower and dropped him mid-chant. The cry turned into a chorus—half fanatics, half terrified men pretending they weren't.

The sound of boots on gravel grew louder.

Valery sprinted through smoke thick enough to chew. The alarms howled across the ridge, red strobes cutting the dark into jerking slices of light. She cleared the last supply crate and nearly collided with a guard crawling backward through the mud, one hand pressed against his thigh, blood pulsing between his fingers.

"Hold still," she said, dropping beside him. A bullet snapped through the air, biting into the crate behind her, spraying splinters. She ignored it, yanked the medical pouch from her belt, and tore it open with her teeth. "Stay down."

The man tried to speak—just noise. She shoved gauze into the wound and wrapped it tight, her heart pounding with every sound of chaos.

Another round cracked past her shoulder, and the crate jumped. A sharp pain shot through her as she fired back, aiming at the shadows moving in the smoke. "You're going to live," she said, even as she felt adrenaline coursing through her veins. "Just keep breathing."

Through the haze, Valery could see movement at the perimeter—three of Harrow's men slipping along the ridge edge, their red-marked coats gleaming in the firelight. She grabbed her radio again. "South line breached!" she shouted. "Three hostiles on the slope—armed heavy!"

Then she raised her rifle and added under her breath, "Not for long."

Another barrage of shots tore through the night as she aimed for one enemy, hitting the leading figure, and the chaos shifted like a tide. Valery's training kicked in, the muscle memory taking over. She wouldn't let herself falter.

"Valery!" Morales's voice crackled in her headset. "Get back to command! You're exposed!"

"Busy," Valery hissed, dragging the bleeding guard deeper behind cover. "He's hit bad. I'm not leaving him."

"Now's not the time to play hero!" Morales shouted, urgency lacing her voice.

"Not playing. He's got half a chance if he doesn't bleed out." She tore the strap from her gear pack and wrapped it tight around the guard's leg, working through the pain that ignited her shoulder.

The guard tried to rise, but she shoved him back down. "You move, you'll make me waste my effort." He tried to protest, but a bullet dropped another enemy soldier nearby, and the impact shook the ground beneath them.

"Stay down," she commanded.

Another shot tore past Valery, this time hitting the crate beside her. "Gotta move," she said. The guard squeezed her hand weakly, fear creeping into his eyes. "You're hurt—"

"Doesn't matter," she said with resolve, rolling onto her good side. She had to keep him alive, had to fight for them both.

Through the chaos, a figure emerged from the smoke—Harrow's second in command, Trapnell, a cruel smile sprawled across his face.

"Didn't think you'd bleed so easy, comm girl," Trapnell taunted, swaggering closer.

"Guess you don't know me at all." Valery stood her ground, determined to push back.

But as she rose to aim her rifle, a burning pain ripped through her shoulder, staggering her backward. Trapnell smirked wider, a predator with his prey in his sights. "You're brave, but bravery only gets you killed."

"Then I'll make sure you know what it feels like." She shot back, twisting her stance, the rifle steady in her grip.

"Valery!" Joe's voice roared through the radio. The familiar surge of resolve blossomed in her chest, drowning out the unintended chaos around her.

A shot rang out from Joe's location—the barrel of his rifle steady and true. The impact hit one of Trapnell's men as Valery aimed carefully.

Heaving a breath, Valery tightened her grip—this was the moment to press forward. But in the chaos, another figure moved to intercept Joe's line of fire, and Valery screamed.

"Joe!"

A sharp yelp followed, but Joe was already on the move as the dust settled. He dropped into the cover of a crate, spinning to return fire. Trapnell glanced over, his confidence wavered.

"Don't think for a minute I won't finish this job!" he bellowed, rushing forward.

"Enough playing!" Joe shouted.

Valery fought through the surroundings, bracing herself before rising to aim once more. "You're going to regret underestimating us!"

As she squeezed the trigger, she felt another strike catch her just below the ribs, forcing her backward. The world dimmed.

"Val!" Joe bellowed, cutting off a figure just before it could charge at her.

"Stay with me!" he shouted urgently, noticing the wound she was trying to ignore.

"I'm fine. I can still fight!"

"Not like this!" he argued, grabbing her by the waist and pulling her with him. "We're retreating! You'll bleed out if you stay exposed."

"I'm not leaving you!" she insisted, but he didn't relent, eyes fierce as he pulled her toward the relative safety of stacked crates.

"Then we'll fight together," he said, voice steady.

Together, they surged forward in unity as the shadows threatened to drag them under. They fought back, refusing to yield in the face of overwhelming odds.

"Keep shooting!" Joe ordered, determination radiating off him.

Valery aimed her rifle again, adrenaline sparking inside her, driven by the need not to falter. "We're not losing this!"

The battle stretched on, the tide of Harrow's forces crashing against them with every strike. They found their rhythm, pushing against the chaos with fierce teamwork.

But as the gunfire continued and the sounds of chaos tore through the night, another explosion rang out. Heavy debris rained down around them as Valery was knocked off her feet.

"Valery!" Joe's voice echoed through the chaos.

Dazed but not broken, she pushed herself up amid the dust, heart racing, reminding herself to focus. "I'm fine," she gasped, gripping the earth with her free hand.

"You're hit!" he shouted, kneeling beside her, concern flooding his expression. "We need to get you patched up!"

"I won't leave!" she shouted defiantly, forcing the pain to the back of her mind.

Valery charged forward with renewed determination, aiming at an advancing adversary in a rush of action. The flames flickered around them, illuminating the chaos as shadows moved in every direction.

"Push forward!" Joe encouraged, his voice ringing with authority. "We need to gain ground!"

With one final push, they pressed against the tide, their resilience shining as they faced down the shadows threatening to consume them. As sweat mingled with rain on their brows, they refused to give in, taking every last flicker of opposition before them.

Finally, amid the din, Valery felt a sudden shift in the chaos. She spotted Harrow's remaining men retreating, realizing they could not break through the defenses. The tide was turning.

"Now!" Joe shouted, rallying the remaining militia as they surged forward, boots pounding the ground. The shadows scattered—a group of Harrow's men pulled back, their resolve dissipating like smoke.

Valery saw the fear flashing in their eyes as they departed, and with it, a sense of victory surged through her. "We did it! They're retreating!" she yelled, adrenaline rushing through her.

The last remnants of Harrow's chaos faded into the dark, and as the initial aftermath of the fight settled, the camp held its breath. Ridge Camp had held against the storm—made it through the endless night.

Soon, the smoke began to clear, and dawn breathed light through the trees, illuminating the aftermath of their struggle. Valery leaned against a crate, breathless and alive, feeling the weight of what they had just faced.

Joe stepped beside her, breathing heavy but a proud smile stretching across his face. "Not too shabby, huh?"

"Not too shabby at all," she replied, heart still racing.

As they looked out over the camp, the reality of their fragile safety began to dawn on them—they had faced shadows and emerged together.

Somewhere in the distance, beyond the ridge, the first hint of sunlight broke through, warming the bruised morning sky. Joe closed his eyes for a brief moment, taking in the sound of their victory, a promise that truth, even amid chaos, could vanquish fears.

Together, they would prepare for whatever came next.

CHAPTER 19
DAWN AFTER THE FIRE

Dawn crept over Ridge Camp quietly, almost like it wasn't sure it was welcome. Thin sunlight filtered through the smoke drifting from the charred patches of earth. The air tasted like burned diesel, metal, and adrenaline sweat. People walked slowly through the wreckage—checking on neighbors, clearing debris, whispering to each other so the morning didn't break under the weight of everything the night left behind.

Joe moved with purpose down the central path—boots crunching over shattered crates, splintered boards, and pieces of red-marked coats left behind by the attackers. His face was unreadable, but his jaw flexed every time he saw a blood smear or a piece of equipment that had been deliberately sabotaged.

Behind him, Valery followed—limping slightly, shoulder stiff, ribs screaming each time she stepped wrong. Her uniform was half-cut open from emergency dressing changes, and her sling hung uselessly at her side because she kept slipping out of it.

"You should be in a cot," Joe said without slowing down.

"You should stop complaining," she shot back. Her voice was thin but sharp enough to stand on.

He stopped and turned, taking in her pale face, clenched jaw, and the way she braced her weight with one hand on a crate.

"Val," he said quietly, the warning right under the surface.

She lifted her chin stubbornly. "Walking. Talking. Breathing. That's all you need from me right now."

He narrowed his eyes. "Add 'not falling over' to the list."

"Then I'm four for four. Relax."

But the second she stepped over a broken pallet, her knee buckled. She didn't hit the ground—Joe lunged and caught her arm, steadying her right before she pitched forward.

Her breath caught in a sharp, involuntary gasp.

He didn't let go. "You're done pretending you're bulletproof. It's getting old."

She winced but didn't look away. "And letting you carry me would be humiliating. So here we are."

Joe stared at her for a long moment—long enough that she thought he might actually drag her back to the med tent. Instead he exhaled slowly through his nose.

"One more fall, and I'm slinging you over my shoulder."

"You couldn't lift me with that bad back," she muttered.

"My back is fine."

"Your ego is worse."

Before he could escalate the argument, Morales jogged toward them, holding a plastic-wrapped scrap of paper. Her hair was soaked with sweat and ash, her shirt half-burned from tackling fires near the generator pit.

"Joe," she said, handing him the packet. "Found this near the east slope."

Joe unwrapped it carefully. Inside was a half-burned sheet torn from a small journal. The handwriting was neat, square, unmistakably military.

Valery leaned closer, breath tight for reasons that had nothing to do with pain. "That's Harrow's handwriting."

The page listed locations in rigid block lettering—storehouse, supermarket, highway depot—and tallies beside them.

Not supplies.

People.

Morales crossed her arms. "We cross-referenced the names at breakfast. None of them are ours. Looks like he recruited them before last night. Maybe even before we landed at Ridge."

Joe's jaw tightened. "He's gathering from everywhere he can reach."

"Not just gathering," Valery murmured. "Organizing."

Joe folded the page. "He's building an army."

Morales nodded grimly. "And last night? It wasn't meant to win. It was meant to measure us."

Valery's fingers curled against the crate beside her. "Then what's the next move?"

Joe tucked the paper into his jacket. When he spoke, his voice was low and dangerous.

"Next move is ours."

The med tent was a patchwork of noise and exhaustion—hisses of pain, clinking metal trays, murmured instructions, and the soft rustle of canvas as people moved between cots. The smell was a bitter mix of antiseptic, burned fabric, and fear that hadn't faded yet.

Amy moved like she always did in crisis—steady, calm, her voice a soft anchor in a room full of frayed nerves. Her apron was streaked with blood that wouldn't be washed out anytime soon, but she didn't seem to notice. She was too busy keeping people alive.

"That's enough sitting up," she told a young guard whose leg was wrapped from hip to ankle. When he tried to rise again, she pushed him back with one hand. "You stand, you bleed. You bleed, I yell. You don't want that."

He sank back with a groan. "Yes, ma'am."

Across the tent, Valery sat on an overturned crate while a corpsman worked on her shoulder. He cut away the dried dressing, revealing angry swollen skin around the bullet hole. When he touched it, she sucked in a breath hard enough that he paused.

"You should've come in last night," he said.

"I was busy," she muttered.

"Still are," he said. "Your pain tolerance is stupid."

Amy walked over, lifting the edge of Valery's sling. "You look like hell, sweetheart."

"Feel better than I look," Valery lied.

The corpsman snorted. "Pretty sure that's impossible."

Amy leaned down to meet Valery's eyes. "You can't fix the whole world if you pass out face-first into the dirt."

Valery didn't smile. "I'm not passing out."

"That's because I'm not done with you yet," the corpsman said. "Bullet didn't hit anything vital, but you're not lifting anything heavier than a radio for a week."

Valery rolled her eyes. "Try telling Joe that."

Amy's expression softened. "Oh, he already knows."

Before Valery could argue, the tent flap opened. The room quieted without being asked—people always looked at Joe the same way after a fight, like he carried the whole ridge on his shoulders even when he wasn't trying to.

He scanned the tent—checking on the wounded, counting empty cots, making sure no new sheets had been pulled over faces since sunrise. His gaze paused briefly at the far end, where a tarp hid four bodies they hadn't had time to bury yet. His jaw tightened.

He stepped toward Amy and Valery. "How many?"

Amy didn't sugarcoat. "Four dead. Eleven seriously injured. Another dozen beat-up but not broken."

Valery sat up straighter. "We held the camp, Joe."

"I know," he said. "But they got further in than they should've."

The corpsman looked up. "She's walking out of here, but not into anything that involves running or bleeding again."

Valery glared. "I can still shoot."

Amy put a hand on her shoulder. "You'll do that from behind a desk today."

Joe folded his arms. "You're on comms until Morales clears you. No field work."

Valery opened her mouth—then shut it again when she saw he wasn't budging. "Fine," she muttered. "But I hate it."

"Good," he said. "Means you'll actually rest."

The tent flap rustled again as Manny and Guido walked in—both filthy, both soot-covered, both wearing expressions that said they had bad news and worse news.

Joe turned to them. "Report."

Manny scratched the back of his neck. "Generator's alive, but barely. They blew two fuel drums, torched the housing, and fried half the wiring. She'll run water pumps and med lights, but that's it."

"No perimeter lights?" Joe asked.

Guido shook his head. "Unless you want the whole grid to go dark. We'll have to go old-school tonight."

Joe nodded. "Fine. Lanterns and torches. Rotate them so nothing runs dry."

Manny exhaled. "People are scared, Joe. They saw the red marks. They heard the shouting. They know it wasn't some random scavenger raid."

Valery shifted, wincing. "We have to tell them Harrow's behind it. They'll hear rumors."

"Oh, we will," Joe said. "But we're not telling anyone how many he's recruited. Or what he stole before the strike."

Amy crossed her arms. "How bad is it, really?"

Joe's voice dropped. "Bad enough he's not working alone anymore. Bad enough he's got families under him now. Bad enough he's planning something bigger than last night."

Manny let out a slow curse. Guido kicked the leg of a cot gently, jaw tight.

Amy stepped forward and put a hand on Joe's forearm. "Then you make sure he doesn't get a third chance."

Joe nodded once. "That's the plan."

Valery met his eyes, her voice quiet but fierce. "So what's our move?"

Joe answered without hesitation.

"We fix what he broke. We secure what he missed."

Then his voice hardened.

"And then we go find where he sleeps."

Outside the med tent, Ridge Camp moved with a kind of stunned discipline. People spoke in low voices, as if anything louder might crack the thin shell of "we're okay" everyone was clinging to.

Joe stepped out first. The light made him squint; smoke still hung in the air like a film. Morales fell into step beside him without a word.

Manny and Guido trailed a few paces back, trading quiet mutters about voltage and pipe fittings.

They walked toward the generator pit.

The earth there was blackened and warped. Two of the fuel drums were scorched open, steel peeled back like someone had tried to open them with a giant can-opener. The generator housing itself was blistered on one side, panels warped from the blast. The smell was acrid and oily.

Donnie was already there, crouched in the rubble with a wrench in one hand and a rag in the other. His jaw was clenched in that "I'm not yelling because yelling won't fix metal" way Joe recognized.

"How bad?" Joe asked.

Donnie didn't look up immediately. He finished tightening a bolt, then stood, wiping his hands on his jeans. "She'll live," he said. "But she's not happy. We rerouted what we could. You've got pumps, med, and a thin line to comms. That's it."

"No perimeter floods," Morales said.

"Not unless you want this thing to die of spite," Manny added. "We're running her at the bare minimum now."

Joe nodded. "Keep it that way. I'd rather be thirsty in the dark than dead with the lights on."

Guido nudged a twisted piece of shrapnel with his boot. "They knew exactly where to hit. This wasn't random."

"Nothing Harrow does is random," Morales said.

Charlie appeared on the edge of the pit, a notebook clutched in both hands, Lenny pressed close to his leg. His eyes were too wide for a boy his age. He tried to smile anyway.

"Dad, I finished the rotation list," he said. "Like you asked."

Donnie took the notebook, flipping through the rows. Names and times were written in careful, crooked columns. "Looks good," he said. "Guard leads can work with this."

Charlie beamed. "Can I help with the fences?"

Joe stepped closer, resting a hand briefly on the boy's shoulder. "You helped plenty," he said. "You and Lenny stay inside the main line today. No wandering past the markers."

Charlie swallowed but nodded. "Yes, sir."

Lenny wagged like he agreed, though his gaze kept sliding toward the burned area.

As the boy led the dog away, Morales watched them go. "He's going to remember last night for the rest of his life," she said quietly.

"So will everyone else," Joe replied. "The trick is making sure what they remember is that we held."

They left the generator pit and headed toward the eastern perimeter, where the attack had started. The ground there looked chewed—boot prints, drag marks, scattered brass. Two of the outer posts were half-collapsed, their supporting stakes shattered.

Morales crouched near a cluster of tracks. "These aren't ours," she said. "Different tread pattern. Lighter boots. They came in low and fast."

Guido pointed with his chin. "And they came close. That's maybe fifty yards from the first family tent."

Joe studied the ground, following the prints with his eyes. "They ghosted the tree line all the way in, hit fuel and power first, then probed the housing line."

"Classic harassment raid," Morales said. "Cause chaos, test defenses, count response time."

"Find the weak spot," Manny added. "Then hit it hard next time."

Guido swallowed. "Comforting thought."

Something caught Joe's eye—a strip of cloth fluttering from a low branch near the trail. He walked over, tugged it free, and turned it in his hand.

It was torn from a coat sleeve. Painted across it, bright and ugly despite the soot, was the red lash.

He held it up.

"Subtle," Guido muttered.

Morales swore under her breath. "He left that on purpose."

Joe folded the cloth once, knuckles whitening around it. "He's done hiding," he said. "He wants us to know he's close."

Manny looked out over the trees, jaw tight. "Then he's either confident or stupid."

"Harrow's never been stupid," Morales said.

Joe tucked the marked cloth into his jacket. "Double posts on the east and north lines," he said. "Two people minimum at every high point, rotating every ninety minutes. Nobody walks perimeter alone after dark. Nobody."

Morales nodded. "Done."

"And tonight," Joe added, "we go full low-tech. Trip wires, cans, bells, dogs if we have to. Harrow knows our hardware. Let him find out we're mean without it."

Guido snorted. "I can make a symphony out of scrap and string."

"Good," Joe said. "I want him to hear us coming if he gets stupid enough to try again."

They stood there a moment longer, the burned ridge at their backs, the shadowed forest ahead. The ground was scarred, but it was still theirs.

"Last night," Morales said finally, "he showed us his reach."

Joe nodded once. "Today, we decide how far we'll reach back."

By the time Joe and Morales reached the command tent, the canvas walls buzzed with voices—quiet, tense, determined. Ridge Camp felt like a shaken hive: wounded inside, guards at the entrances, smoke drifting in slow spirals from the areas still smoldering. The attack had changed something fundamental. The air felt sharper. People walked differently—heads up, eyes scanning, hands never far from tools that could double as weapons.

Inside, the long table was already crowded.

Manny stood with a grease-stained notebook, flipping pages with quick, restless fingers. Guido sat backward in a chair, forearms draped over the backrest, tapping a wrench against his knee in an impatient rhythm. Donnie leaned over a crude map sketched in pencil and charcoal. Amy hovered near the entrance, arms crossed—her apron still streaked with dried blood.

Valery slipped in behind Joe a moment later, sling hanging loose at her side like a formality she resented. Her face was pale, but her eyes were sharp, scanning the room for any hint of bad news.

When Joe stepped to the head of the table, the tent fell quiet.

Morales spread out the map—creased, smudged, edges curled from being handled too roughly in too many long nights. "We need to rework the perimeter," she said. "East slope is a soft entry. They used it. They'll use it again."

Donnie nodded. "We'll reinforce it with two raised platforms. One here—" he tapped the ridge lip, "—and one here. Overlapping sight lines. I'll build both by tomorrow morning."

Manny looked up from his notes. "Panels on the generator are holding, but it's a patch, not a fix. We need proper wiring and new plating from Arnold. Otherwise we're gambling every time someone breathes near the housing."

"No generator means no pumps," Amy said. "No pumps means half this camp dehydrates within two days."

Joe nodded once. "Arnold run goes on the board for the next forty-eight hours. But not today. Today is about securing ground."

Valery shifted, ignoring the pain that flickered across her face. "He's not done. Harrow isn't someone who tests a wall once. He tests until he finds the crack."

Guido lifted his wrench. "We found a crack all right. He slipped through it like a rat on roller skates."

"Which means," Manny said, "we need to plug it. And fast."

Joe placed both hands on the table. "We're not reacting anymore. We're organizing."

He pointed to the map, sweeping his finger across the tree line.

"Harrow's been busy. We know he has at least a dozen fighters. Probably more. He's collecting families too. Maybe willingly, maybe not. Either way, last night wasn't his opening move—just the loudest one so far."

Amy leaned forward. "What do you mean 'collecting families'?"

Joe hesitated—not because he didn't want to tell her the truth, but because it sat wrong in his mouth. "We found signs in the woods. A temporary camp. Bedding for kids, food remains, small footprints."

Morales added quietly, "He's moving civilians. Not just combatants."

Amy's hand went to her chest. "Why? What could he possibly want with families?"

Guido answered darkly, "He wants followers. People who owe him. People who fear him."

"People who will fight for him," Valery said softly.

Joe nodded. "He's building a force—not just with guns, but with desperation."

Manny flipped his notebook closed. "What's the plan?"

"Two phases," Joe said. "One: secure Ridge Camp so last night never happens again. Two: locate Harrow's base. Find out how many people he's got, how armed they are, and what he's planning."

Morales crossed her arms. "Recon run?"

"A small one," Joe confirmed. "Tomorrow morning, we follow the trail he left. He was sloppy last night. Footprints, markings, even a carved symbol in a tree. He wants us to know the general direction."

Guido muttered, "Real considerate of him."

Valery stepped closer to the table. "You need comms support."

"You're injured," Amy snapped immediately.

Valery didn't break eye contact with Joe. "I can sit in the command tent. Monitor radio. Track any transmissions from his band. I won't be useless."

Joe gave a small, grateful nod. "Good. I need you listening. Morales, Manny, Guido—you're with me."

Morales's expression didn't change, but she shifted her stance—centering herself like someone preparing for a long fight.

"And what happens," Amy asked carefully, "if you find Harrow?"

Silence hung heavy for a breath.

Joe exhaled slowly. "Then we watch. We count. We map. And we come home."

Guido raised a brow. "No punching?"

"No punching," Joe said.

Manny lifted a hand. "No shooting either?"

"No shooting."

Morales finally spoke. "So... just marking territory? Like wolves?"

Joe managed the ghost of a smile. "Exactly."

Valery leaned back, a shadow passing behind her eyes. "This isn't over," she whispered. "Not even close."

Joe nodded. "That's why we stay smarter than him."

Outside, the wind shifted—bringing the scent of charred earth, diesel, and something colder.

A promise of the next fight.

Dawn sharpened into full daylight, but Ridge Camp still felt wrapped in the bruised quiet after a storm. The fires were out, the smoke thinner, yet the air carried that metallic aftertaste of fear. People moved with purpose, but their eyes lingered on shadows longer than before.

Morales stood at the ridge entrance as Joe, Manny, and Guido returned from the tree line. She'd been watching the woods like she expected them to spit out ghosts. When she saw the plastic dinosaur in Joe's hand, her posture changed—just slightly, just enough.

Below, camp life pulsed unevenly. Donnie and his crew hammered new bracing onto the fuel shed. Carlos and Jin—two of the ten military volunteers who had rotated into security—hauled scrap metal to reinforce the east barricade. Hannah and Luke, the two med-support volunteers, carried fresh bandages into the tent while Amy reorganized a shelf like she could will order back into the world.

It looked like rebuilding, but everyone knew it was bracing.

Joe walked down the slope, and conversations faded as people watched him pass. Not out of fear—out of want. They wanted answers. They wanted certainty. They wanted him to tell them it was over.

He couldn't.

Morales followed at his shoulder. Manny and Guido flanked him, quiet for once, boots sinking into the damp soil as they approached the central fire.

Valery sat on a crate nearby, radio tablet open across her knees. Her sling was clipped back into place only because Amy yelled at her until she complied. She wasn't listening to the radio—she was listening to the silence between signals. When she saw Joe, something unreadable flickered across her face.

"Any chatter?" Joe asked.

"No broadcasts from Harrow's band," she said. "Just standard static. The calm part is what worries me."

"Good," Joe said. "Let him think we're licking wounds."

Guido kicked a stray rock into the fire pit. "We kind of are licking wounds."

"We're allowed," Manny muttered, brushing pine needles off his sleeve. "We got hit hard."

Joe nodded. "We did. But we held. That matters."

Morales scanned the faces gathering around them—families, guards, engineers, cooks, mechanics—the full heart of Ridge Camp. "They need to hear that from you."

Joe took a breath, stepped up onto one of the new logs Donnie had set as seating for the kids' learning circle. The crowd quieted, slowly, like a tide receding.

He didn't shout. He didn't need to.

"We survived last night," Joe said, voice steady. "We fought back. We protected each other. That's who we are."

A murmur rippled through the camp—affirmation, exhausted but real.

"Harrow hit us because he thinks fear will break us," Joe continued. "But he doesn't understand something important. Fear doesn't break us. It focuses us."

He lifted the small plastic dinosaur, holding it up for the closest groups to see.

"We found signs of families in the woods. People he's been moving around like supplies. People following him because they don't know we're here—or because they're scared to do anything else."

Several people in the crowd stiffened. A woman near the back covered her mouth. Donnie looked away, jaw tight.

"We're not letting him grow stronger," Joe said. "We're not waiting for him to strike again. Tomorrow morning, we track him. Quietly. Carefully. We learn what he's built. And then…"

He let the words settle like dust after an explosion.

"…we take it apart piece by piece."

The reaction wasn't cheers. It wasn't applause.

It was a deeper sound—a collective, quiet exhale of a people who'd decided they weren't prey.

Amy stepped forward, resting a hand on Joe's back for a moment before moving to address the crowd. "Get water. Get food. Check on your neighbors. Keep your kids close. Stay smart. We're strong because we do this together."

Behind her, Charlie stood with Lenny clutched to his chest, the dog's ears twitching like he understood every word.

Valery watched Joe with something fierce and unspoken in her eyes. She didn't argue about being benched. She didn't joke. She simply nodded once—a soldier's promise.

Morales clapped Joe on the shoulder. "Recon at first light. No exceptions."

Joe nodded. "We finish it before he does."

The wind shifted across the ridge, carrying away the last curls of smoke.

Ridge Camp wasn't healed.

But it was awake.

And angry.

And ready.

CHAPTER 20
THE HARDENING

Morning came without softness. Ridge Camp moved like a machine that had slept with one eye open. The attack was behind them, but the tension stayed in the air, sharp and clear, guiding every step people took across the ridge. No one laughed. No one talked about anything except what needed fixing or what needed guarding. It was the first morning in months where the entire camp acted like a militia instead of a community.

Joe met Morales at the command tent just after sunrise. She already had the maps spread out and three clipboards stacked in front of her.

"Training starts today," she said. "Everyone who can stand gets a place."

Joe nodded. "Let's get it moving."

Morales walked outside and blew the metal whistle Donnie had rigged for her. It cut across camp fast. Within minutes people gathered—new volunteers, guards from last night, older teenagers who wanted to help, a handful of former fleet workers who'd never held a rifle but refused to hide anymore. Morales counted them fast.

"Thirty-two," she told Joe. "More than we expected."

"Fear motivates," Joe said.

She didn't argue.

Valery arrived next, sling tight, face pale, but focused. She stayed by the comm table they'd dragged outside so she could run radio checks and monitor nearby frequencies. Amy had made her promise not to lift anything heavier than a pencil, but Valery's eyes were everywhere, following the movement of every guard and volunteer like she was tracking a puzzle only she could see.

Donnie and Manny rolled out the equipment cart they'd thrown together overnight—rubber training knives, wooden rifles, spare radios, whistles, ropes. Guido followed carrying three first-aid bags and a bucket of sand in case someone passed out. He grumbled the whole way, but he didn't stop moving.

Joe stepped forward and raised his voice. "Listen up! Last night wasn't the worst of it. Harrow is still out there, and he has families with him. That means he has food, shelter, and reasons to fight harder. So today we start preparing to take the initiative. No more waiting."

The group nodded. Some looked scared, some determined—but all of them listened.

Morales pointed to the first group. "You four—rifle safety and reload drills. You three—defensive formations. Back row—medical basics with Amy. Manny and Guido, you're with me on physical drills."

She turned back to Joe. "You get the new recruits. Give them basics they can actually use."

Joe stepped in front of six volunteers—three men, two women, and a teenage boy who insisted he was seventeen even though he looked fourteen.

"None of you are soldiers," Joe said plainly. "That's fine. You don't need to be soldiers. You need to follow orders, keep your head clear, and understand basic survival when bullets start flying."

He handed each of them a wooden rifle.

"Rule one—don't aim at anything you're not prepared to shoot. Rule two—don't freeze. Freezing gets you dead. Rule three—if Morales gives an order, you follow it even if you don't understand why."

One of the men lifted the wooden rifle and held it awkwardly. "Feels weird without any weight."

Joe adjusted the man's stance. "Then hold it until your arms remember it. Tomorrow you get the real one."

In another corner, Morales ran the more advanced group like she'd been born doing it. She corrected stances, adjusted footwork, shoved people sideways to see if they'd hold balance. Two recruits fell. She made them get up and try again.

Manny led rope drills—climbing, tying emergency restraints, pulling victims out of danger zones. He wasn't gentle, but he was clear.

Guido handled the first-aid station, barking orders at anyone who touched a bandage wrong.

Amy supervised him with a raised eyebrow. "Don't yell unless the wound's real," she told him.

He grinned. "I can yell preemptively."

"No," she said.

Inside the perimeter, people not training rebuilt—hammering platforms, securing water lines, repairing the scorched generator housing.

Ridge Camp wasn't mourning anymore. It was preparing.

And that was exactly what Joe wanted.

Training spread across the ridge like a second sunrise. The air filled with the thud of boots, the snap of wooden rifles, the sharp commands Morales barked across the camp. Sweat hit the dirt. People stumbled, corrected themselves, tried again. Every mistake they made now was one less they'd make when bullets were real.

Joe moved from group to group, checking stances, correcting grips, making sure nobody lost fingers to enthusiasm. The teenage boy—Noah, finally admitting he was fifteen—kept tripping over his own feet.

"Slow down," Joe told him. "You're not racing anyone."

"I'm trying!" Noah said, red-faced.

"No," Joe corrected, "you're rushing. Rushing gets you shot. Plant your feet. Look first. Move second."

He repositioned the boy's hands on the wooden rifle and nodded. "Try again."

Noah took a breath, stepped forward, and this time didn't fall. The grin on his face was brief, but real.

Across the training yard, Manny blew his whistle and pointed at two recruits tangled in a rope line.

"No, no, no—if you tie a granny knot like that in the field, the only thing you'll be rescuing is your dignity after Morales yells at you."

Morales shouted from ten feet away, "I already am yelling at them!"

Guido, bandana soaked with sweat, squatted next to a volunteer struggling to apply a pressure wrap to a dummy arm. "Gentle! This is a person, not a salami."

The volunteer loosened their grip.

"Not that gentle," Guido snapped. "You're not petting it."

Amy walked by, swatted Guido's shoulder with the back of her hand, and handed the volunteer a fresh roll of gauze. "Ignore him. He was raised by wolves."

"Very medical wolves," Guido muttered.

On the far edge of camp, Valery sat under the shade of a tarp, headset on, monitoring all frequencies. Amy checked on her hourly, ensuring she stayed seated and didn't wander. Valery pretended to obey. She hated being sidelined, but she understood the reason—her entire left side ached every time she breathed too deep.

Still, her mind was sharp.

Every time she glanced up, she watched how the recruits moved. How they followed orders. Who froze. Who thought too much. Who thought too little.

When Morales jogged over for a status check, Valery said quietly, "Group Three is too slow in reload drills. Group One's left line keeps opening when they pivot. And Noah—he steps with his toes instead of his heel. Fix that or he'll sprain something."

Morales stared at her. "You're not even facing them."

"I don't have to," Valery said. "I can hear their patterns."

Morales blinked once. "You're terrifying. Stay terrifying."

"I intend to."

Joe called the first break after an hour. People collapsed on crates, panting, clutching canteens like they were gold. The younger ones sprawled flat on the dirt. A few older men leaned on each other, laughing despite themselves.

But even during the break, there was no drift into comfort. Eyes kept sweeping the tree line. The memory of last night's fire still burned too hot to forget.

Joe took a swig of water, then clapped his hands once. "All right! Form up! We're starting drills with movement under pressure."

Groans rose like a choir.

"Don't complain," Morales snapped. "If you're going to hit the dirt, you might as well learn how not to die in it."

She sent her group to the slope, where she had set up a small obstacle course between brush and crates. Volunteers crawled, ducked, scrambled, swore, and occasionally face-planted. Morales didn't soften —not even when someone fell hard.

"You fall," she barked, "you get up."

One young woman hesitated after tripping. Morales knelt beside her.

"You scared?"

The woman nodded, embarrassed.

"Good," Morales said. "Fear is information. Listen to it. Move anyway."

She hauled her up and sent her back through the course.

Joe watched from a distance. This was what he needed—people waking up to the reality of what came next. Not panic. Not despair. Readiness.

Donnie joined him, wiping sweat from his brow. "We're getting there," he said. "Gonna take time, but we're getting there."

Joe nodded. "They'll be ready."

Donnie gave him a long look. "Will you?"

Joe didn't answer immediately. His gaze swept the camp—the volunteers pushing through drills, the wounded resting under shade, Valery watching everything with quiet fire.

Finally, he said, "I have to be."

Donnie nodded once. "Then we're good."

The whistle blew again, sharp and demanding.

Break was over

Training resumed.

And Ridge Camp kept hardening.

After the second round of drills, the sun climbed higher, burning away the morning haze. Sweat slicked every face on the ridge. Even the breeze felt sandpaper-hot. Still, nobody quit. Not today.

Joe gathered the next wave of trainees—those who'd handled the basic rifle work without dropping the weapon or themselves. He walked them to the shaded edge of the ridge for the next lesson.

"Cover and movement," Joe said, pointing down the slope. "One person shoots. The other moves. You keep each other alive. If you move alone without cover? You die alone."

The group stiffened, taking the words seriously.

He paired them off. Two men who didn't like each other. A mother of two with the ex-electrician from Deck Four. Noah with a retired botanist who'd once been too afraid to step onto the floating pier without holding someone's sleeve. Today he held a wooden rifle with both hands steady.

Joe raised his voice. "First pair, go!"

The electrician laid down a steady stream of mock fire while the woman moved to the next crate, stumbled, caught herself, then dove behind cover. When she lifted her wooden rifle again, her eyes were fiercer.

"Good!" Joe called. "Next!"

The botanist surprised everyone with his speed, sliding into position smoother than any of them expected. Noah nearly tripped again, but when Joe barked, "Fix it!" the boy planted his next step cleaner.

Joe felt it then—a flicker of something long buried. Not hope. Not pride.

Readiness.

A hum of movement behind him signaled Morales approaching with her squad of volunteers. She was dust-covered, scraped, and looked like she'd fought the obstacle course herself.

"They're stiff," she said, voice low. "But getting sharper. Give me three days and they'll move like they actually want to survive."

Joe nodded. "They will."

Morales scanned the group. "You're thinking."

"I'm always thinking."

"Too much, usually," she said.

Joe didn't deny it.

Across the ridge, Manny climbed onto a crate and whistled loud enough to make three volunteers jump.

"All right, rope rats!" Manny called. "We're doing emergency drags. Nobody faints or I drag you back using your face!"

Guido rolled his eyes. "Such motivation."

"You want to teach it?" Manny asked.

"No," Guido said immediately.

They had the volunteers practicing rescue pulls—looping rope under arms, dragging bodies behind cover, hauling weight uphill without tripping. One man dropped the rope twice, panting, face flushed.

"Faster!" Manny shouted.

"I'm trying!"

"Try when you're baking bread. Move when you're saving lives!"

Amy arrived with a bucket of water and two cups, handing them out where needed. She looked tired, but relentless.

"Small sips," she warned. "If you chug and puke, I'm not mopping it."

Guido grinned. "She's so nurturing."

"Keep talking," Amy said. "I'll find more chores."

Not far away, Valery rose from her seat, sling hanging loose. Amy shot her a warning glare.

"Sit," Amy said.

"I'm just stretching."

"You stretch sitting."

Valery huffed and sat again. Amy didn't buy it.

Joe approached. "How's the signal board?"

"Quiet," Valery said. "Too quiet. Either Harrow's regrouping, or he's hunting. Neither option is comforting."

"Nothing new on the coded bands?"

"Same pattern. Three shorts, one long. Same spacing. Same intervals. He's broadcasting confidence."

Joe frowned. "That's what bothers me."

Valery leaned back, watching the training yard. "He's watching. Somewhere. Maybe thinking we won't be ready in time."

Joe followed her gaze. "Let him think that."

"He won't for long."

"No," Joe said, "he won't."

The third round of drills began. Morales led formation movement —shifting positions, rotating lines, tightening defense. She corrected people with taps, nudges, and hard truth.

"You hesitate, someone dies," she said, steady and even. "Don't think—react with control."

A woman flinched at the weight of the wooden rifle. Morales pulled her aside.

"You scared of the weapon?" Morales asked gently.

"Yes," the woman admitted.

Morales nodded once. "Good. Fear keeps your hand steady. Now hold it correctly."

Joe watched the transformation—how fear turned into resolve, how trembling hands steadied under instruction.

Guido shouted, "Medic drill!" and half the volunteers dropped to the ground while the other half scrambled to assess fake wounds.

"Too slow!" he barked. "Bleed-out happens in minutes, not next week!"

A volunteer tried to lift an unconscious dummy the wrong way.

"Back straight!" Guido yelled. "You're not made of spare parts!"

Joe allowed himself a breath. Not relief—never that—but something close to acknowledgement.

They were getting there.

Ridge Camp wasn't an army. Not yet. But it was learning.

Learning to fight.

Learning to move.

Learning to harden.

And Joe intended to make sure they reached the finish line before Harrow did.

Afternoon heat pressed down on Ridge Camp like a hand meant to test its strength. Instead of buckling, the camp pushed harder. The drills grew louder, sharper, more focused. Every thud of boots against dirt echoed like a vow.

Joe gathered his core fighters—Manny, Morales, Guido, and ten of the military volunteers who'd stepped forward after the attack. These ten weren't Navy reactor techs or tug operators. They were former infantry, security, and aviation gunners—men who'd known violence long before the Firestorm.

Joe addressed them simply, no theatrics. "You're the backbone. When we go after Harrow, you're the ones leading the push. Today we drill discipline."

One of the volunteers, a tall ex-Army specialist named Jaxon, nodded. "Orders, Commander?"

"I'm not a commander," Joe said. "But you follow my lead."

"Yes, sir."

Joe looked them over—faces hardened by survival, not authority. They weren't fleet officers. They weren't polished. But they knew how to hold a line.

He pointed to Morales. "You're taking first squad."

She jerked her chin. "Copy."

"Manny, you're second."

Manny groaned. "I hate second."

"Then move faster," Morales said.

"Rude."

Guido smirked. "I'll support whichever team cries first."

"I vote Manny," Jaxon said.

"Traitor," Manny muttered.

But the exchange eased the tension the fighters all carried.

Joe clapped once. "Form up!"

The squad leaders moved their groups into triangular formations—tight, efficient, two covering one. Joe walked between them, correcting posture and spacing.

"When you advance," he said, "you move like an animal with three legs—one limb moves while the other two anchor. You never expose your flank. You never charge blindly. You don't win fights by being brave. You win by being smarter than the men shooting at you."

Jaxon's squad nailed the formation instantly. Morales tested them by shoving the rear man. He compensated without falling out of pattern.

"Good," she said. "Now again. Faster."

They did.

Manny's squad... slightly less graceful.

"Stop drifting left!" Manny shouted at one man.

"I'm not drifting!"

"You're leaning like a dying goat!"

Guido cupped his hands. "Does anyone know how goats die?"

"Quiet," Manny snapped.

"You're welcome."

Still—after the third attempt, the squad tightened. Their movements clicked into sync, not perfect, but workable.

Joe nodded. "Better. Again."

In the center of camp, Valery monitored their movements from her communication post. She no longer flinched at every burst of shouting or crash of bodies hitting dirt. She studied every pattern, noting weak points, calculating corrections.

Amy approached with another round of water jugs. "How's your shoulder?" she asked.

"I've had worse."

"You've complained louder."

"Not recently," Valery muttered.

Amy smirked. "Because you know I'll sedate you."

"And miss all this?" Valery asked, gesturing toward the training yard. "Not a chance."

Amy placed a firm hand on Valery's good shoulder. "At least pretend to rest."

"I'll rest when Harrow's buried," Valery said quietly.

Amy swallowed. "Just don't let that bury you too."

Valery glanced toward Joe—running drills like a man preparing his own replacement.

"He won't let it," she whispered.

Beyond them, near the lower edge of ridge, Donnie worked with several volunteers to reinforce the watch platforms. Idiot-proof, he called them. High enough to see danger, low enough that falling wouldn't break anyone's neck. Charlie and Lenny brought him nails and wooden pegs with military solemnity.

"You are my best crew," Donnie told them.

Charlie puffed up with pride.

Lenny sneezed on a nail.

Another section of camp was dedicated to med training. Amy and two corpsmen drilled volunteers in wound packing, tourniquets, splinting, and emergency drags.

"Pressure here," Amy said, demonstrating. "If you hesitate, they bleed out. If you panic, they bleed out. If you pass out—Guido will mock you for the rest of your life."

Guido shouted from across the ridge, "Truth!"

A teenager paled. Amy steadied him. "Don't worry. You'll get used to the blood. Most of it won't be yours."

The boy swallowed hard. "That's supposed to help?"

"No," Amy said. "But it's true."

Back at the formation yard, Morales tested her squad with simulated suppression fire—popping caps from a training launcher. The volunteers hit the dirt instantly, spread correctly, returned fire with wooden rifles.

Joe smiled faintly. "They're getting it."

Morales grinned. "Oh, they're terrified of me. Terror motivates."

Manny jogged over. "My group hasn't cried today. That counts as improvement."

"Your group cried yesterday?" Guido asked.

"Shut up."

Joe lifted his hand. "Fall in!"

Both squads formed into two parallel lines. They were winded, dripping sweat, covered in dirt—

—but they stood straighter than they had that morning.

"Listen up," Joe said. "This isn't to turn Ridge Camp into soldiers. It's to keep your families alive."

The line stiffened.

"You're fighting men who follow orders without conscience. Men who have families of their own, but they still choose Harrow. They won't hesitate. Neither can we."

Silence. Hard, heavy, ready.

He looked from one face to the next.

"This is just the start. Tomorrow we drill close quarters. Day after, coordinated movements. We're not meeting Harrow halfway. We're beating him to his next move."

Morales stepped up beside him. "And if he tries anything before that? We're ready."

Joe nodded once. "Dismissed."

The squads broke with a ripple of exhausted relief.

Training wasn't just working.

It was transforming them.

Making them sharper.

Faster.

Harder.

Ridge Camp was no longer a settlement.

It was becoming a force.

By late afternoon, the drills slowed—not because anyone wanted to stop, but because bodies finally started to shake and misstep from fatigue. Morales blew her whistle once, short and sharp.

"Stand down!" she called. "Water, food, ten minutes. Then review."

The order rippled through the ridge. People sagged where they stood, some dropping to sit right in the dirt, others leaning against crates or tent poles. Sweat streaked through soot on faces. Hands shook when they took cups from the water line, but nobody spilled much. They were past shaky now. They were learning control.

Amy walked through the resting lines with Charlie and two teenagers carrying buckets. "Small sips," she reminded them. "If you chug and throw up, I'm not cleaning it."

Guido lifted his cup. "You're all heart, Amy."

"Drink," she said. "Less talking, more recovery."

Morales made a slow walk across the ridge, watching how people sat, how they breathed, how quickly they steadied. Joe joined her, arms crossed, scanning like a man taking mental inventory of every potential weak link—and every potential leader.

"They're tired," Morales said.

"They need to be," Joe answered quietly. "Not broken. Just tired enough to remember how much last night cost us."

Morales kicked lightly at the dirt. "We're close to having something real here. A unit. Not just bodies who know where the trigger is."

Joe nodded. "I saw three of the new recruits cover each other without being told. Two days ago they would've run in opposite directions."

"Fear bonded them," Morales said.

"Training keeps them alive," Joe replied.

Across the clearing, Manny had collapsed on a crate like a man whose soul had stepped out for air. Ashley handed him a cup of water and shook her head. "You taught ropes like you were trying to strangle someone."

"That's how you know I care," Manny muttered, wiping sweat from his brow.

Valery, still anchored at the communications table, watched every movement on the ridge with sharp, unblinking focus. She wasn't in the drills, but she wasn't resting either. The sling lay across her lap now instead of holding her arm up—it was the closest she'd come to admitting she hurt.

Joe walked toward her. "You holding up?"

She didn't look away from the radio readings. "I'm sitting still. Isn't that what everyone wanted?"

"Not what I want," Joe said. "I want you healed."

"You'll get that eventually," she said. Her voice was level, but her eyes flicked up to meet his. "But you'll get information faster."

Joe didn't argue. She was right. She always was when it came to comms.

She slid a small notebook toward him. "Frequency logs from the last twelve hours. No sign of Harrow's beacon. Either he's lying low or he moved."

"Moved," Joe said. "He tested us. Now he's planning."

Valery tapped her pen restlessly. "Then we train harder."

Before he could answer, Morales blew the whistle again. Everyone stood—aching, stiff, but standing.

"All right!" Morales called. "Final drill. Pair up. Defensive stance only. No attacks, no dodges, no falls. Just hold your line."

Groans rose from every direction, but people obeyed. Even the teenagers straightened their backs and lifted their makeshift rifles.

Joe watched them settle into two long rows—faces serious, stances shaky but improving.

He stepped forward and raised his voice. "Listen up! Harrow hit us once. He'll try again. Next time he'll bring more than torches. Maybe more rifles. Maybe more men. Maybe families he forced to fight for him."

Silence spread, heavy and sharp.

Joe continued, voice steady. "When he comes, we're not meeting him as victims. Or survivors. Or civilians. We meet him as a united front—with discipline, with teamwork, with control."

Morales walked the line like a drill sergeant forged from wildfire. "Hold steady. Every second you stay balanced now is one less second you bleed later."

People braced themselves. Feet planted. Shoulders squared. Lines tightened.

It wasn't perfect.

It wasn't even pretty.

But it was becoming something strong. Something dangerous.

Joe stepped back beside Morales. "Day one," he said. "And they already look different."

Morales nodded. "Give me a week?"

She didn't have to finish. Joe already knew.

Ridge Camp wouldn't just defend itself.

It would strike back.

And Harrow wouldn't be ready for what this ridge was turning into.

CHAPTER 21
INTO THE TREE LINE

The morning hit Ridge Camp like a cold slap. No fog, no softness, just sharp air and the smell of burned diesel drifting off the east slope. People moved fast, purposeful, carrying ropes, rifles, tool kits—anything that would help rebuild what Harrow tried to break. Today wasn't about fear. Today was about preparation. The attack had shaken the camp, but it hadn't cracked it. If anything, it welded the people tighter.

Morales stood in the center of the ridge with a clipboard in one hand and a grease-pencil map in the other. She barked orders like she was building a fortress with nothing but her voice. "Perimeter teams, rotate every ninety minutes! Spotters stay on the platforms and keep your eyes open! If something moves and you don't recognize it, call it in before you blink!"

Joe stood beside her, double-checking the recon packs he planned to take into the tree line. He had stripped them to essentials—water, radios, med kits, blades, spare ammunition. No noise-makers, no flares. Everything had to be silent. Everything had to be fast.

Valery sat near the main radio mast, her sling on, headset pressed to one ear, logging frequencies and adjusting the gain with her good hand. She wasn't complaining, but anyone with eyes could see the pain sitting just beneath her breath.

Morales leaned toward her. "You good?"

Valery didn't look up. "I'm not getting shot today, so yeah. Fantastic."

Morales smirked. "Stay sharp."

"Always," Valery answered.

Amy emerged from the med tent carrying a box of bandages and antiseptic wipes. She passed Joe and touched his arm gently—the small grounding kind of touch only a mother could get away with. "You come back with all ten fingers, all ten toes," she said. "I don't have time to sew you together today."

"I'll behave," Joe promised.

Donnie jogged up from the water line carrying a wrench twice the size of his forearm. Charlie followed with Lenny sprinting circles around his ankles.

"Pressure's stable," Donnie reported. "Filters are holding. Give me two hours and the tank supports'll be reinforced."

"Good," Joe said. "We need everything locked down before nightfall."

Manny and Guido arrived next carrying the last of the recon gear. Manny handed Joe a compact radio. "Short range only. Won't bounce off anything. Clean signal."

Guido ducked his head to peer inside Joe's pack. "Food? Water? Your sense of humor?"

Joe zipped it shut. "Just the important things."

Guido held up a small coil of wire. "Trip sensor. In case you need something that screams when you don't."

Joe added it to the pack without comment.

Behind them, the militia trainees gathered again—tired, bruised from yesterday's drills, but ready.

Morales stepped forward. "Today's focus is infiltration patterns and coordinated movement. No live fire. No heroics. Just discipline." She clapped her hands and pointed. "Pairs! Move!"

People scrambled.

Manny passed Ashley a wooden rifle to use for formation drills. Their fingers brushed briefly—her cheeks flushed, Manny cleared his throat and pretended a sudden interest in the dirt—but neither said anything.

Guido noticed and raised one eyebrow so high it practically left his skull.

Manny shot him a glare.

Guido grinned. "Looks like somebody's got—"

"Say it and I'll tie your boots together," Manny muttered.

"Promises, promises," Guido murmured.

Ashley didn't hear—they'd already moved into position beside the other trainees—but a tiny smile tugged at the corner of her mouth.

Joe ignored the exchange and stepped up onto an overturned crate.

"Listen up!" The crowd snapped to attention. "Recon team leaves in thirty minutes. Morales has command until we return. Your job is simple: strengthen the ridge. Tighten the perimeter. Drill until you're too tired to lift a rifle. Because when the time comes to move on Harrow's camp, I can't have anyone freezing."

The trainees nodded. The older volunteers nodded too. Even the teens straightened their shoulders.

Joe continued, "We're not marching out today. We're not fighting today. Today is about readiness. Today is about discipline. Harrow isn't going to hit us the same way twice. So we stay ahead of him."

He paused, scanning the camp—253 people rebuilding their home after last night's firestorm.

"We didn't choose this fight," Joe said. "But we sure as hell finish it."

Morales blew her whistle. "Drill stations! Move!"

The ridge exploded into controlled motion. People sprinted to assignments. Hammers rang. Rifles clicked. Radios chirped.

Ridge Camp wasn't afraid.

Ridge Camp was sharpening.

Joe checked the straps on his pack one more time. Then he tightened his grip on his rifle and looked toward the tree line where Harrow had vanished the night before.

"Let's get to work," he murmured.

And the day began.

Thirty minutes later, Ridge Camp watched its hunters leave.

Joe, Morales, Manny, Guido, and ten handpicked militia walked single file past the last tent line and into the shade of the trees, rifles low, packs snug, eyes scanning. No speeches. No drama. Just boots, breath, and the quiet clink of gear.

Valery's voice crackled softly in Joe's earpiece. "Radio check. Team One, confirm."

"Team One copies," Joe murmured. "Signal's clean."

"Roger that," she replied. "You're on a tight leash. You go quiet more than five minutes, I start yelling."

Morales smirked. "She will too."

"I know," Joe said.

The forest swallowed the camp noises fast. Within twenty yards, the hammering, the shouts, the clank of metal were gone, replaced by wind through branches and the faint drip of last night's rain sliding off needles. The ground was soft, the soil dark and churned in places where Harrow's people had already walked.

Manny knelt at the first cluster of tracks. "Same pattern as yesterday," he said quietly. "Boot size, gait, direction. They weren't hiding."

"They wanted speed," Morales replied. "They got it. We get information."

Guido scanned the tree line, squinting. "I hate how pretty it looks out here," he muttered. "Trees, birds, murderous lunatics. Real postcard stuff."

One of the militia—a tall woman named Reyes—hid a nervous smile. The others stayed quiet, eyes wide, taking it all in.

Joe pointed ahead. "We follow the heaviest path but we don't step in it. I want our trail as clean as possible. Reyes, your rear. Torres, you're on our left line. Kim, you're right. Call anything that looks wrong."

They moved in a loose diamond, disciplined but flexible. Every crackle of twigs drew a glance. Every shifting shadow got a second look.

After ten minutes, Morales lifted a hand.

The column froze instantly.

She pointed with two fingers at the ground ahead.

There, half-hidden by leaves, was a bare footprint beside a boot print. Smaller. Narrow. Toes splayed.

"Not just soldiers," she murmured.

Joe's jaw clenched. "Families."

"Or hostages," Guido said.

Manny shook his head. "Hostages don't leave prints that deep. That's someone who's been walking like this for a while."

One of the younger militia men, Davis, swallowed hard. "Why would they follow him? After what he did at sea?"

"Because he offers something," Joe said. "Safety, power, a story where they're not the ones dying. Scared people listen to whoever shouts loudest."

He stepped around the tracks and moved on. Nobody stepped in them. Nobody wanted to feel that weight under their boots.

The farther they went, the clearer Harrow's route became—branches cut, brush kicked aside, a strip of worn earth winding like a scar across the slope. At one point the trail split, then rejoined, like he'd sent a small group ahead and pulled them back once the path was clear.

Morales studied the split. "He's probing terrain," she said. "Not just wandering. He's thinking in patrols."

"Yeah," Guido muttered. "Hate it when the enemy's competent."

A few minutes later, Manny raised a hand. "Hold."

They stopped behind a clutch of rocks and low scrub. He pointed ahead, where the ground dipped slightly. The soil there looked wrong—too smooth, too clean, like someone had brushed it with a hand.

"Trip line?" Reyes whispered.

"Maybe," Manny said.

Joe stayed low, eyes narrowed. He pulled a small stick from the ground and gently swept it across the dip.

The stick snagged.

Fine fishing line shimmered when the light caught it, barely visible. It ran from one sapling to another, leading to a rusted tin can filled with rocks, balanced on the edge of a higher branch.

"Noise trap," Guido said softly. "Cheap but effective."

"Would've given us away," Morales added. "And told them we can't see past our boots."

Joe looked back at the group. "This is why we train," he said. "You don't rush ahead because you're scared or eager. You move smart. You move alive."

He carefully lifted the can free and lowered it to the ground without a sound, then snapped the line with two fingers. "Leave it," he said. "No sign we were here."

They pushed on.

The forest thickened, the light thinning into green and gold shards. Somewhere above, a crow called once, harsh and lonely. The air smelled like damp earth and old leaves—a world that didn't care about their war.

Joe's radio clicked once. Valery's voice came through, low and even. "Team One, status."

"Still moving," Joe answered. "You're clear on this band?"

"Nothing but background hiss," she said. "No new beacons. No chatter. He's quiet."

"Means he's busy," Morales muttered.

"Means he's planning," Joe replied.

They reached the small clearing where they'd found the temporary camp yesterday—tarps, ash, scattered bedding now gone. The fire pit was cold, stamped out. The red lash symbol remained carved deep into the tree.

Someone had added a second stroke beside it.

Fresh, wet, darker.

Guido stared at it. "That wasn't there yesterday."

"No," Joe said quietly. "It wasn't."

Morales' eyes swept the trees. "He knows we're following."

"Good," Joe said. "Then he also knows we're not blind."

Manny crouched at the edge of the clearing, fingers sifting soil. "Pressure marks here… and here. They stood in a circle. At least fifteen, maybe twenty people. Some smaller prints. Kids again."

Reyes' voice shook a little. "They're living like this?"

"For now," Joe said. "Desperation makes a lot of things seem normal."

He dug into his pack and pulled out a small grease pencil. On the back of his glove, he jotted rough numbers. "Headcount estimate: twenty. Weapon count unknown, but assume overarmed."

"Assume they're always overarmed," Morales said.

Joe nodded. "We go another half mile, then we turn back. This isn't the day we bump their main camp. We're mapping routes, nothing more."

"Copy that," Valery said in his ear. "I'm tracking your timestamps. If you go one step over your limit, I'm sending Morales' angry twin as backup."

Morales snorted. "I don't have a twin."

"Exactly," Valery said. "Imagine if you did."

Despite the tension, Manny smirked. The sound of Valery's voice brought a faint ease into his shoulders he didn't bother to hide.

Ashley's voice came faint over the shared channel from camp, background noise he hadn't expected: "Radio test from Ridge. We're running drills on frequency two. Don't worry, I won't break your toys."

Manny rolled his eyes but couldn't stop the ghost of a smile. "You break it, you fix it," he muttered under his breath.

"What was that?" she asked.

"Nothing," he said. "Just making sure you're not touching my calibration settings."

"Relax," she replied. "I like my generators not exploding."

Morales gave Manny a sideways look. "Focus," she mouthed.

He cleared his throat. "Focused."

The recon team pressed on until the trees began to thin and the ground rose toward a rocky outcrop. Beyond it, faint in the distance, lay the suggestion of movement—nothing clear, nothing close—but enough that Joe lifted a fist and dropped to one knee.

"Stop here," he whispered.

The others sank into the brush, rifles ready, breaths slow.

Somewhere out there, beyond the next rise, Harrow's world waited.

Today wasn't the day to walk into it.

Today was the day to learn how.

The recon team held position in silence, the forest wrapped tight around them like a closed fist. Nobody moved. Nobody breathed loud. Joe scanned the ridge line ahead—just shadows and trees, but something about the air felt… watched. Not seen, not stalked, but observed with quiet calculation. Harrow's style.

Morales shifted beside him, her voice low enough to barely exist. "No movement. No silhouettes. Nothing on the wind."

"Doesn't matter," Joe whispered. "We're close enough for them to hear us if we sneeze."

Guido exhaled slowly. "Great. And I had pepper powder on my breakfast."

Manny elbowed him. "Try surviving five minutes without commentary."

Guido's lips twitched. "Impossible. I'd die."

Joe motioned them backward with two fingers. No talking. No cracking branches. No sudden noise. They eased away from the ridge line, step by slow step, until the trees swallowed the slope again and the sense of being watched faded like a hand letting go of their collars.

Once they were deep enough, Morales finally spoke. "They have sentries out there. Had to."

"Yeah," Joe said. "But they didn't take the bait. They didn't reveal themselves."

"Smart," Guido said. "Annoying, but smart."

Joe knelt and drew a rough map in the dirt with a fallen twig. "This ridge here—too open. They've got high ground beyond it. Probably a lookout post. We circle wide next time."

"Or we lure them out," Morales said.

"Not yet," Joe replied. "We don't know their numbers."

Manny tapped one section of the drawn path. "Based on yesterday's tracks? Twenty, maybe thirty people."

Guido shook his head. "And maybe forty more we didn't see."

Joe didn't respond because they all knew it was possible.

He erased the map with his palm. "We turn back."

They headed down the slope, careful and quiet. The forest was warming now, sunlight threading between branches in bright, narrow spears. Manny pushed aside a thicket of brush and frowned. "Hey, Joe —look."

A ribbon of fabric hung from a low branch. Gray, frayed, torn.

Morales lifted it. "Kid's shirt."

"Fresh tear," Manny said. "Edges aren't weathered."

"Last twelve hours," Joe estimated. "Maybe less."

Guido rubbed the back of his neck. "So families are still on the move."

"Which means Harrow's consolidating," Morales added. "Pulling everyone closer to his new camp."

Joe's jaw tightened. "Yeah. And we're running out of time."

They kept moving, weaving between trunks until they reached the wide bend of the creek. The water shimmered under the morning light, soft and deceptively peaceful. Moss coated the rocks. A frog croaked somewhere upstream.

Joe paused, crouched, checked the muddy bank.

Footprints. Deep ones.

"Two adults," he murmured. "Carrying weight."

Manny crouched beside him. "Heavy weight. Supplies?"

Morales shook her head. "Or a child."

Silence hit them like a drop in air pressure.

Then Joe stood. "We head back now. No delays."

They moved faster on the return—still quiet, still deliberate, but with purpose threading each step. Even Guido stopped cracking jokes, which meant the gravity had fully settled in.

Halfway back, Joe's radio clicked once.

Valery's voice whispered through, calm but edged. "Team One, be advised—camp reports movement east of your trail. No visuals. No sound. Just disturbed brush picked up by a perimeter patrol."

Morales stiffened. "Distance?"

"Half a mile from your last ping. Could be animals. Could be watchers."

Joe said, "Copy. We'll be eyes-up."

Then another voice came over the channel—Ashley's. "Just to be clear, we're not dealing with bears, right? Because I'm not prepared for bears today."

Manny sighed loudly. "Ashley—stop talking."

She snorted. "Relax, Manny. I'm just trying to lighten—"

A burst of static cut her off before she said lighten the mood, but Manny's tiny smirk said he'd heard enough.

Guido nudged him. "She's flirting with you, genius."

Manny flushed. "No she's not. She's—she's distracting the channel."

"Uh-huh," Guido said. "Distracting you, specifically."

Morales smacked them both lightly on the arms. "Eyes forward. You can gossip when we're not two minutes from someone's ambush."

They resumed their careful descent. Through the trees, Ridge Camp began to reappear—smoke from the cooking fires rising in thin threads, the distant clink of hammers, voices calling orders. Civilization breaking through wilderness.

When they stepped out past the last pine, Valery was already at the perimeter with her radio slung across her good shoulder, watching them like she'd been rooted there since dawn.

"You're early," she said.

"We had enough," Joe replied. "We didn't push the line."

Her eyes raked over the group, counting heads with military precision. "Good."

Guido dropped onto a crate dramatically. "You miss us?"

"I missed silence," she said. "Camp's been buzzing since you left. No one likes waiting."

Joe stepped closer so only Valery and Morales heard him. "We saw fresh movement. Families. Kids. And Harrow marked the clearing again."

Valery's face went still. "He knows you're tracking him."

"Yeah," Joe said. "And he's inviting us."

Morales cracked her knuckles. "Then we accept it. On our terms."

Joe nodded once. "We start planning tonight."

Guido flopped back, staring at the sky. "At least tell me there's food first."

Before Joe could reply, Ashley jogged up with a radio coil in her hands. She glanced at Manny—quick, shy, trying to play it off as nothing.

"You, uh… dropped this earlier," she said.

Manny blinked. "No, I didn't."

"Well, you did now," she said, handing it to him.

Their fingers brushed. Manny froze like an idiot. Ashley turned away too fast, cheeks red.

Guido whispered loudly, "Oh boy."

Morales sighed, "Not now."

Joe looked toward camp again, expression steady and grim.

Harrow was waiting.

And now, Ridge Camp had seen enough.

Time to move from defense to preparation—for the war they all knew was coming.

Afternoon training hit Ridge Camp with the force of a second wind—hard, fast, and loud. The recon team barely had time to grab water before Morales dragged Joe to the temporary training field. The volunteers waited in two uneven lines, sweaty, bruised, and determined, wooden rifles hanging awkwardly at their sides.

"Round two," Morales barked. "No excuses. No quitting. If you faint, Guido drags you to Amy. If you throw up, try not to do it on my boots."

Guido raised a finger. "New rule: if you puke near me, I'm filing for reassignment."

"No one cares," Morales snapped.

Valery stood at the comm table again, but now she leaned on it with her good arm, field notes arranged in neat rows. She wasn't on the field, but she wasn't resting either. She tracked each drill with a sharp, calculating stare—studying weaknesses, memorizing who hesitated, noting whose hands shook.

Joe stepped forward and clapped once. "Formation drills first. Pair up."

The recruits scrambled.

Manny led a rope course on the left side—drag lines, obstacle crawls, improvised stretcher hauls. Ashley had been assigned to help him, mostly because she had working ribs and sharp instincts for knots and pulls.

"Okay, rookies!" Manny called. "If you can't drag a full-grown adult thirty feet, how the hell are you going to pull someone out of a burning tent?"

A volunteer raised her hand. "We won't be near burning tents, right?"

Manny didn't even blink. "Tell that to the people from last night."

Ashley cut a rope and tossed him the end. "You're being too intense again," she said under her breath.

"It's war training," Manny muttered. "Intensity is appropriate."

She smirked. "So is not making them cry."

"I'm not—" Manny started, but Ashley already stepped away, shouting, "Okay, line up! Don't listen to Manny. He had caffeine for breakfast."

Guido overheard. "All of us had caffeine for breakfast."

"Yeah," Ashley said, "but you need it to function. Manny just becomes a menace."

Manny shook his head and pretended he wasn't secretly admiring the way she took command.

Across the field, Joe corrected stances, adjusting hands and feet, shifting shoulders with gentle but firm pressure. He could tell who had natural reflexes and who wasn't cut out for the front line. That was fine. Not everyone needed to be front.

"Morales," Joe said quietly, "we need three tiers. Advance, support, perimeter."

She nodded. "Already started the list."

She pointed with her chin at a trio who kept forgetting to chamber imaginary rounds before firing. "Back line—support only. They freeze every time someone yells."

"Agreed," Joe said.

"And those two?" Morales indicated a pair moving through cover drills like they'd been born in tactical gear. "Advance."

"Absolutely," Joe said.

Valery coughed once loudly. "Or," she called, "you could test them under pressure before making the list."

Morales narrowed her eyes. "Define pressure."

Valery raised her good arm and hit a switch.

An airhorn detonated from somewhere behind the tents.

Half the recruits shrieked.

Three dropped their rifles.

One fell over completely.

Morales froze mid-step. "Valery."

Valery shrugged. "Realistic simulation."

Joe fought a smile. "I'm not against it."

Ashley clapped, laughing. "Do it again!"

"No," Manny groaned. "Please don't."

But Valery was already adjusting the volume.

"Val," Joe warned.

She held up one finger—just one—and pressed the button again.

The second blast sent one recruit sprinting in the wrong direction, another flat on his stomach, and Manny—and this happened purely by accident—straight into Ashley's arms.

They froze.

Guido's jaw dropped. "Oh my sweet God."

Ashley blinked up at Manny. "You okay?"

Manny stepped back so fast he tripped over his own rope coil. "I—yes—I mean—no—I mean—fine—totally fine—never better."

Ashley looked like she wanted to laugh but held it in.

Morales slapped the airhorn off. "Valery, if you touch that again, I'm taking the battery out with my teeth."

Valery raised her hands. "Noted."

Training resumed—louder, sharper now that everyone's nerves were shot. The airhorn incident had a side effect: no one froze anymore. The fear was already shaken out of them.

Joe watched as one recruit belly-crawled through the mud without complaint. Another practiced reloading until her hands stopped shaking. Manny coached a young man through a drag test until the kid's face went red but didn't quit. Ashley steadied a rope line for two teenagers who couldn't tie knots to save their lives—literally.

Ridge Camp wasn't a militia.

Not yet.

But it was becoming one.

And that shift—slow, painful, necessary—was exactly what Joe needed before they made their next move.

As the sun dipped lower and the shadows stretched longer across the training field, Morales called for final drills. "One more set! Then water, then regroup!"

Joe glanced toward the tree line they'd explored that morning. His jaw tightened. "Harrow's watching us," he murmured.

"Good," Morales said. "Let him watch."

Joe nodded.

"Because when we're ready..." Morales added, eyes narrowing, "he won't see us coming."

Training ended the way all hard days should: with sweat, bruises, and silence. Not defeat—just bones-deep exhaustion that meant people had pushed themselves past what they believed they were capable of. Morales dismissed the recruits with a curt wave. "Wash up, hydrate, eat something that isn't fear," she said. "We're back at it at dawn."

The volunteers scattered—some toward the water barrels, others toward the mess tent where the kitchen crew was already ladling out stew. A few curled up on crates, heads tilted back like they might sleep right there in the dirt.

Joe remained on the field a moment longer, hands on his hips as he watched them go. For hours he'd been caught in the rhythm of training, of fixing stances, teaching reloads, sharpening instincts. Now his brain returned to the clearing in the woods. To the footprints. To the toy dinosaur still tucked in his jacket pocket. A reminder of what Harrow was doing. Who he was using.

Morales stepped beside him, shaking dirt from her gloves. "They're not ready," she said bluntly.

Joe nodded. "Not yet. But they're better than they were this morning."

"They'll need to be better than that," she said. "If we're going to hit Harrow's camp and get everyone back alive."

Joe didn't argue. Couldn't. His jaw flexed once before he exhaled. "Tonight we rest. Tomorrow we build them into something that can't be shaken."

Morales snorted. "Rest? You planning to teach the entire camp how to do that too?"

He allowed a small smile. "Maybe."

She bumped her shoulder against his lightly. "Get cleaned up. You smell like fear and hard decisions."

"And you smell like scorched rope," he countered.

"That's called leadership," Morales said, turning toward the path. "Try it sometime."

As she walked off, Joe caught sight of Valery stubbornly trying to stack crates one-handed. He moved toward her before she could pretend she wasn't struggling.

"I thought Mom said nothing heavier than a pencil," he said.

Valery froze mid-lift, then glared. "This crate is emotionally lightweight."

He took it from her hands easily. "Sit."

"No."

"Val."

She huffed and dropped onto a crate like a rebellious teenager. "Fine. Five minutes. But I'm not fragile."

"I know," he said quietly. "That's the problem."

Valery glanced at him then, eyes tired but fierce. "You think I'm going to fall apart?"

"Never," Joe said. "But I think you bleed. And I'd like you to stop proving that point."

She looked down at her sling. "I hate sitting out."

"I know," Joe said. "But right now I need you alive more than I need you out there."

Her breath caught—not from pain this time, but from something softer, deeper.

She broke eye contact quickly. "Don't get sentimental. It makes me itchy."

Joe smirked. "Wouldn't dream of it."

Across the way, Manny and Ashley cleaned up the rope course. Manny tried to rewind a rope coil but managed to tangle it around his own arm.

Ashley took it from him gently. "If this rope had feelings, you'd be arrested."

"I was having a moment with it," Manny said defensively.

"Yeah," she laughed, "a traumatic one."

He watched her hands as she worked—quick, sure, efficient. "You're good at this," he said, almost surprised.

"Being useful? Yeah, I try."

"No," he said, rubbing the back of his neck. "I mean… you make things feel less overwhelming."

Ashley paused, warmth touching her expression—real, soft, unguarded. "You do that too, Manny."

He flushed instantly. "I—uh—okay—I mean—cool—good."

Ashley grinned and walked off before he could further implode.

Guido, coming up behind Manny, muttered, "Smooth as sandpaper dipped in vinegar."

"Shut up," Manny said, but he smiled anyway.

As dusk bled across the ridge, camp life settled into a quiet hum—pots clattering, murmured conversations, the faint yip of Lenny chasing moths near the tents. The day had been brutal, but the evening carried something new: resolve. Real, collective, heavy resolve.

Joe stood on the overlook above the tree line, the last of the sunlight fading behind him. Morales joined him again, arms crossed as she surveyed the forest.

"Recon at first light," she said. "Teams are already prepping packs."

Joe nodded slowly. "We'll take six. Quiet. Light. No heroics."

Morales arched a brow. "That last rule applies to you too."

He didn't deny it.

Instead, he reached into his pocket and pulled out the toy dinosaur. Morales stared at it, her expression darkening.

"Souvenir?" she asked.

"Warning," he said.

She exhaled long and slow. "We take him down soon."

"Yes," Joe said, gaze hardening as the forest deepened into darkness. "Very soon."

The wind shifted across Ridge Camp, carrying the scent of pine, smoke, and something colder beneath it—anticipation.

They weren't just prepared now.

They were committed.

And Harrow had no idea what was coming.

CHAPTER 22
INTO THE DARK

Morning dragged itself over Ridge Camp gray and reluctant, as if it wanted no part in what was coming. The sky hung low, heavy with clouds that hadn't decided whether to break or brood. Nobody spoke above a murmur. Nobody joked. The air tasted like metal and anticipation. Today wasn't a rebuilding day. Today was the day they prepared for the strike.

Joe stood inside the command tent, staring at the map like it might finally talk if he glared hard enough. Morales hovered over his shoulder, tapping her pen against the rough-drawn terrain. "Three entry points," she said. "North slope, east ravine, or the creek line. Creek is quietest but the mud will slow us."

Joe nodded slowly. "North slope gives us height. East ravine gives us cover."

"And Harrow knows that," Morales muttered. "He'll have eyes on both."

Valery sat at the radio table, sling cinched tight across her ribs, hair tied back, expression hard enough to cut rope. Her fingers worked across the dials and switches without hesitation. She was pale—still hurting—but she wasn't sitting this out. "East ravine has the least chatter," she said without turning around. "I've been scanning all morning. No movement on the lower frequencies."

Joe glanced over. "You were supposed to rest."

"I did," she said. "For a whole ten minutes."

Morales smirked. "Practically a vacation."

Valery shot her a look. "I'll send you a postcard next time."

The tent flap rustled and Manny and Guido entered, each carrying gear bags and a stack of neatly rolled rope. Manny looked like he'd slept three hours. Guido looked like he'd slept upside down in a tree. "Okay," Guido said, dropping the ropes. "We brought the essentials: rope, radios, duct tape, morale issues, and Manny's bad jokes."

Manny shoved him lightly. "My jokes are a coping mechanism."

"They're a threat to national security," Guido replied.

Joe ignored their bickering—mostly because it was the only sound in the tent that didn't feel like fear wearing boots. "Spread everything out," he said. "We check and double-check. Nothing goes wrong because someone forgot batteries."

Valery interrupted without looking up. "Check them again anyway. I'm not trusting Guido with batteries."

Guido pointed at her. "One time. One time the flashlight died in a tunnel and suddenly I'm the reason we're all blind."

"You tripped into a wall," Manny said.

"It was dark," Guido hissed.

Joe rubbed his forehead. "Can we stay on task for five minutes?"

Morales snapped her fingers. "Alright, children. Focus."

They did.

Morales slid the map closer to the lantern. "So—team of ten. We move at dusk. Silent, staggered formation, no radios unless someone's bleeding."

Joe added, "Valery stays here with the comms team. She'll monitor every band we've got."

Valery stiffened but didn't argue—not out loud. Her eyes tracked Joe, sharp and unreadable. She knew this part mattered. The bench wasn't punishment. It was protection.

Manny began sorting gear into piles—blades, compact radios, spare magazines, cable ties. Guido checked each radio twice, then smacked one against his palm when it sputtered. "This one's moody," he said.

"Makes sense," Manny said. "It belongs to you."

Ashley slipped into the tent quietly, her arms full of folded bandages and antiseptic cloths. She handed them to Manny, their fingers brushing briefly—too briefly—but she still flushed. Manny pretended he didn't notice, which fooled absolutely no one.

"Got the med supplies," Ashley said softly. "Amy's packing more, but she wants to make sure you all know you're not allowed to come back with extra holes."

Guido raised a brow. "Define extra."

Ashley gave him a deadpan stare that could peel paint. Manny chuckled under his breath.

She lingered a moment longer—long enough for her eyes to flicker to Manny again—then excused herself and slipped away.

Guido elbowed Manny. Hard. "You're welcome."

"I swear to God," Manny muttered.

Morales coughed once—loudly. "Back to work."

Joe knelt beside the map, tracing a line with his finger from Ridge Camp to Harrow's territory. "We go in quiet," he said. "We gather intel. We find his numbers, his escape routes, his supply lines. No engagement. Not yet. Tonight is reconnaissance only."

Morales nodded. "Tomorrow we plan. After that?"

Joe's eyes darkened, the decision already carved into his bones. "We end this."

Valery spoke softly from the radio table, her voice steady but edged with something fierce. "He won't see you coming."

Joe looked at her—really looked—and nodded once, a silent promise passing between them.

Morales slapped the last clipboard closed. "Then that's it. Today we train. Tonight we scout. Tomorrow… we make the decision Harrow's been forcing on us since the moment he broke out of that brig."

Joe stood, tightening the straps on his vest. "This is the last calm before the storm. Use it."

The tent fell quiet—not afraid quiet, but the charged, electric quiet of people about to walk knowingly into danger. Nobody asked if they were ready. Ridge Camp didn't have the luxury of questions anymore.

They were doing this.

And by nightfall, the path to Harrow's end would already be lit.

Word spread without anyone needing to say it aloud: Joe was picking the team.

By midmorning, a knot of people waited near the edge of the training yard—guards, ex-military, a couple of hard-eyed civilians who'd seen too much to back down now. They stood straighter when Joe approached with Morales, Manny, and Guido flanking him.

Morales didn't waste time. "If you're here because you think this is glory, leave," she said. "If you're here because you want revenge, leave. If you're here because you understand this is ugly work and some of you might not come back—stay."

Nobody moved.

Guido muttered, "Well, that's comforting," under his breath.

Joe stepped forward. "This isn't the big strike. Not yet. Tonight is recon. We go in, we look, we listen, we count, and we leave

Ridge Camp throbbed with controlled chaos through the afternoon. It wasn't loud—not the kind of noise that came with panic or confusion—this was different. Focused. Rhythmic. The steady cadence of a community sharpening itself like a blade.

Joe spent the first hour double-checking supply loads. Morales spent it threatening to strangle anyone who misplaced their gear. Manny spent it elbow-deep in the weapons crate, muttering about "idiot-proofing" the firing pins. Guido spent it taste-testing rations "for poison," which nobody believed but everyone tolerated because it kept him out of trouble.

The chosen twelve trained together away from the main group, in a strip of open clearing carved between two lines of pines. The ground there still smelled faintly of smoke from the attack, but it was stable—solid footing for practice.

Joe ran them through silent signals first.

Hand up—freeze.

Two fingers—eyes on target.

Palm down—drop.

Fist to chest—confirm visual.

Fist up—pull back immediately.

Some of them learned fast. Ortiz barely needed instruction. Harper mirrored Joe like she'd been born to it. Singh moved like a man who'd been waiting for something to fight for.

Others… not so seamless.

Briggs tripped over his own bootlaces twice, swearing both times. Tran kept mixing up the palm-down drop with the palm-forward halt. Nolan kept pushing his glasses up with the back of his hand because he refused to let go of the signal he was practicing.

But they were trying, and that mattered more than elegance.

Morales paced behind them like a wolf, correcting posture, grabbing shoulders, shoving people into cover positions. "You don't lean," she snapped at Phelps. "You lean, you die. Straight spine. Solid core. Commit to your stance."

He corrected instantly.

Joe watched her work with a faint nod. She'd been his right hand since the flood, and she didn't know how essential she'd become. She probably never would. Morales wasn't built to need praise—she was built to keep people alive.

Meanwhile, Manny and Guido worked with the second cluster—checking weapons, running dry-fire drills, testing trigger discipline. Manny handled technique. Guido handled nerves.

"Breathe," Manny told Harper. "In through the nose, out through the mouth. If your pulse spikes, your aim drifts."

Guido added, "And if you shoot me in the ass again, I'm haunting you."

Harper smirked. "I didn't shoot you in the ass."

"It grazed me," he said. "Emotionally."

Manny shook his head. "Ignore him."

Ashley approached with a crate of spare mags and neatly packed med strips, her braid falling over her shoulder. She set the crate down beside Manny. "Inventory's squared," she said. "And for the record, Guido, you can't be emotionally grazed. You barely have emotions."

Guido gasped dramatically. "I am wounded."

Ashley didn't look at him. "You'll live."

Manny's lips twitched.

She caught the almost-smile and paused. Not long. Just long enough for her heartbeat to skip and restart in a different rhythm.

"Try not to break anything," she said quietly, handing Manny the list.

He took it—and for a moment their fingers brushed.

Ashley pulled her hand back a half-second too late.

Manny didn't comment. But the small, flickering grin he tried to hide said enough.

Joe noticed. Morales noticed. Valery noticed from across the clearing and raised one eyebrow high enough to hit orbit.

But no one said a word.

Not yet.

—

By late afternoon the drills intensified.

Joe led the twelve into the tree line for silent movement exercises. They practiced stepping over roots, forming staggered lines, freezing at signals. They crawled, crouched, rolled behind cover, shifted position without speaking. Morales corrected mistakes with sharp taps to the back of the knee or the shoulder.

Guido tripped on a branch.

Manny muttered, "We are doomed."

Guido hissed, "I'm building tension. It's character arc."

"Your character arc is falling," Manny replied.

They kept going.

Again.

And again.

And again until sweat soaked shirts and breath steamed in the cooling air.

Joe didn't let them stop. Not when they got tired. Tired was when real mistakes happened.

He moved among them silently, correcting, tightening, shaping.

Preparing.

Because by nightfall, this wasn't training anymore.

It was survival rehearsal.

—

When the sun finally dipped behind the ridge, Morales called a halt.

"Eat. Hydrate. Check gear. After this, no talking. We move in silence until we come back." She pointed her whistle at Briggs. "Especially you. You breathe like a tractor."

Briggs opened his mouth—then wisely closed it.

Joe walked to the ridge edge, watching the sky bleed into early dusk. The trees beyond the slope darkened like something waiting. Something listening.

Harrow was out there.

Watching.

Breathing.

Preparing.

Joe tightened his jaw.

Tonight, we see you.

Morales stepped beside him. "They're ready."

"No," Joe said softly. "But they're as close as they can be."

Behind them, Valery adjusted the band of her headset, her eyes sharp despite the sling and exhaustion. "Radio check in ten. I'll be on all channels. Don't make me call search-and-rescue. I hate that."

Guido whispered to Manny, "She scares me."

Manny nodded. "Good. Fear keeps you alive."

Ashley stood near the med table watching Manny, unaware her expression had softened.

Ridge Camp moved around them—quiet, tense, alive.

Tonight they would step into Harrow's shadow.

Tomorrow, depending on what they found, they would plan how to drag him out of it.

Night spread over Ridge Camp like ink poured across the sky—slow, heavy, swallowing color until only lantern-glow and ember-light held back the dark. The attack had burned away whatever innocence the camp once had. Now every person—every family—moved with sharper purpose, as if the ridge itself understood there were no safe dawns left unless they made them.

Joe gathered the twelve at the far edge of camp where the ground dipped into the forest path. There were no torches here, no lanterns—

only shadow and the thin silver wash of moonlight. Morales stood beside him, rifle slung, jaw set. Manny, Guido, Ortiz, Harper, Singh, and the others formed a tight semi-circle.

Valery limped into view, ignoring the sling entirely. "Comms check in two minutes," she said. "Silent channel only. If you hear me say 'cloud bank,' that means pull back immediately. No exceptions." Her eyes flicked dangerously toward Guido, who immediately mouthed, "I wasn't going to say anything," even though he absolutely had been.

Ashley arrived carrying small medical packs. She handed one to each volunteer without fuss, her voice steady, her hands sure. When she reached Manny, he took his pack—then hesitated.

She looked up at him. "You come back with all your limbs."

"Which ones can I live without?" he asked lightly.

She didn't smile. "None."

He swallowed. Hard.

Ashley turned away quickly, but her ears were pink.

Valery, from ten feet away, muttered, "Oh for God's sake," under her breath.

Morales blew a thin breath through her nose. "Focus up. If Harrow's got scouts on the ridge, they'll see your hormones before they see your rifles."

Several people choked on their own spit in surprise.

Ashley fled back toward the med table.

Manny stared at the ground like it suddenly contained the secrets of the universe.

Joe didn't comment—he just clapped his hands once, sharp, pulling everyone's attention tight.

"All right," he said. "Listen up. Tonight is recon, not confrontation. We find their trail. We find how close they are. We find out what we missed last time. That's it."

He scanned each face, making sure they were truly hearing him.

"No one fires unless I fire first. No one breaks formation. No one plays hero. If anyone goes down, we all pull back. We do not push deeper. Understood?"

A murmur of agreement rose—quiet, unified.

Morales stepped forward, voice low but firm. "Pair up. Ortiz with

Singh. Manny with Guido. Harper with Briggs. Nolan with Tran. Phelps with me. Joe leads. Stay close. Stay alert."

They formed their pairs quickly.

Valery handed Joe a final radio pack. "I'll give you a ping every ten minutes," she said. "If you don't answer within three, I'm sending Morales's entire secondary team after you."

"I'll answer," Joe said.

"You better." She hesitated—just long enough that Joe noticed the crack in her voice before she masked it. "Don't underestimate him."

"I never do."

Valery nodded once and stepped back—but the look she gave him was pure steel.

"Go get the bastard."

They slipped into the forest like a shadow stretching long and thin. Joe took point, moving with a predator's slowness—calculated, precise. Morales followed right behind, her steps near-silent on the damp earth. The rest spread in staggered formation, eyes scanning, breaths low.

The woods were colder than the ridge. Quieter too. The kind of quiet that didn't feel natural.

Manny whispered, "Why is it always creepy at night?"

Guido whispered back, "Because evil doesn't clock out at sunset."

"Shut it," Morales hissed.

Silence resumed.

Joe stepped over a downed log and crouched, touching the churned soil. The trail was fresh—hours at most. The prints were deep. Fast-running. Multiple groupings. Two sizes—adult and small.

Children.

His chest tightened.

Morales saw the set of his jaw and said nothing, but her grip on the rifle shifted.

They moved further in, following the faint line of crushed underbrush. A piece of fabric caught Joe's eye—cotton, torn, smeared with mud. A child's shirt sleeve.

He tucked it silently into his belt.

Behind him, Manny muttered to Guido, "You feel that?"

"What?"

"The air. Feels like it's… waiting."

Guido shivered. "Yeah. Hate it."

The deeper they went, the thicker the trees grew—towering pines and gnarled oaks twisting together above them, forming a canopy that swallowed the moonlight whole. The air smelled of damp moss, old water, smoke, and something else…

Movement.

Joe lifted one hand—a flat halt command.

Everyone froze instantly.

Morales slid to his side. "What do you see?"

Joe didn't answer yet.

Not movement.

Not sound.

Something missing.

The forest had gone dead silent around them. No wind. No insects. Nothing.

Morales's voice was barely a breath. "We're in someone's space."

Joe nodded.

Behind them, Ortiz shifted her footing just enough to break a twig.

CRACK.

Every rifle lifted at once.

Every breath caught.

No attack came.

But Joe felt the hairs on his arms rise.

"They're close," he whispered. "Watching."

Morales's jaw flexed. "Fall back?"

"Not yet."

They moved another thirty yards—slow, deliberate.

And then they saw it.

A small clearing.

Dead fire pit.

Footprints everywhere.

Tarps. Torn bedding. Rope. Ash scattered in circles.

And carved into three different trees were the same symbol—

A slashed red lash.

Fresh.

Joe exhaled once, shaky.

Morales murmured, "He camped his families here."

"And he left this," Joe said softly, touching a carving. "For us."

Guido whispered, "It's a message."

"No," Joe corrected. "It's bait."

Behind them, Manny suddenly stiffened.

"Uh, Joe?" he whispered. "We've got company."

Joe turned.

Two silhouettes stood far off between the trees.

Watching.

Not attacking.

Not running.

Just… waiting.

Morales raised her rifle.

Joe lifted a hand. "Hold."

The silhouettes stayed motionless.

Then, as silently as smoke, they stepped backward into the trees and vanished.

Like they were never there.

Manny whispered, "What the hell was that?"

Joe didn't answer at first.

He finally said, "Scouts."

Morales nodded. "Watching. Measuring us."

Joe stepped back from the clearing.

"Retreat," he said. "Slow. Eyes open. Don't engage."

They pulled out carefully, the clearing shrinking behind them as the shadows swallowed it again.

Not a word was spoken until they saw the faint glow of Ridge Camp ahead.

Only then did Joe breathe fully.

Morales glanced at him. "You know what that was, right?"

"Yeah," he said quietly.

"A warning," she said.

"No," Joe murmured. "An invitation."

Ridge Camp looked almost peaceful when they broke out of the trees—a cluster of lanterns on the ridge, fires banked low, silhouettes moving between tents. From this distance you could almost pretend it was just a settlement on a hill and not a target someone had already marked.

Joe didn't let himself pretend.

They crossed the last stretch in silence. Valery was waiting at the perimeter line, headset crooked around her neck, sling still ignored. She scanned their faces fast—counting heads, measuring wounds.

"You're late," she said, voice tight.

"We're alive," Joe answered.

Her eyes flicked over the team again, as if checking the math. No one was missing.

"Report," she said.

"Inside," Joe replied. "All of you—grab water, then command tent in five."

Morales peeled off to dismiss the patrol pairs. Manny and Guido drifted toward the tool racks, the adrenaline wearing off and the bone-tiredness digging in. Ortiz, Harper, Singh, and the others headed straight for the water drums.

Valery fell into step beside Joe as he crossed camp. "You look like hell," she said.

"We met the neighbors," he answered.

"And?"

"And they wanted us to see them."

She swore under her breath. "Of course they did."

Inside the command tent, the air felt thicker, like it always did when too many decisions were about to be made at once. The map table waited, scarred and crowded with notes. Amy stood near the back with a mug of coffee she hadn't touched. Donnie leaned over a crate that doubled as a bench. Lenny lay at his feet, tail thumping once when Joe walked in.

Morales pushed through the flap second, followed by Manny, Guido, and Valery. The others stayed outside—this was core-circle business.

Joe planted both hands on the table. "We found one of his camp sites," he said. "Temporary. Families. Kids. Signs of at least four family units moving with him. Eight to ten armed adults. Maybe more."

Amy's face tightened. "Children? With him?"

"Yeah," Joe said. "He's not just building a squad. He's building a following."

Donnie swore softly. "Coward's insurance. You shoot at him, you hit civilians first."

Morales crossed her arms. "He carved his symbol into the trees. Fresh. Three of them in one clearing. Like a brand."

Valery's jaw clenched. "He wants us to know it's his ground."

Guido leaned back against a pole. "We also had two of his scouts watching us. Not far off. Didn't fire. Didn't run. Just stared until we saw them, then ghosted."

Manny added, "Felt like they were sizing us up. Counting rifles. Counting nerves."

Amy looked from one face to the next. "So what are we saying? That he's inviting us in?"

Joe met her eyes. "Yeah. He is."

The tent went quiet for a long beat.

Finally Morales said, "We can't ignore that. He's daring us to come into his reach. If we wait, he picks the time and place. Again."

Valery tapped a knuckle on the map, right where the northern tree line began. "He's testing our patterns. Our response times. Our rotations. He'll keep hitting the soft spots until there aren't any people left to train."

Amy set the untouched coffee down. "Then what do we do? We go after him with children in tow? We start something we can't contain?"

"We don't start anything," Joe said. "He already did. We just decide whether we fight on his terms or ours."

Donnie spoke next, voice slow, measuring. "We can't let him keep raiding outposts and dragging survivors into this. Every person he pulls to his side is one less pair of hands here—or worse, one more rifle pointed at our people. This doesn't stay small."

Guido nodded. "He's building his own little empire of desperate. The longer we wait, the more solid it gets."

Manny rubbed the back of his neck. "Question is: can we hit him without leveling his followers in the process?"

"That's the line," Joe said. "We take him, not the families."

Valery's eyes narrowed. "You really think he'll stand still and let you pick him out of the crowd?"

"No," Joe said. "But he's arrogant. He likes speeches. He likes being seen. Battle day he'll put himself where everyone can watch him. He always has."

Morales looked at him sideways. "You're talking like you already know how this ends."

Joe met her gaze, no softness in his voice. "We all know how this ends. It's just a matter of when."

Amy stepped closer to the table. "So when?"

"Not tonight," Joe said. "Not tomorrow. We need one more day. Tomorrow we lock down the camp. We rotate drills with real ammo. We brief the militia on exactly what we're walking into. Day after that, at first light—we move."

Two days. It sounded both too far and too close.

Valery studied him. "I'm coming," she said.

"No," Joe answered.

She opened her mouth.

He cut her off. "You're staying on comms. Eyes and ears. You're the only one who can track multiple bands and keep the whole picture in your head while we're in the mud. I need you here."

She hated it. Every line of her face said so. But after a long breath, she nodded once. "Then I want every movement reported. Every five minutes. If you go dark, I will drag everyone with a pulse out there to find you."

"Deal," Joe said.

Morales frowned at the map. "Twelve going out, plus the ten militia we've been shaping."

Amy's head snapped toward her. "Twenty-two?"

"We can't hit an armed camp with less," Morales said. "Not if we want to have a chance of winning without turning it into a slaughter."

Donnie's voice was quiet but firm. "I'll stay and hold Ridge Camp. Someone has to keep water flowing and lights on if this goes long."

Amy added, "And someone has to keep the wounded breathing when it's over. I'm staying."

Guido snorted. "Lucky us."

No one laughed, but the corner of Joe's mouth twitched.

He straightened. "Here's the line. In forty-eight hours, we take a strike team into Harrow's territory. Objective one: identify his camp layout and heavy weapons. Objective two: remove Harrow from the equation. If we can peel his people away in the process, we do it. If we can't, we still cut off the head."

Manny exhaled slowly. "And if it goes bad?"

"Then we fight our way back to this ridge," Joe said. "Or we die buying them another day up here. But we do not get taken alive into that camp. Understood?"

Every person in the tent nodded. No one looked away.

Valery's voice softened, just for a second. "You really think we can pull this off?"

Joe thought of the carved symbols. The plastic dinosaur. The toy still in his pocket. Of Charlie's determined little face. Of Lenny's bark during the night attack. Of Ridge Camp standing in the half-light—battered, scorched, but unbroken.

"Yes," he said. "Because we don't have a choice."

He looked around the tent one more time. Morales. Valery. Manny. Guido. Mom. Dad. The people who had dragged him through every fire so far.

"Get some rest," he said. "Tomorrow we drill until we can't see straight. Night after that, we end this."

They started to file out—Morales to adjust watch rotations, Guido to check the outer trip lines again, Manny to shore up the generator housing a last time.

Valery lingered at the flap. "Joe?"

He looked up.

"When it comes down to it," she said quietly, "and it's him or you..."

"It won't be him," Joe said.

For a heartbeat she looked like she might say something else.

Instead she just nodded once, that same soldier's promise from the night after the attack.

"Then don't miss," she said, and stepped into the dark.

Outside, the ridge wind rose, rattling canvas, tugging at ropes. Ridge Camp did not sleep easy that night. It sharpened itself—on fear, on anger, on the thin edge between survival and extinction.

Two days.

Then the dark wouldn't be something they waited through.

It would be something they walked into on purpose.

CHAPTER 23
THE LAST RECKONING

The sky wasn't even gray yet when they gathered at the edge of Ridge Camp. Just a smear of lighter black over the trees, the kind of hour that belonged to ghosts and bad ideas. Joe checked each face in front of him and found neither. Just tired people who'd decided they were done being hunted.

Ten militia volunteers stood in a rough line—men and women with worn boots, mismatched jackets, and borrowed rifles. Behind them, Morales, Manny, and Guido adjusted straps and checked mags like this was just another job. It wasn't. Joe could feel it in the air. This was the hinge. One way or the other, tonight changed everything.

Valery's voice crackled soft over his shoulder from the comms table they'd dragged out under the tarp. "Scope's clear so far," she said. "No radio chatter on Harrow's bands. You've got a clean window."

She had her sling on, but Joe could tell she hated every second of being stationary. The headset looked like a leash.

"Keep listening," Joe said. "If he sneezes, I want to hear it."

"You will," she replied. "And Joe?"

He glanced back.

"Bring them home," she said. She didn't mean Harrow.

Amy stepped up then, apron already on like she was heading straight back to med when this was done. Her eyes were red but steady. She put a hand flat on Joe's chest, right over his heart, like she was checking for something.

"You come back upright," she said. "I'm too old to start over training another son."

He huffed a small breath that almost counted as a laugh. "Yes, ma'am."

She squeezed once, then moved to Manny, straightened his collar, and kissed his cheek. "Don't let him be stupid," she told him.

"No promises," Manny said. "But I'll try."

Donnie came last, a wrench still hooked in his belt out of habit. He pressed a sheathed knife into Guido's hand. The grip was wrapped in worn black tape.

"In case things get stupid," Donnie said.

Guido looked at it, then at him. "You say that like stupid isn't the plan."

"Try not to die," Donnie answered. "I don't feel like training another you, either."

"Now that hurts," Guido said, tucking the knife at his back.

Joe slung his rifle, checked the sidearm at his hip, then looked over the team one more time. Ten militia, Morales at his right, Manny and Guido anchoring the line. They weren't an army. But they'd never needed to be.

"All right," he said. "We move quiet. We move smart. No one fires unless I or Morales call it. This is recon and removal, not a screaming contest. Harrow goes down, his men stand down, and we walk out with whoever he's been holding."

"Or?" one of the militia asked quietly.

"Or we improvise," Joe said. "But the goal is clean."

Morales blew out a breath. "You heard him. Last chance to change your mind."

No one moved.

"Good," she said. "Let's go."

They slipped into the trees single file, leaving Ridge Camp behind with only the faint clink of gear and the soft pad of boots on dirt. The sounds of camp faded fast—hammer taps, low voices, a child's laugh—and then there was only forest.

Except the forest was wrong.

After a few hundred yards, Joe realized what was bothering him. No birds. No insects. No wind threading the branches. Just a flat, dead quiet that made every breath sound too loud.

"This is bad ground," Manny muttered under his breath. "Feels like the world's holding its breath."

"Then don't give it a reason to exhale," Guido said.

Joe raised a fist, and the line halted automatically. He crouched by a fallen limb, fingers brushing a nearly invisible filament stretched between two saplings. Fishing line. Just high enough to catch a boot.

Morales knelt beside him, eyes narrowing. "Trip wire," she said. "Scavenged."

"Outer ring," Joe said. "We're close."

He snapped the line gently, coiled it, and pocketed it so no one behind him would catch it by accident. As he moved forward, he saw more signs—branches broken in a deliberate pattern, not storm damage; scuffed patches of dirt where boots had turned in place; cigarette ash ground into the soil in a small circle where sentries had stood too long.

"Harrow's boys are patrolling," Morales whispered. "Regular intervals."

Joe adjusted the column with a series of hand signals. Two militia peeled off to the flanks, one dropped back to cover their rear. The rest tightened up, less space between each body, more control if they had to hit the dirt.

Manny shifted his pack and muttered under his breath, "Could really use Ashley yelling at me about this being a bad idea."

The thought was quick, uninvited, and gone—but it left his gut tight. He hoped she was still asleep, dreaming about anything but him walking into somebody else's war.

"Eyes up," Morales said quietly. "Save the romance for after we don't die."

Manny shut up.

They kept moving, deeper into territory that felt like it belonged to someone else. The air picked up a faint metallic tang, like old engines and rusted steel. Somewhere ahead, something clanked—rhythmic, distant, like metal on metal.

"Gate," Guido breathed. "Or he's moving equipment."

Joe dropped to a knee, raised his hand, then tapped two fingers against his ear. Valery's voice came faint through the small radio at his collar.

"Still no broad transmissions," she said. "But I'm getting occasional short-range bursts. Local chatter. You're near the edge of his net."

"Copy," Joe murmured. "We're in his yard."

He signaled down. Everyone flattened, melting into brush and shadow. A moment later, voices drifted through the trees—two men, laughing too loud, words carried on the wrong kind of confidence. Outer scouts.

Morales met Joe's eyes and tilted her head forward slightly. Not yet. Wait.

Joe listened, counting their steps, their pauses, the way one's boots scraped more than the other's. Pattern, weight, range.

The team lay still, every muscle tuned. The world seemed to narrow to the sound of those approaching voices and the cold, simple fact underneath all of this:

They were close.

Closer than they'd ever been.

And the next move would decide who walked back out of these trees.

The scouts passed so close Joe could smell the cold grease on their rifles. One of them spit into the dirt, cursing about the cold. The other kicked a stone, bored out of his mind. Once their footsteps faded, Morales made a tight chopping motion and the team moved again—this time slower, deeper into the hush of the woods.

Thirty more feet and the trees thinned just enough to give them a partial view of another clearing. Not the main camp. A staging area. Rough shelters made of pallet wood and tarps ringed a dying fire pit. Piles of scavenged goods lay stacked under a tarp—buckets, old tools, half-full fuel cans, crates with faded "FEMA" stamps.

Manny surveyed it with a grimace. "He's been stripping every ruin within walking distance," he whispered.

Guido kicked a clod of dirt, then caught himself and muttered a curse.

"Families were here," Morales said quietly.

She was squatting by a patch of flattened ground—prints too small for adult boots, a discarded woven bracelet, the corner of a child's drawing half buried in leaves.

Joe felt something cold settle in his ribs. "When?"

"Yesterday. Maybe this morning," Morales answered, touching the dirt. "This spot is still warm from bodies."

Joe scanned the tree line. "Then where are they now?"

Manny pointed to the far side of the clearing. "Tracks heading north. Lots. Harrow moved them."

Joe gritted his teeth. "We're close."

Voices snapped through the air—sharp, irritated, unmistakably human. Everyone dropped low again. Two men stood near a stack of crates, half arguing, half whispering.

Joe motioned for quiet and signaled the militia to fan out. They moved into position like they'd been doing it for years, not hours.

The two men didn't notice.

"He wants everyone ready by sundown," one snapped. "Says it's time for the message."

"We're already ready," the other replied. "The hell else does he want?"

"He wants Ridge Camp to learn who's in charge," the first hissed. "He said tonight we show them what leadership looks like."

Morales' eyes sharpened.

Joe raised one eyebrow: Go?

She nodded.

Three militia circled wide, silent as the thought of regret. Joe crept from behind a tree, using brush as cover. Manny mirrored him from the opposite side. Guido moved in a drunken zigzag that somehow made no sound.

The two men kept talking, clueless.

"Harrow's pissed the last strike didn't break them. Says he'll do it himself this time."

"What? He's going there?"

"He's going to make them kneel, that's what he said."

Joe had heard enough. He made a fist-clench gesture—take them.

It happened fast and clean.

Manny swung behind the bigger one, arm snapping around his throat, locking in. Guido swept the legs out from the other and clamped a hand over his mouth before he could squeak. Morales moved like a flash, pulling zip ties from her vest and securing wrists behind backs in seconds.

Joe stepped forward and pressed a knee to the bigger man's spine. "Quiet. Breathe wrong and we end you."

Both men froze.

Joe jerked his chin at their camp. "Search it."

Manny and two militia dug through the crates. Morales checked pockets. Guido rummaged in a bag with unnecessary enthusiasm.

Quick results piled up fast.

A torn notebook with crude maps of the valley.

A list of names under the title LOYAL FAMILIES.

A page with supply counts.

And another sheet with a big, ugly number:

26

Beside a stick-figure rifle.

Manny held it up, jaw tightening. "Twenty-six armed men," he whispered.

Guido let out a low whistle. "Hell of a headcount."

Morales looked to Joe. "If he's got twenty-six with guns, he thinks he's a warlord."

"He's about to find out he's not," Joe said, pocketing the map. "We move."

Before they left, the smaller captured man choked out a muffled plea behind Guido's hand—something desperate, unintelligible.

Guido paused. "He's trying to say something."

Joe shook his head. "We can't risk noise. They stay gagged and tied until we're done. Then we bring them back."

Morales gave a sharp whistle—two notes, low. A signal to advance.

They crossed out of the staging area and into the deeper, darker woods beyond it.

The trees thickened. The air grew colder.

And somewhere ahead, faint through the brush, smoke curled upward in thin serpent trails.

Harrow's true camp.

Manny swallowed hard. "Well. This is where the fun starts."

Morales elbowed him lightly. "Keep your voice down."

Joe didn't turn, but his reply was ice-edged steel. "From here on, nobody breathes loud. We're in his shadow now."

And the shadow was getting darker with every step.

They reached the ridge overlooking Harrow's main camp just as the sun began pressing weak gray light into the treetops. Joe crouched behind a toppled log, scanning the ground below. The others knelt beside him, breaths held without needing to be told.

What lay beneath them wasn't a camp—it was a wound carved into the earth.

Fire pits smoldered lazily, the smoke curling around makeshift shelters built from pallets, tarp scraps, and twisted sheet metal. Old military crates had been turned into tables, barricades, or crude benches. The red lash symbol—Harrow's twisted brand—was painted across tree trunks in ugly streaks, like the forest had been clawed by something rabid.

Families huddled near the center. Children clutched each other or hid under blankets. Adults kept their eyes down, shoulders slumped in submission.

The militia around them froze. Hard men, trained or seasoned, went silent in a way that had nothing to do with stealth.

Morales whispered, "He's using them as shields."

Joe said nothing. His face didn't shift, didn't flinch, didn't betray the fury burning behind his eyes. But Morales saw it anyway—the kind of quiet rage that meant Harrow's hours on this earth were numbered.

Two armed guards paced near a larger shack—the only structure that looked intentionally built. Light flickered inside. A shadow passed across the window—broad shoulders, slow gait. Harrow.

Manny swallowed hard. "That's him. Has to be."

Guido flicked his eyes around the camp. "Lot more people than we thought."

Joe's voice was low but lethal. "They're not fighters. They're hostages."

A rustle came from the far end of the ridge. Joe whirled—only to stop when he saw her.

A woman. Early thirties. Dirt streaked across her face. She had a toddler pressed against her chest.

She saw Joe, Morales, Manny, Guido—saw the drawn weapons, the small formation of militia behind them.

She didn't scream. She didn't look afraid.

She lifted one trembling finger to her lips.

A silent plea:

End this.

Please.

Joe nodded once—barely a motion—then gestured to his team.

Spread out.

Quiet.

No sudden moves.

Down by the fire pit, Manny froze suddenly, breath catching in his throat.

A little girl was staring at him from behind a stack of crates. Her hair was tangled. Her cheeks streaked with soot. In her small hand—clutched tight—was a cracked plastic dinosaur.

The same kind Joe had found in the woods.

Manny's eyes widened.

Morales grabbed his sleeve. "Focus."

He forced himself back, jaw tight.

Joe swept two fingers to the left: Morales, take flank.

A point right: Manny with two militia to guard the families and extract hostages first.

Three taps on the ground: Guido and three more circle wide behind the command shack.

Joe himself moved straight toward the heart of the camp, staying low, every muscle drawn tight.

They made it halfway before a board snapped under Guido's boot.

A sharp crack echoed like a gunshot through the trees.

The camp stirred.

Harrow's men jerked awake, heads lifting.

A guard stepped out of the shack with a half-eaten ration bar in hand.

His eyes widened when he saw Joe.

Morales breathed, "Here we go…"

The guard reached for his rifle.

Morales dropped him before he could shout, one clean swing of her arm knocking the weapon aside as she slammed him into the mud.

Joe surged forward.

The families gasped but stayed silent. They had been waiting for this. Hoping for this.

And now—

The tension snapped.

Everything in the camp tightened like the last breath before a scream.

Morales whispered, "On your count, Joe."

Not loudly. Not desperately.

Just ready.

Joe didn't take his eyes off Harrow's shack.

He inhaled once, steady.

The air tasted like smoke and fear.

He could sense Harrow behind that door—like a storm waiting to break.

Joe raised his fist.

Held it.

Then dropped it.

A silent command.

The world shifted.

The fight began.

Chaos erupted the instant Joe's fist dropped. The militia surged from the tree line in controlled bursts, two teams sweeping left, two sweeping right, rifles raised but muzzles angled away from the clusters of terrified families.

Morales went in fast and clean, slamming her shoulder into the nearest guard before he could fire a wild shot into the crowd. The man hit the dirt with a thud, Morales' knee pinning his wrist until Guido darted in and zip-tied him with a practiced pull.

On the other side of the camp, Manny threw himself behind a crate as bullets ripped through the air. Harrow's men fired recklessly, panicking at the sudden assault. A group of children screamed and scattered. Manny scrambled forward, grabbed two of them by the shoulders, and hauled them behind a broken pallet wall.

"Stay down—don't move!" he barked, then motioned for two militia volunteers to cover them.

The militia pushed Harrow's right flank while Morales signaled her left team onward.

Joe moved straight for the command shack. Gunfire cracked behind him; silhouettes darted between tents; the smell of dirt, sweat, and burning trash filled the air.

Joe's boots hit the ground in hard, controlled strides.

Then the shack door exploded open.

Harrow burst out like a man born from violence. Broad shoulders, jaw clenched, rifle slung across his chest, knife strapped to his thigh. His eyes found Joe instantly, and something feral lit inside them.

"Protect the loyal!" Harrow roared to his men. "Kill the rest!"

Several of his fighters hesitated—half-starved fathers, trembling teenagers—but the lieutenants barked orders of their own and shoved them forward.

Joe lifted his chin in a slow, deadly acknowledgment.

Harrow grinned like he'd been waiting for this.

He swung his rifle up.

Joe dove aside—just as Harrow's shot cracked through the air.

Behind Joe, Guido slammed into one of Harrow's lieutenants, smashing the man's face into a rock with a sickening crack.

"That's for blowing up our fuel drums, jackass!" Guido hissed as the man went limp.

Manny dragged another group of children into cover, shouting over the noise, "It's okay! It's okay! Stay low!"

Joe rolled to his feet in one smooth motion—just in time for Harrow to crash into him.

They slammed together hard, fists and elbows flying. Joe's forearm blocked a punch; Harrow's knee drove toward his ribs. Joe twisted his weight, slammed his shoulder into Harrow's chest, and they both hit the dirt.

The knife flashed.

Harrow drew it in a single fluid motion, steel glinting. He brought it down toward Joe's ribs with a brutal, practiced thrust.

Joe caught Harrow's wrist mid-strike. Their muscles strained; teeth clenched; dirt sprayed as they writhed for leverage.

Harrow snarled through clenched teeth, "You can't protect them, Joe! You never could!"

Joe's jaw tightened. "Watch me."

Harrow jerked violently, nearly breaking free, but Joe twisted hard. Harrow's wrist cracked with a sickening pop. The knife dropped.

Harrow roared and lunged for it with his free hand—fast, too fast for a wounded man.

Joe had no choice.

He drew his sidearm mid-roll.

Harrow reached the knife—

just as Joe fired.

One round. Clean. Final.

The bullet hit Harrow square in the chest.

His body jerked once, staggered, then crumpled backward into the dirt.

For a moment he lay blinking at the sky, breath rattling, confusion flickering across his face like he couldn't quite understand the ending he'd earned.

Joe stood over him, chest heaving, mud streaked across his clothes and blood across his knuckles.

Harrow's lips moved.

No sound came out.

Joe stared down at him without triumph—only truth.

"This ends now."

Harrow's eyes glazed. His chest fell once more, then didn't rise.

Across the camp, Harrow's remaining fighters saw their leader fall.

One by one, rifles lowered. Fear slipped into their eyes, then surrender, then—unexpectedly—relief.

A young man barely twenty dropped his weapon first, tears streaking through grime. "We're done," he whispered.

Another man raised his empty hands. A woman fell to her knees, sobbing.

The fight drained out of the camp like air from a ruptured lung.

Morales grabbed the last resisting guard and slammed him face-first into the dirt, shouting, "Enough!"

The militia closed in, not with cruelty, but with precision. Disarming. Restraining. Containing.

Joe holstered his weapon slowly, breathing hard, sweat stinging his cuts.

It was over. For real this time.

Behind him, Manny emerged from behind the crates with a little girl clinging to his leg.

Guido limped out of the shadows, wiping blood off his forehead and muttering, "Not bad for a morning."

And from the families—timid, traumatized, shaking—came the first sound of hope.

A single exhale.

A whisper of disbelief.

Then, softly—sobs of relief.

Joe didn't smile. He didn't celebrate.

He just looked at Harrow's body one more time.

The nightmare was finished.

Now came the work of piecing together what remained.

Silence didn't fall all at once. It settled slowly, like dust after a demolition. Harrow's men lowered their weapons one by one, hands trembling, eyes flicking between Joe, Morales, and the lifeless body on the ground.

A few still clutched rifles with white-knuckled desperation, but when Morales leveled her stare at them—cold, sharp, unmovable—they dropped them with a clatter.

One man sank to his knees.

Another covered his face with both hands and shook.

A woman stared at Harrow's body as if she expected him to rise again and strike them for weakness. When he didn't, she whispered, "Thank God," like a prayer she hadn't dared say in months.

Morales raised both hands, voice steady. "It's over! No more fighting. No more threats. You're safe now. Stay where you are."

The militia stepped forward, rifles lowered, movements controlled and deliberate. They weren't here to punish; they were here to contain.

Manny jogged to the far side of the camp where several children huddled behind a pallet wall.

"Hey," he said softly, sliding down beside them. "It's okay now. Can you stand?"

A little girl—mud streaked across her cheeks, holding tight to a battered toy dinosaur—peeked past his shoulder.

"Is he gone?"

Manny nodded slowly. "Yeah. He's gone."

She launched forward and wrapped her arms around his neck, nearly knocking him backward. Manny froze for a moment—surprised, breath caught—then held her gently, eyes stinging.

Guido limped past him with one hand pressed to his ribs. "If you fall over, I'm not carrying both of you," he muttered.

The girl blinked at him.

Guido sighed and softened—not much, but enough. "He's fine, kid. Tough as a rusted bolt."

"Thanks, Guido," Manny said dryly. "Touching."

Guido pointed at him with his good hand. "You're welcome."

Morales moved through the camp like she'd been born to command aftermaths.

"Secure weapons. Check every tent. Prioritize medical needs. Nobody moves alone," she ordered.

The militia spread out in pairs. Former fighters—shaking, starved, broken—put their hands up without resistance. Most didn't even look at their captors; they looked at the families nearby, as if begging forgiveness.

Joe stood over Harrow's body for a long moment. Not out of vengeance. Not satisfaction. Just quiet confirmation. The shadow Harrow cast over the valley was gone. Finally gone.

Joe knelt, reached down, and closed the man's eyes.

"No more," he whispered.

Then he stood and clicked his radio. "Ridge Camp, this is Blue Leader."

Valery's voice came through immediately—tight, breathless, vibrating with restrained desperation.

"Joe—Joe, talk to me."

He exhaled. "It's done."

Silence.

Then—Valery exhaled a shaky laugh. "Bring our people home."

Joe's gaze drifted across the camp.

Families emerged from shadowed shelters, stepping slowly as if afraid the ground might vanish beneath them. A mother rushed forward, scooped the girl from Manny's arms, and held her tight enough to bruise.

"Mom," the girl sobbed into her shoulder.

Manny stepped back, swallowing hard.

Guido slapped Manny lightly on the arm. "Try not to look like you're gonna cry. It ruins my reputation."

Manny shoved him weakly. "Shut up."

Across the clearing, two militia volunteers guided a group of dazed men out of a makeshift cage built from scavenged pallets.

"He kept us here," one whispered. "Said Ridge Camp would enslave us if we ran."

Morales heard him and shook her head. "You're free now."

Joe moved from group to group, checking eyes, checking wounds, checking that the terror was fading—not instantly, not fully, but enough to breathe again.

When he reached the families at the back of the camp, a young woman stepped forward. Dirt streaked her face; her hands shook.

"You… you came for us."

Joe shook his head. "For all of you."

Her lip trembled. "I told my son someone would. I told him someone still cared."

Joe didn't speak. He didn't need to.

Behind him, Morales signaled. "We're ready to move."

The militia gathered the surrendered weapons—rifles, makeshift knives, crowbars wrapped in cloth—and dumped them into a single pile.

Manny found the knife Harrow pulled on Joe lying in the dirt. He picked it up, turned it over once, then hurled it far into the trees without a word.

Guido grinned approvingly. "Good arm."

"Better aim," Manny murmured.

Morales organized the freed families into groups—wounded first, children next. She assigned militia escorts to each cluster.

"Nobody walks alone. Nobody at the back. Keep your eyes open but your weapons down."

Joe stood at the front. The dinosaur toy lay half-buried near his boot. He picked it up, brushed the dirt off, and handed it to the little girl Manny had saved.

Her eyes brightened, small fingers curling around it. "Thank you," she whispered.

Joe gave a single nod.

Then he lifted his rifle, checked the line, and motioned forward.

"Let's go home."

They began the long walk back to Ridge Camp—families protected, armed men disarmed, Harrow's tyranny ended. Behind them, the broken camp grew smaller and smaller, swallowed by the trees that had once hidden horrors.

Ahead, Ridge Camp waited like a light on a dark coastline.

And Joe walked at the front—not proud, not triumphant—just steady. The kind of leader who carried the weight so others didn't have to.

Harrow was dead.

But the work of rebuilding had only just begun.

———

CHAPTER 24
THE NEW DAWN

The column of rescued families stepped through the ridge gap just as the morning sun broke over the tree line, turning the camp's smoke trails into soft gold. The people of Ridge Camp, already gathered in a wide semicircle, fell into a hush. No fear. No staring. Just quiet welcome.

Joe walked at the front with Morales and Manny, guiding the newcomers across the final stretch of rock. When his boots hit camp soil, he paused and exhaled long and low. Behind him came the exhausted families—thin, bruised, frightened, clutching small bags and smaller children. They hesitated at the edge of Ridge Camp as if afraid to hope.

Valery pushed through the crowd the moment she saw Joe, sling tight around her healing shoulder. She didn't run—she could barely run anyway—but she moved fast enough that Manny muttered, "She's gonna tear that stitch," under his breath.

Joe stepped toward her without thinking.

She reached him first, stopping just inches away. For a heartbeat, neither spoke. Then Valery breathed out, "Took you long enough," her voice rough but steady.

Joe let out a tired laugh. "Traffic was hell."

She leaned in and wrapped her uninjured arm around him. He pulled her close—not holding her back, not stopping her, just holding her like he'd been waiting to breathe again. Then he pressed a brief kiss to her forehead, subtle and soft.

"Good to see you standing," he murmured.

"Good to see you in one piece," she replied. "Mostly."

Their moment ended when the crowd surged slightly, the rescued families pausing as if unsure they were allowed to keep walking. Joe and Valery stepped apart—professional again—but the warmth between them didn't fade.

Immediately, Ridge Camp moved.

Not in a rush. Not in a swarm. In a tide of gentle hands and soft voices.

Women came forward first, offering water cups and blankets. Men moved aside, creating clear walking paths. A man with a braided beard gave a hesitant father a nod, then took his heavy pack without a word. A young pair of cooks handed out warm broth to the older survivors who looked half-frozen.

One Ridge Camp mother stepped up to a trembling woman holding a toddler and whispered, "You're safe now. Come with me. I'll show you where your little one can rest."

The woman burst into quiet tears, but she followed.

Charlie, who had been bouncing lightly on his toes all morning, rushed forward with two other camp kids at his side. Lenny barked cheerfully beside them, tail spinning.

Charlie puffed his chest and said to the newcomers' children, "Wanna see our fort? It's not really a fort, but we pretend it is. You can come. We've got sticks."

The kids—nervous, stiff, unsure—looked to their parents, who nodded. Slowly, the children stepped forward. Charlie grinned, grabbed the nearest boy's sleeve, and led them off toward the ridge path. Lenny trotted proudly after them as if escorting royalty.

Amy climbed onto a crate to make herself seen. She didn't shout; she didn't need to. Her voice carried calm like a hand easing pain.

"All right, everyone," she said. "If you're injured, or if your children are tired or sick, come with me to the medical tent. We'll take care of you immediately. No lines, no waiting."

Several families stepped forward at once, relief softening rigid faces.

Morales stepped up beside Joe and addressed the rest. "Everyone else, follow me. I'll show you where everything is—water, food, sanitation, sleeping tents. No pressure, no rush. You set your own pace."

The newcomers listened—really listened. Their fear loosened.

As the families moved deeper into camp, embraced from every side, Valery glanced at Joe.

"You did good," she murmured.

He shook his head. "We all did."

She smiled slightly. "Don't get modest now. I was starting to like the confident version."

He chuckled under his breath. "Let's get them settled first."

And together, they walked forward into the new morning—leading the rescued families toward a place that finally felt like home.

The flow of people moved slowly but steadily across Ridge Camp as Morales led the first group deeper inside the settlement. The rescued families glanced around with guarded curiosity, still unsure whether this new place was refuge or another kind of prison. Joe walked with them, keeping a respectful distance, letting them take the space they needed.

The camp itself had shifted into a quiet, purposeful rhythm. People worked and watched at the same time—lifting crates, guiding children, handing out cups of warm broth, all with the same gentle focus. The faces were tired, soot-smudged, and bruised from the night before, but this morning carried something softer. Relief. Maybe even pride.

Manny noticed a small girl staring at him with huge brown eyes. Her hair was tangled, one sleeve torn. She clutched a battered toy dinosaur—the same model as the one Joe had found days ago. Manny froze for a second, his throat tightening. Then he crouched down, careful not to look too intimidating.

"Hey," he said softly. "Cool dinosaur."

She blinked up at him, hesitated, then whispered, "His name's Flip."

"Flip, huh?" Manny nodded like this was royal information. "Good name. Strong. Fast."

The girl didn't smile exactly, but her shoulders loosened a little. Her mother stepped forward and touched Manny's arm in silent thanks before guiding her daughter toward the medical tent. As they walked off, Manny looked to Guido, who nudged him with an elbow.

"Kid likes you," Guido muttered.

"Yeah, well," Manny said, rubbing his neck, "Ashley says I'm good with kids."

Guido smirked. "Oh yeah? That what she said?"

"Shut up," Manny grumbled—but he didn't hide the small smile that crept in anyway.

A group of Ridge Camp men approached the rescued men, offering firm handshakes, clapping them on the back, guiding them toward the tool racks. "You'll get to work if you want," one of them said. "Not today. Tomorrow. Today's for breathing."

A man with a mechanic's hands—grease permanently lining the creases of his fingers—lifted his chin at Donnie. "I heard you got machinery running up here?"

Donnie nodded. "Generator, pumps, some rigs."

Reed Lawson stepped forward, "I used to work aircraft maintenance. Mechanical systems. Gearboxes. Hydraulics."

Donnie gave him a wide, tired, sincere grin. "We'll take every set of hands we can get. Come by tomorrow morning. I'll set you up with Manny. And bring your boy—Charlie'll lose his mind if he meets another kid who likes engines."

For the first time since arriving, the man smiled.

Behind them, Amy guided a cluster of frightened mothers toward the med tent. "Come sit, rest your backs, let's check the kids first," she said gently. "If you need blankets, food, or quiet, we've got all three." She handed one woman a warm cloth for her baby and touched another woman's shoulder reassuringly. "You're safe here. No one's rushing you."

Joe watched her work, gratitude softening his face.

Near the south end of camp, Morales continued her tour. "This path goes to water collection and filtration," she explained. "This one leads to sanitation. We've got bathing tents separated by families and age groups. Kitchen's this way. You'll always find someone there during daylight."

A man raised his hand hesitantly. "Do we...need to sign anything? Or...register?"

Morales shook her head. "No bureaucracy here. Just people. Once you've eaten, slept, and breathed, we'll gather volunteers. Until then? Your only job is to rest."

They kept moving until the group reached the center of camp. Ridge Camp had been rough around the edges since its first day, but the core was solid—huts in neat rows, tents reinforced against wind, workstations carefully spaced, paths swept clear of hazards. It wasn't pretty, but it was organized. It felt like safety.

Children ran past carrying sticks, pinecones, and one very excited Lenny who kept yapping as if announcing everyone's arrival. The rescued children watched them with wide eyes. Then Charlie ran back over, waving frantically.

"Hey! You guys wanna come see the lookout post? Morales says I can show people if I don't touch anything sharp!"

One of the new boys—twelve or maybe thirteen, tall and wiry with curious eyes—stepped forward. "Can I come?" he asked quietly.

Charlie blinked, then grinned like lightning. "Yeah! Come on!" He grabbed the boy's sleeve and pulled him along, Lenny galloping behind them. The other kids followed, shy but eager, the tension melting as they disappeared up the ridge path.

Valery watched all of this, leaning slightly on her uninjured side. "They're adjusting faster than the adults," she murmured.

"Kids always do," Joe replied. "They don't carry the guilt. Just the memories."

Valery looked at him sideways. "You carrying some?"

Joe didn't answer at first. Then he said quietly, "Not anymore."

She nodded once—a simple acceptance.

The last of the newly freed families were guided gently into camp by the Ridge volunteers. Some stumbled from exhaustion, others wept silently, and a few looked around as if searching for traps that weren't there. But step by step, the weight on their shoulders eased.

A new chapter had begun.

And Ridge Camp was ready to hold it.

Amy climbed onto a crate near the center of camp, wiped her hands on her apron."Okay, listen up for a minute," she called, voice warm but carrying. "If you've got children sixteen and under, I want you to bring them over here. We're going to show them where everything is so they don't have to guess."

There was a pause while the new families looked at one another, uncertain. Then one mother nudged her teenage son forward. Another woman lifted a toddler onto her hip and stepped closer. A man with a girl clinging to his shirt joined them, face drawn tight with fatigue.

Amy softened her tone. "Nobody's in trouble. Nobody's getting separated from their parents. We just want your kids to know where the food is, where the med tent is, where they can play without getting run over by a cart."

That broke some of the tension. Slowly, a knot of children and teenagers formed in front of her—some from Ridge Camp, some new. Charlie stood near the front with the wiry, dark-haired boy who'd walked in with the aircraft mechanic.

Amy smiled at the group. "All right, field trip time. Parents, you're welcome to come along if you want, or you can grab ten minutes to sit down and breathe. Your choice."

A few of the newcomers' eyes filled suddenly at that. Ten minutes to breathe felt like a luxury.

Charlie shot his new friend a quick grin. "Come on, Ben. You gotta see the kitchen. Mom lets me steal extra bread sometimes if I pretend I'm working."

Ben huffed a quiet almost-laugh. "You're not supposed to tell people that."

"I trust you," Charlie said solemnly, like that settled it.

Amy hopped off the crate. "Okay, troops. First stop—food. Follow me."

The pack of kids and a handful of parents trailed after her. As they walked, she pointed things out. "That big tent there? That's the kitchen. If you're hungry and it's daylight, somebody in there can help you. You don't have to ask permission from anybody wearing a red mark, or any mark at all. You just show up and say you're hungry."

A little girl with tangled curls tugged on her sleeve. "Even if it's not mealtime?"

Amy glanced down and nodded. "Especially then."

They moved past the cook area where a couple of Ridge volunteers stirred pots thick with stew. One of the cooks, a broad woman with tattooed forearms, spotted the group and winked at the kids. "Any friend of Amy's gets a ladle scrape later," she said. "Don't let me forget."

That won a few real smiles.

Amy led them on. "Those tents with the blue flags? That's med. You get hurt, scraped, scared, you come there. You don't need to be bleeding to ask for help. And if you see someone else hurt, you bring them."

At the entrance, Dr. Elise looked up from a supply crate. Her hair was tied back, sleeves rolled to her elbows, hands already stained with antiseptic. Her gaze swept the knot of kids with a quick, gentle assessment.

"I'm Dr. Elise," she said. "I fix bodies and help brains untangle themselves. We'll have a quieter space set aside for anyone who needs to talk in private. That goes for grown-ups too."

One of the older teenagers from Harrow's group shifted uncomfortably, eyes dropping to the ground. Elise didn't press. She just nodded to Amy. "I'll be here."

"Med, check," Amy said. "Next—sanitation. You're all going to love this part."

Groans and snickers rippled through the Ridge kids. Ben leaned toward Charlie. "She serious?"

"Oh yeah," Charlie said. "She talks about clean hands like some people talk about winning the lottery."

They passed the row of privacy screens and wash barrels where volunteers showed a couple of new mothers how to use the makeshift showers—gravity-fed water lines, soap slivers in a tin, hooks for hanging clothes. It wasn't luxury, but compared to Harrow's camp, it might as well have been a spa.

"Boys on this side, girls on that side," Amy said, pointing. "No spying, or Morales will make you dig latrine trenches for a week. Understood?"

A chorus of "Yes, ma'am" answered her.

Up the slope, Morales had gathered the new adult arrivals under a shade tarp. She waited until Amy's group passed by before raising her voice.

"Everyone who feels up to walking, I'm going to show you the lay of the land," she said. "This is not a work assignment. This is so you don't get lost and don't have to ask a guard where the bathroom is at three a.m."

A few men chuckled weakly.

Morales pointed toward the north edge. "Water lines and tanks are there. If something goes wrong, you'll hear Donnie yelling about it before anything explodes. South side is gardens—small, but growing. We'll be expanding those. West is storage. You'll have supervised access once we get you assigned to teams."

One of Harrow's former men—bearded, gaunt, eyes hollow—raised a hand. "You…really want us working? After what…after what some of us did?"

Morales held his gaze for a long moment. "If you did something that needs addressing, we'll address it. But most of you were following orders you were scared to disobey. That ends here. You want to earn your place? You show up ready to work and ready to protect the people next to you. That's the only line we're drawing right now."

The man swallowed hard and nodded.

"We're not assigning jobs today," Morales continued. "You've been walking under somebody else's rules for too long. Take one day to remember your people, not tools. Tomorrow, there'll be a sign-up board by the main fire. Mechanics, builders, cooks, med support, patrol, teaching—if you've got skills, we'll use them. If you don't, we'll train you."

She glanced toward where Amy's group of kids moved past the garden rows. "You want to belong here? Start by wanting them safe."

Heads turned toward the children. Something shifted in the air—small, fragile, but there.

Joe watched from near the edge of the training yard, arms folded loosely, not looming, just present. Valery stood beside him, one hand resting lightly against his arm.

"You're good at this," she said quietly. "The whole 'build a town out of wreckage' thing."

"Was kind of hoping I'd never need that particular skill set," he answered.

She huffed a soft almost-laugh. "Too late now."

They stood in silence for a moment, watching Amy lead the kids toward the practice field where older teens ran simple drills with sticks and obstacle courses. Ridge Camp and the newcomers were mingling already—hesitant at first, then more naturally, like threads beginning to weave together.

"Think they'll stay?" Valery asked.

Joe's gaze tracked Ben and Charlie arguing cheerfully about the best way to brace a ladder.

"If we do this right?" he said. "Yeah. I think they might actually start to live here."

Not just survive.

Live.

The sun inched higher, turning the smoke-stained sky a pale gold as the camp settled into a nervous rhythm. Ridge Camp had never been this full—new faces, unfamiliar voices, children weaving between tents like startled deer slowly remembering how to run again. It wasn't chaos, not exactly. It was becoming something. A beginning.

Morales finished the last leg of her adult tour and paused near the new water barrels. The group around her—twenty-seven adults, some gaunt, some exhausted, all wary—stood quietly waiting for whatever came next. She placed her hands on her hips and looked them over with a soldier's eye, but her voice stayed soft.

"You don't owe us your whole story today," she said. "Hell, you don't owe us anything today except honesty when it matters and respect for the people around you. That's it. Basic human decency."

A few heads lowered. Others lifted as if surprised by the simplicity of it. Morales gestured back toward camp. "Your families are safe here. Your kids are safe. No one's going to threaten them or use them to make you fall in line. You're not property. You're not leverage. And if anyone here treats you like you are, you tell me."

Her voice didn't rise, but the promise in it was iron.

A woman clutching a thin child to her chest blinked fast, like she wasn't used to softness anymore. "What if… what if we don't know what we're good at?" she asked. "Harrow didn't let us… choose much."

Morales gave her a slow nod. "Then you try things until something fits. That's how we all started."

A ripple went through the crowd—a small shift, more air in the lungs, more space in the shoulders.

Behind them, Manny and Guido hauled a crate toward the supply tent, arguing as usual.

"It's not crooked," Guido insisted.

"It is absolutely crooked," Manny shot back. "Look at it. It's leaning like it partied too hard last night."

"It survived an explosion, two recon trips, and my uncle's stupid chili. It's fine."

"Your standards," Manny said, "are subterranean."

Ben—still close to Charlie—paused to watch them bicker. "Do they fight a lot?" he asked.

Charlie shrugged. "Yeah. But they like each other."

Ben blinked. "Really?"

"Oh yeah," Charlie said. "It's how they bond."

Ben's smile was small but real, the first all morning.

Amy returned from her tour of the children, wiping sweat from her brow. "All right," she said, scanning the clusters of people. "If you're hungry or tired, kitchen tent's open. If you need water, barrels are over there. If you need space, we've got quiet corners. If you need medical, go see Dr. Elise. And if you need a hug… Valery probably won't give you one, but I will."

"Hey," Valery called dryly, "I give excellent hugs. Selectively."

Amy winked. "Selective is fine. We've got options."

Her tone coaxed out a handful of quiet chuckles. Not many—but enough to ease the air.

As people dispersed—guided by volunteers, greeted by strangers who didn't stay strangers for long—Joe stepped into the center of the flow, taking everything in. His shoulders stayed squared, but his face had softened into something thoughtful. Calculating, but hopeful.

Valery moved beside him, staying close but not hovering. Her arm hung in the sling, but her posture stayed proud. "You okay?" she murmured.

"Not sure yet," he answered honestly. "Ask me in an hour."

"Okay," she said. "In an hour I'll ask if you're still pretending you're fine."

He shot her a sideways look. "Pretending?"

"You're the king of pretending."

"I'm not the king of anything."

She nudged him lightly with her elbow—her good one. "Fine. You're the local duke of pretending."

Joe huffed a laugh through his nose. "Better."

Amy appeared beside them, hands on her hips. "You two done flirting or do I need to get a broom and sweep the air between you?"

Valery choked on a laugh. Joe muttered, "Mom."

Amy lifted a brow. "Don't 'Mom' me. Someone needs to keep things moving before half these people collapse from nerves."

Joe sobered, glancing out over the crowd. "They're scared," he said quietly.

"Of course they are," Amy replied, her voice softening. "They've been controlled, starved, threatened. They don't know what to do with kindness yet. They'll adjust."

"Yeah," Joe said softly. "I know."

Amy followed his gaze to the children weaving between tents—Ridge Camp kids showing newcomers where to find the berry patch or how to climb the old driftwood stump that everyone pretended was a castle. Tiny bursts of ordinary life beginning to spark.

"You did good, Joey," she said. "All of you did."

He didn't answer, not right away. His jaw worked, emotion flickering beneath the hardened surface.

Finally he exhaled. "Let's hope it's enough."

Valery leaned slightly into him—not clinging, not overstepping, just a quiet presence anchoring him. "It will be," she said.

Around them, Ridge Camp and the rescued families blended—shy introductions, careful smiles, hands extended in welcome. Little by little, fear loosened its grip.

A community was forming.

Not perfect. Not peaceful.

But real.

And that was enough.

By afternoon, Ridge Camp pulsed with new motion—like a heart learning a new rhythm. Volunteers guided families toward empty huts and tents, pointing out shaded spots, fire pits, the wash line, anything that might make a strange place feel less like a battlefield and more like a home. No one rushed them. No one barked orders. Everything moved with a quiet patience that seemed to surprise the newcomers.

Joe stood near the supply tent with Morales, watching the flow of people. "We're going to need more shelter by tonight," he said.

"Already got Donnie and three volunteers cutting lumber," Morales replied. "But we're out of nails. We'll have to get creative if we want frames up by dark."

Joe smirked faintly. "When have we ever not gotten creative?"

Across the camp, Manny crouched beside a wagon, showing two men how to reinforce the wheels with scrap metal. The newcomers watched him like he was demonstrating a magic trick. One finally asked, timidly, "This won't break?"

"Only if you treat it like a mule," Manny said. "Be nice and it'll live longer than all of us."

The men laughed—quiet, uncertain, but real. Manny grinned, satisfied, then nearly tripped when Ashley brushed past him with a bundle of blankets in her arms.

"Watch your feet," she teased softly.

He straightened, clearing his throat too fast. "I—yeah—I was just—uh—helping."

"Mm-hm," she said, amused. "Then help me carry these to the north huts."

"Sure. Yeah. I can do that."

Morales watched him go and shook her head. "That boy's doomed."

"Let him be doomed," Joe said. "He deserves something good."

A small voice tugged his attention. "Are you Joe?"

He turned to find a boy of about ten standing there—thin, dusty, clutching a cloth dinosaur like it was his heartbeat. Ben hovered behind him, offering encouragement with a nervous half-smile.

Joe knelt. "I'm Joe. What's your name?"

"Colton," the boy said. "My mama said… my mama said you're the man who fought the bad one. The one who—"

He stopped, breath snagging.

Joe placed a steady hand on his shoulder. "You're safe here, Colton. No one's going to hurt you again."

Colton nodded, though his eyes stayed watchful, uncertain. "Mama said I should help," he whispered. "Whatever you need help with."

Joe's chest tightened. "You don't owe us anything. Not today. Just rest, okay?"

Colton hesitated. Then he whispered, "Okay," and slipped his hand into Ben's. The boys walked off toward the other children, where Charlie greeted them with a wide grin and immediately began explaining how to build the "best secret fortress in the woods."

Amy approached then, wiping clay from her hands. "We've got a little triage line forming. Nothing catastrophic, mostly exhaustion and dehydration, but several of them have old injuries nobody treated right."

Joe nodded. "Tell me what you need."

"A second pair of hands once dinner's going," Amy said. "Valery already volunteered, sling or not."

Joe snorted. "Of course she did."

Amy's expression softened. "You should go talk to her. She's been pretending she's fine for hours."

Joe lifted a brow. "You're telling me to check on someone else's mental health?"

Amy flicked his shoulder. "I'm an excellent example of compartmentalized emotional avoidance, thank you very much."

He smirked. "That's one way to put it."

She gave him that mom-look—the one that could slap sense into a brick wall. "Go."

He went.

Valery stood near the comm table, adjusting the antenna with her good hand while a newcomer teen watched, fascinated. "No, not like that," she said. "If you angle it wrong, you'll catch static from the ridge line. Try it again—slow."

The boy tried. The static vanished.

Valery smiled. "See? Told you. Not magic. Just science."

Joe stopped beside her. "Recruiting already?"

She gave him a sideways look. "Camp doesn't run itself."

"You should sit."

She raised an eyebrow. "You should stop saying that to me."

He let out a slow breath. "Fair. Just… proud of you. That's all."

Her expression softened, subtle as a shift in wind. "Right back at you."

A lull settled between them—comfortable, quiet. Around them, Ridge Camp thrummed with rebuilding energy. People who had been prisoners hours ago were now learning, talking, testing safety like a door they weren't sure would stay open.

"You know," Valery said, breaking the silence, "watching all of this… it almost feels like hope behaves like fire."

Joe looked at her, puzzled but intrigued. "Fire?"

"It needs oxygen. Fuel. Someone to guard it. Someone to feed it. And if you put it in the right hands… it spreads."

He absorbed that, mouth curving slightly. "Yeah. I guess you're right."

Valery nodded toward the families settling into their spots. "They'll heal. Not fast, not neatly. But they will."

"And us?" Joe asked before he could stop himself.

She paused, eyes steady. "We're already healing. We just don't notice it because we're too busy surviving."

Joe exhaled, deep and grounding. "Yeah."

She nudged him lightly with her elbow. "Also, stop looking like you're carrying the whole world alone. You have a team, remember?"

He smirked. "The team that keeps getting injured, yelling at me, or flirting badly?"

"Exactly," she said. "It's perfect."

Ahead of them, Morales climbed onto an overturned crate. "Listen up!" she called, voice ringing across camp. "If you're sick, scared, tired, hungry, or confused—which should be all of you—head to the center circle in ten minutes. We'll do another walk-through and help you get settled."

People nodded, murmuring thanks, drifting toward the open area.

Joe looked over the growing village—walls half-built, tents fluttering, smoke rising in thin spirals. It wasn't neat. It wasn't fully safe. It wasn't peaceful.

But it was theirs.

A new start.

A new dawn.

CHAPTER 25
FAULT LINES

Ridge Camp woke slowly the next morning. The light came in pale and thin, hitting the tents and the smoke from the cook fires. People moved carefully, like they weren't sure if they were allowed to relax yet. The camp was quiet, but not calm. Everyone was trying to figure out what the new day meant now that Harrow was gone.

Joe walked down the main path with tired eyes and stiff shoulders. He nodded at people as he passed them. They nodded back. No one expected him to talk yet. He was still processing the night before—fighting, rescuing families, leading them home, and then trying to sleep while adrenaline still burned under his skin.

The rescued families stood near the long tables, unsure of what they were supposed to do. They clutched their bags and their children like someone might still try to take them away. Ridge Camp residents offered blankets, food, and directions in quiet voices. It helped, but the tension was still there. Too many new people. Too much fear left over. It would take time.

Morales approached with a clipboard tucked under her arm. She didn't bother with a greeting. "We've got problems."

Joe let out a long breath. "How many?"

"All of them," she said. "Sleeping space is tight. The food line jammed twice. Two guys almost fought over a pot of boiling water.

And someone tried to walk into the storage tent because he thought supplies were free-for-all."

Joe rubbed his forehead. "Tell me nobody got hurt."

"No," Morales said. "Ashley yelled at him before anything happened."

They both looked toward the med tent. Ashley stood there with her arms crossed, her stare sharp enough to cut steel. The man she yelled at earlier avoided her completely. Manny stood nearby pretending to fix a rope on a crate, glancing at her every few seconds like he wasn't sure if she was impressive or terrifying.

Morales followed Joe's gaze. "Don't even think about trying to manage that situation."

"I wasn't," Joe said.

"Sure," she replied, unconvinced.

Before Joe could argue, Evan Calder walked up. He was tall and thin, with grease on his hands and a face that looked like it hadn't had a good night's sleep in years. "Morning," he said in a flat voice.

Joe nodded. "How's the equipment team doing?"

"Mixed," Calder said. "Some people know what they're doing. Some barely know which end of a hammer to hold. We'll sort it out."

"What do you need?"

"Shelters," Calder said. "We need more of them by tonight. Half the newcomers' tents are falling apart. And we're out of nails."

Joe raised an eyebrow. "Already?"

"We built half a town in one day," Calder said. "Nails disappear fast when you're trying to keep a roof from falling on someone's head."

Joe sighed. "Can we switch to wooden pegs?"

"Manny already said he can make them," Calder said. "He also said it would be 'fun.' That comment worries me."

Joe let out a tired laugh. "If Manny says something is fun, it's usually a bad sign."

"Exactly," Calder said. "But we don't have another option."

Joe nodded. "All right. Tell Manny to start shaping pegs. I'll send Guido to help."

Calder gave him a look. "Guido? Helping? You sure?"

"No," Joe said. "But send him anyway."

Calder shook his head and walked off.

Morales checked her clipboard again. "You ready for the rest of the problems?"

"No," Joe said. "But go ahead."

"Good," she replied. "Because we haven't even gotten to the water system, the food stores, or the patrol schedules yet."

Joe stared at the sky for a moment like he was asking it for strength.

Another day had started. Ridge Camp was alive. Ridge Camp was messy. And Joe was responsible for all of it.

Joe followed Morales toward the storage tent, where several newcomers stood waiting in a loose, uncomfortable cluster. They whispered among themselves, looking anxious and confused. The problem was obvious the moment Joe arrived—two stacks of crates had been pulled apart, and several items lay scattered on the ground.

Morales crossed her arms. "They said they were hungry and thought they could take whatever they needed."

One of the newcomers, a man in his thirties with a sunken face and tense shoulders, stepped forward. "We weren't trying to steal. We just... we didn't know the rules."

Joe nodded calmly. "There's no punishment. But we need order so nothing runs out. Talk to the cooks. They'll get you fed. No one here goes hungry."

The man's shoulders slumped in relief. "Thank you."

Morales gave Joe a small nod. That was one crisis handled.

As Joe and Morales reorganized the crates, Guido strolled over looking half-awake and half-annoyed. "Something explode yet?"

"Not yet," Joe said.

"Then it's too early for me," Guido muttered, yawning.

Morales pointed at a pile of loose tools. "Start stacking those properly. And if you break anything, I'll make you do all the latrine digs."

Guido blinked at her. "You know threatening me doesn't actually get me to work faster."

"Yeah," Morales said. "But it makes me feel better."

Joe hid a small smile, then carried a crate outside and nearly bumped into Ashley. She was standing with a tight posture, arms crossed, her jaw clenched. Her eyes were scanning the crowd like she expected trouble.

"You okay?" Joe asked.

Ashley shifted her weight. "Just keeping an eye on them. People get scared in new places. Scared people make bad choices."

Joe studied her closely. She was steady, but her shoulders were tense, like she was braced for something to go wrong. "Take a break if you need it."

She shook her head quickly. "If I sit still, my brain starts picking old memories apart. Working's better."

"Fair enough," Joe said gently. "But you don't have to do everything today."

"I know," she said quietly, though it didn't sound like she believed it.

Manny jogged up holding a bundle of carved wooden pegs like they were trophies. "Got the first batch. These things are solid. I can make a few hundred if we keep the wheels turning."

Ashley glanced at the pegs, then at Manny. For a moment, something softened in her expression. "Good job," she said.

Manny tried to play it cool. He failed miserably. "Well, you know… I'm basically the Michelangelo of woodworking. If Michelangelo made tent parts."

Ashley raised one eyebrow. "Right."

Manny grinned, cheeks slightly red, and hurried off toward the half-built shelters. Ashley watched him go, her fingers brushing the small bolt charm on her wrist before she caught herself and stopped.

Joe didn't comment. He wasn't blind, but he wasn't going to embarrass her either.

A loud whistle cut the air. Donnie waved them over to the water line, where a group of new volunteers had gathered. "We've got a problem," he said. "Pressure in the gravity tanks isn't even. It'll cause a backup if we don't fix it now."

Morales groaned. "Of course it will."

Donnie pointed to the group. "I need two people who aren't afraid to climb."

Ben, the boy who had bonded with Charlie, nudged the aircraft mechanic—his father, Reed. "Dad, you can do it."

Reed hesitated only a second before stepping forward. "All right. What do you need?"

"Follow me," Donnie said. "And hope you aren't afraid of heights."

Joe glanced around the camp. People were moving—working, learning, helping, stumbling, trying again. Tension ran under the surface, but so did determination.

Ridge Camp wasn't perfect. But it was trying.

And Joe could feel the fault lines shifting—cracking in some places, healing in others.

It wasn't going to be easy. But they were building something real.

Morales led Joe toward the north end of camp where a small crowd had gathered around the half-built shelters. The newcomers stood awkwardly at the edges, unsure whether to help or stay out of the way. Ridge Camp volunteers hammered, tied ropes, and adjusted beams, but the lack of nails slowed everything down. Wooden pegs were useful, but they took time.

Manny stood in the middle of the chaos holding a mallet and shouting instructions with surprising authority. "Line up the pegs before you hammer! If you angle them wrong, the whole frame goes crooked and I'm not rebuilding it twice!"

Guido was beside him, hammering badly and complaining loudly. "You say 'angle them wrong' like that means something to me!"

"It means," Manny said, "stop hitting things like you're angry at the ground!"

"I *am* angry at the ground!" Guido snapped.

"Why? What did the ground do?"

"It tripped me earlier!"

Manny stared at him. "That was a tree root."

"Same difference," Guido said.

Ashley walked past them carrying a stack of rope coils. "You two argue more than old married people."

Manny tripped over his own words. "We—what—no—we're not—he's not—"

Guido folded his arms. "Honestly, I could do better."

Ashley rolled her eyes and walked off. Manny watched her go, then elbowed Guido. "Don't joke like that."

Guido shrugged. "Relax. She only has heart eyes for one person and it's not me."

Manny turned pink and nearly dropped his mallet.

Joe stepped in before Manny combusted. "How many shelters finished?"

"Two," Manny said. "But we can get five more done today if we keep going."

Several newcomers overheard him. One man stepped forward slowly. "We can help," he said. "We were builders before… everything happened."

Another man nodded. "I worked construction. I can set beams."

A woman added, "I can measure and cut wood."

Manny blinked at them. "Seriously? You're actually volunteering?"

They nodded, cautious but sincere.

Joe stepped closer. "Good. Work with Manny. He'll show you what he needs."

Manny straightened, suddenly serious. "Okay. Yeah. Let's do it. Grab those boards and follow me."

The newcomers eased into the workflow—awkward at first, then smoother as their old skills resurfaced. They lifted beams, tied knots, checked levels, and asked Manny questions. Manny answered each one with careful patience, slowing down when someone looked confused.

It was the first time Joe had seen the newcomers focus on something other than fear.

He moved on, checking the edges of camp. The patrol team stood near the ridge line, reviewing the new watch schedule. Since Harrow's group was gone, the danger wasn't immediate, but no one wanted to assume safety.

Charlie came running down the slope with Lenny barking happily behind him. "Joe! Joe! Come look!"

Joe braced. Charlie's version of excitement could mean anything from "found a bird" to "accidentally discovered a new valley."

"What is it?" Joe asked.

"Tracks!" Charlie said proudly. "Fresh ones!"

Morales groaned. "Please don't be more escaped Harrow guys."

"No," Charlie said. "Animal. Big animal. Bigger than a deer. Could be food!"

Morales pinched the bridge of her nose. "Or it could be something that wants *us* for food."

Charlie shrugged. "Everything's food for something."

Joe sighed. "Show me."

Charlie led them up a narrow trail toward a patch of soft dirt near the trees. Large prints—wide, heavy—pressed deep into the mud. Morales crouched down and examined them.

"Bear," she said. "A big one."

Charlie's eyes lit up. "Can we hunt it?"

"No," Joe said immediately.

"But—"

"No," he repeated. "We're not sending anyone after a bear. Not today."

Charlie kicked a rock. "Fine."

Lenny barked at the tracks, then backed up nervously.

Morales stood. "We'll set up noise deterrents along the ridge. Last thing we need is a bear wandering into the food area."

"Agreed," Joe said.

They turned back toward camp, where people were starting to talk more than whisper and move without flinching. The shelters were going up. The water system was getting fixed. Kids were running. The newcomers were working.

It still wasn't easy. Tension remained everywhere. But something was shifting.

Ridge Camp was holding.

By midday, Ridge Camp felt like two camps trying to merge into one. The older residents moved confidently—checking water lines, hauling lumber, teaching newcomers how to secure tent ropes. The newly rescued families moved slower, careful with every step, flinching whenever someone raised their voice or moved too quickly near them. That mix of hope and fear tangled through the air, making everything feel both promising and fragile.

Joe walked toward the cook line where volunteers handed out bowls of thick stew. A small commotion was growing near the end of the line. Two newcomers were arguing with a Ridge Camp volunteer, their voices rising fast.

"I've been standing here longer," one man said, frustration tightening his face. "My family hasn't eaten since yesterday."

"You can't cut the line," the volunteer insisted. "Everyone's tired. Everyone's hungry."

Joe stepped between them. "Hey. Slow down. Nobody here gets ignored. But we have rules so everyone gets fed."

The man opened his mouth to argue again but stopped when he saw the calm in Joe's face. Not anger. Not judgment. Just steady control.

Joe motioned to the cook. "We'll get these two fed now. Then back of the line after that. Fair?"

The men exchanged a look, then nodded. "Fair."

The cook handed them bowls, and the tension eased. A small crisis, solved before it could grow teeth.

Joe moved on and found Valery sitting near the radio mast adjusting the tuning dial with her good hand. Sweat covered her forehead, but she was focused.

"You should take a break," Joe said.

"Not until the frequency stabilizes," Valery replied. "We keep getting short bursts of interference."

"From what?" Joe asked.

"I'm not sure. Could be leftover equipment at Arnold. Could be atmospheric changes. Could be someone out there trying to reach anybody who's listening."

Joe crouched beside her. "You think survivors are calling in?"

Valery gave a small shrug. "Maybe. But we need the setup stable before we can pick up anything clear."

Ashley approached holding a stack of blankets and a coil of rope. Her expression was tight, but her hands were steady. "Valery, you need water?"

Valery offered a grateful smile. "Yeah. Thanks."

Ashley handed her a cup. Manny caught up moments later with a wooden peg stuck in his hair and sawdust all over his shirt. "Ashley, you forgot—oh." He noticed Joe. "Hey. Everything good?"

"Depends," Joe said. "Why do you look like a lumberjack exploded on you?"

Manny sighed. "Guido tried to help. His definition of 'help' is questionable."

Ashley couldn't hide a small smile. "He dropped a beam."

"On me," Manny added. "On my head. Which, yes, explains the peg lodged in my hair."

Valery reached up and gently pulled the peg out. Manny flushed bright red.

Ashley watched all this carefully. Her posture relaxed a little, but only a little.

A shout rose near the garden area. Morales' voice followed, sharp and irritated. "Nobody move!"

Joe and the others hurried over. A group of newcomers stood near the fence, staring at something on the ground. One of the Ridge Camp gardeners pointed at a metal canister—half-rusted, dented, but unmistakable.

"Found it buried under the soil," the gardener said. "Old chemical container. Could be toxic."

Joe's stomach tightened. The last thing they needed was contaminated soil.

Ashley stepped forward, eyes sharp. "I can check it. Carefully."

"No," Joe said. "We don't know what's inside. We'll do this measured."

Morales nodded. "Agreed. We isolate it, mark the area, and dig around it slowly. No one touches it until we know exactly what it is."

The newcomers murmured nervously.

Valery arrived, leaning slightly on her good side. "Everything okay?"

"Possibly," Joe said. "But we're taking precautions."

It wasn't a crisis. Not yet. But it was a reminder.

Even safe ground could hide danger.

The chemical canister was roped off within minutes. Manny and Guido worked together to set a perimeter while two of the newcomers

—both former farmers—stepped forward to suggest how to test the soil without risking contact. Their quiet confidence helped steady the group, and soon Morales had a small team assigned to dig around the edges instead of touching it directly.

"Slow and steady," she ordered. "Nobody rush this."

As the work began, Joe stepped back and scanned the camp. Even with the scare, people kept moving—carrying water, setting up shelters, helping one another without being told. The newcomers watched the older residents and copied what they did, learning the rhythm by imitation.

For the first time since they stepped onto this ridge, Ridge Camp felt like a place where life could actually continue.

Valery approached him again, a little pale. "Radio's clean for now," she said. "I left it running on auto-scan."

"Did you drink water?" Joe asked.

"Maybe," she said.

He raised an eyebrow.

She sighed. "Fine. I will."

Joe shook his head, amused despite the stress.

Nearby, Ashley stayed posted near the rope line around the canister. She kept her shoulders squared and her eyes sharp, but the tightness in her jaw betrayed how close she still lived to her past. Manny joined her, wiping sweat from his forehead.

"You okay?" he asked.

"Yeah," she said quickly. Too quickly.

"You sure?"

Her expression softened a little. "I'm… trying."

"That's enough," Manny said quietly.

She looked at him in a way that said she believed him—maybe for the first time.

Guido stomped over, hands covered in dirt. "Okay, so we're officially calling that thing 'Rusty the Death Can,' right?"

Manny rolled his eyes. "We are not naming it."

"I already did," Guido said. "It's important to label threats."

Ashley cracked the faintest smile.

Joe turned when he heard footsteps behind him. Amy approached, wiping her hands on a towel. "Med tent is running steady. Most of the injuries are from Harrow's camp. Old bruises, untreated infections, a few cracked ribs. Nothing we can't handle."

"Good," Joe said. "Anything you need?"

"A third hand and a miracle," Amy answered. "But we'll manage."

Valery smirked. "She always manages."

Amy pointed her towel at her. "You take care of your shoulder. Or I'll sew you to the chair."

Joe snorted. Valery muttered, "I'd like to see you try."

The afternoon settled into a steady pace. Work crews moved lumber down the slope. Children ran between tents with Lenny chasing after them. Manny and the aircraft mechanic— Reed Lawson —worked side by side replacing broken wagon wheels. Reed's quiet skill drew a crowd, and Manny introduced him to anyone who wandered near.

"This guy knows aircraft mechanics," Manny explained proudly. "He can fix anything with gears."

Reed shrugged shyly. "I can fix most things. Not everything."

"He's modest," Manny told Ashley.

Reed smiled at her. "Good to meet you."

Ashley nodded politely but stayed close to Manny. Manny didn't seem to mind.

By late afternoon, Morales returned from the garden site. "Chemical canister's sealed," she reported. "Soil around it is safe. We'll dispose of it carefully."

A wave of relief passed through the camp.

Joe exhaled. "Good. That's one less thing to worry about."

Morales looked toward the distant shimmer of water where the fleet rested offshore—Savannah, Halcyon, the tugs, the skiffs. "We still have reactors to shut down, supplies to salvage, fishing lines to set, winter to prepare for..." She blew out a breath. "We're going to be busy."

"We always are," Joe said.

Valery stepped to his side. "At least now we have people to help."

"And people to protect," Joe added.

Ashley and Manny were now helping a group of newcomers build hut frames. Manny cracked a joke that made Ashley hide a smile behind her hand. Reed offered them a tool and was pulled into the conversation like he'd always belonged there.

Guido shouted at someone about using the "good rope," Donnie yelled back about not wasting supplies, and Charlie sprinted past carrying two sticks like swords while yelling, "Defend the ridge!"

Joe took it all in—the chaos, the rebuilding, the fear, the cautious joy—and felt something rare settle into him.

Promise.

As evening approached, Morales called out, "Gather by the fire in ten minutes! We'll go over tomorrow's work, then eat."

People drifted toward the center of camp.

Valery touched Joe's arm. "We're doing it."

"Doing what?" he asked.

"Building something that lasts."

Joe nodded, looking out at the growing sanctuary around them. "Yeah," he said quietly. "And this is only the beginning."

CHAPTER 26
STORM ON THE LAKE

The fog lifted off the lake in slow patches that morning, drifting like torn cloth across the water as Ridge Camp gathered at the shoreline. The air was cold enough to sting, and the sky held that heavy gray look that warned storms could form without much warning. But today wasn't about comfort. Today they needed answers—how deep the water was, where the fish were, and whether the tugboats could handle real runs now that the storms of early fall had settled.

Kite and Marlin bobbed lightly at the end of the floating pier, their hulls scratched but solid. These boats had survived the ocean, the storms, and Harrow's sabotage attempts, but they hadn't been tested in Tennessee waters yet. Their paint was chipped, their engines loud, but they were still dependable.

Joe stood at the end of the pier with Manny, Ashley, Guido, and Reed Lawson—the quiet aircraft mechanic who had turned out to be surprisingly good with boats. Valery stood onshore near the comms table with a headset, ready to coordinate.

"Engines first," Joe said. "Let's make sure they're not going to stall halfway across the lake."

Reed ducked into Kite with confident movements. "Fuel lines look clean. Filters are clear. It's old, but it'll run."

Guido smirked. "Sounds like me."

"Except you don't run," Manny shot back.

"You want to swim today?" Guido asked, raising a brow.

"Not particularly."

"Then hush."

Joe didn't bother interrupting them. Their bickering was background noise—steady, familiar, and strangely comforting.

Ashley stepped onto Kite with that careful steadiness she always used around water. "Should we test the sonar first or wait until we're in deeper water?"

"Deeper water," Joe said. "We need a clean reading. If fish are anywhere, they'll be down near the old valleys."

The air smelled like cold metal and lake weeds as the boats powered on. The engines coughed once, then steadied into a low hum. Reed adjusted a valve, nodded, and signaled thumbs up.

"Looks good!" he called.

Onshore, Valery's voice crackled through the radio. "Comms check. Kite, give me something."

Ashley picked up the handheld. "Kite copies."

"Marlin?" Valery asked.

Guido clicked his radio on. "Loud and annoying, like always."

Valery's sigh was audible. "Copy that."

Joe stepped onto Marlin with Manny while Ashley, Reed, and Guido climbed onto Kite. The plan was simple: run parallel paths down the lake and see what the sonar caught. If the water held fish, Ridge Camp would finally have a dependable source of protein heading into winter.

"Let's move," Joe said.

A small skiff crew had launched earlier to map the shallows on the north side of the lake, staying closer to shore while the tugboats handled the deeper runs.

Marlin pushed away from the pier first, gliding across the calm water. Kite followed close behind. The lake surface was glassy, broken only by the steady churn of their engines. The cold hit them immediately—sharp wind, colder spray—but everyone held steady.

"Reading?" Joe asked.

Manny adjusted the sonar screen bolted to Marlin's control panel. Lines flickered. Shapes appeared faintly.

"Small clusters," he said. "Nothing big yet."

"Could be baitfish," Joe said. "Keep going."

On Kite, Ashley watched her screen while Guido scanned the water with binoculars.

"Seeing anything?" Reed asked.

Ashley zoomed in on a cluster of shapes. "Maybe. Could be submerged debris. Could be fish. Hard to tell."

"Go southwest another two hundred yards," Valery advised from shore. "You're getting close to one of the deeper pockets on the map."

Both boats adjusted course.

Then—finally—Manny's screen lit up in thicker shapes.

"Joe," he said, pointing. "There. That's fish. A whole school."

Joe leaned in. "Valery, mark this spot."

Already ahead of him, Valery responded, "Marked. That's a promising zone."

Joe took a long breath of cold air. For the first time in weeks, the future felt survivable.

They had fish.

They had boats.

And for an hour, everything felt steady.

But storms never warned twice.

Valery kept her eyes on the lake as the wind shifted from steady to unpredictable. It slapped cold spray across her face even though she stood back from the shore. Her shoulder still hurt, but she ignored it. She leaned over the radio table and pressed the transmit button. "Kite, Marlin, check in. I need your status right now."

Static cracked through the speaker before Kite finally answered. "Kite here. Winds are getting rough. Whitecaps everywhere. We're heading back in. Marlin's trailing us."

Valery didn't like the hesitation in his voice. "How far behind?"

"Half a mile, maybe. They slowed down to help Skiff Two. That one's taking on water."

Valery stiffened. She turned toward Morales, who was helping secure a tarp near the camp center. "We've got trouble on the lake!"

Morales jogged over. "How bad?"

"Skiff Two's in trouble. Marlin is assisting, but the storm's picking up fast. Look at the ridge line."

Morales followed her gaze upward. The sky had gone from silver to an ugly, bruised gray. A long cloud mass pressed low over the ridge, moving fast.

"Storm's early," Morales said.

"Storms don't care about schedules," Valery answered.

Morales shook her head. "What's the plan?"

"We get them home," Valery said, flipping switches. "Kite is already returning. Marlin needs direction. The skiff won't make it on its own."

She stepped closer to the shoreline as waves slammed against the rocks. Spray soaked her boots. Ridge volunteers rushed to secure tents and anchor loose equipment as the wind picked up.

"Marlin, come in," Valery said. "What's your position?"

The captain answered, sounding strained. "Skiff Two has water up to the floorboards. We're tethered to them. Trying to pull them in, but the wind's pushing us sideways."

Valery gripped the radio harder. "Cut the tether."

"Negative."

"Captain," she said, voice sharp, "if you stay connected, this storm will flip both vessels. Cut loose, circle back, and guide them instead. Kite can intercept."

Morales muttered under her breath, "He won't want to cut it."

"He has to," Valery said.

She keyed the mic again. "Marlin, cut the tether. That's an order."

A long silence followed. Valery imagined the crew staring at the rope, afraid cutting it meant abandoning the skiff. Finally, the captain answered. "Copy. Cutting now."

Valery exhaled.

Morales nodded. "Good call."

Valery lifted binoculars. Kite was making steady progress, bouncing over the choppy water. Marlin, farther out, blurred in the spray. The skiff bobbed alone now, small and vulnerable.

"Hold together," Valery whispered.

A strong gust shoved her sideways. Morales caught the edge of the table before it flipped. Shouts rose as a pot rolled across the stones.

"Storm line's hitting early!" someone shouted.

Valery didn't look away from the lake. The wind tore across it now, flattening the water into streaks and lifting white foam into the air.

"Kite, accelerate," Valery said. "You're moving into rougher water."

"Already in it," the pilot replied. "Visibility dropping."

"Keep talking to me," she said. "Don't go silent."

Morales touched her elbow. "You're running this operation."

"Someone has to," Valery said. "I'm not losing anyone today."

Thunder rolled across the valley. The wind howled harder, bending the trees.

"Marlin, status," Valery called.

"Tether cut. Skiff drifting. We're trying to circle. Hard to stay steady."

"You're not alone out there," Valery said. "Keep the skiff in sight. Kite's almost on you."

Another burst of thunder shook the air. The storm arrived all at once, slamming into the lake with violent force.

Valery stood firm, dripping, boots planted against the wind.

"Everyone hang on," she whispered, voice steady and fierce. "We're bringing them home."

The storm hit the camp hard enough that several people had to grab tent poles to keep them from snapping. Rain turned the shoreline into slick mud within seconds. Valery hunched over the radio, shielding it with her body as wind whipped the tarp overhead. "Kite, where's the skiff?"

Kite's pilot answered between jolts of static. "We've got visual. They're drifting fast. Hold on—trying to get close."

Morales was already directing volunteers to form a line along the rocks. "We're throwing ropes the second they're in range. Keep kids away from the water!"

Joe arrived at a run, soaked and breathing hard. "Update."

"Skiff Two's taking water and drifting," Valery said. "Kite is almost to them. Marlin is circling."

Joe wiped rain out of his eyes. "Anyone in the water yet?"

"Not yet," Valery said. "But they're close to losing that boat."

Morales pointed toward the lake. "Look!"

A wave rolled over the skiff's bow, slamming it sideways. One of the men aboard lost his balance and nearly fell out before grabbing a rail. The others bailed water with buckets, but it wasn't enough.

Kite fought its way closer. Two crewmen threw a rope, but the wind snapped it aside.

"Again!" Valery barked into the radio.

Kite threw a second line. This one hit the skiff's bow, and one of the crew tied it around a metal cleat.

"Good," Valery said. "Hold steady. Start towing them in."

"They're heavy," Kite's pilot grunted. "Storm's fighting us."

A sudden crack echoed from the lake—something wooden breaking. Guido cursed under his breath. "The skiff's hull can't take this."

He was right. A large wave hit, and the skiff lurched violently. The rope pulled tight—and then snapped.

The skiff spun, half-submerged now. A crate slid across the deck, slammed into the rail, and tipped over the side. It bobbed once, twice, then disappeared beneath the surface.

"Damn it!" Manny shouted. "That was one of the new supply crates!"

Joe didn't hesitate. "Divers on standby!"

Manny, Guido, and Reed—ran toward the gear tent to grab wetsuits and tanks. The storm had soaked everything, but diving gear didn't care about weather.

Valery stepped to Joe's side, hair plastered to her face. "We can't send them out there until the skiff's secured."

"We won't," Joe said. "But that crate had replacement tools and part of the medical inventory. We can't lose it."

Kite made another attempt at the skiff. This time, they approached from the windward side, sliding in dangerously close. Two of their crew leaned over the rail, muscles straining as they hooked a metal grappling pole to the skiff's bow.

"We've got them!" the pilot yelled.

Morales signaled the shore team. "Get ready! That line's going to come fast!"

A group of volunteers braced themselves. Guido returned in a wetsuit, tank strapped to his back, and stopped beside them. "Tell me when."

"You'll get your turn," Morales muttered.

Kite revved its engine, fighting the storm. Slowly, painfully, the skiff inched closer to shore. The waves kept slamming it sideways, but the grappling pole held.

Joe cupped his hands and shouted over the wind, "Pull! Now!"

The shoreline crew heaved on the rope as soon as it reached them, boots slipping on the wet rocks. Inch by inch, the skiff drifted toward land.

Finally, a large wave shoved it far enough that the hull scraped against the rocky bottom.

Joe moved first. "Secure it!"

Valery ran beside him, ignoring her shoulder. People rushed forward, tying ropes around the skiff and anchoring them to trees and boulders.

The men inside looked exhausted but alive.

One of them pointed out toward the deeper water. "The crate… it sank right before the skiff hit the last wave."

Joe nodded. "We know. Divers are going after it."

Guido pulled on his mask. Manny checked his tank. Reed adjusted his fins.

The storm still raged, but the crate held too much to abandon.

Joe stepped back and let the wind whip at his jacket. "All right," he said quietly. "Let's get it back."

The storm eased just enough for visibility to improve, but the lake was still rough and unpredictable. Manny, Guido, and Reed moved toward the waterline, tanks strapped to their backs, masks ready. The wind shoved at them in irregular bursts, forcing them to brace each step. Joe walked with them partway. "Only go if you can keep control. No hero moves."

Guido snorted. "Control is my middle name."

Manny muttered, "Pretty sure it's 'Trouble.'"

Reed adjusted his regulator and added, "Just keep talking. Makes it easier to pretend I'm not terrified."

Joe gave him a steady nod. "Fear keeps you sharp. Just don't let it make your choices."

Behind them, the shoreline crew secured the rescued skiff fully, tying down every rope and checking knots twice. The storm had weakened, but it still tossed occasional sheets of rain across the rocks.

Valery stayed at the radio table, giving updates to Kite and Marlin. "Hold position until we finish retrieval. Do not attempt another pass."

"Kite copies," came the reply.

Morales scanned the water. "Where's our entry point?"

Guido pointed at the churning edge about twenty feet south of the skiff. "Deep enough and the rocks thin out. We'll drop there."

Ashley jogged over with a bundle of extra line and tossed it beside Manny. "In case you need another tether. Don't go down too far."

Manny gave a crooked grin. "Yes, ma'am."

She rolled her eyes, but her voice softened. "Come back up."

Guido elbowed Manny lightly. "Nice of her to worry. Must be the charm."

Manny hissed, "Shut up," but his ears were pink.

Ashley took a step back, crossing her arms. The storm whipped a gust of wind across the rocks, carrying wet grit with it. A wave slammed against the shoreline, spraying everyone. She turned to move out of the splash zone—and her foot hit a slick patch of dark algae hidden between two stones.

Her boot skidded.

Her balance vanished.

She let out a short gasp as the world tilted.

Manny reacted before thought could catch up. He lunged forward, dropping his tank-heavy weight in a controlled slide, catching her arm with both hands. His momentum pulled him down to one knee, but he held her upright.

Ashley froze in his grip, eyes wide.

Manny's voice shook slightly. "Got you. You okay?"

Ashley swallowed hard and nodded. "Yeah… yeah. Just slipped."

He still held her arm, fingers wrapped tight.

She noticed.

He noticed her noticing.

Morales coughed loudly. "Maybe finish the flirting after the storm, huh?"

Ashley stepped back, flustered but smiling. "Right. Sorry."

Manny cleared his throat. "Uh… okay. Diving. Yes. Diving."

Guido grinned like a man who'd just witnessed a major plot twist. "Yep. This is going in the camp gossip."

"Touch anything and I drown you," Manny muttered.

"Romance threats. Love it," Guido said.

Ashley's voice softened again as she met Manny's eyes. "Be careful."

He nodded once, more serious than before. "Always."

Guido clapped Manny on the shoulder. "Let's go before we melt from all this emotional warmth."

Reed stood ready, mask on, fins in hand. "Storm's calming. Visibility still bad but manageable."

Manny slipped his mask over his face. "All right. Keep formation. Crate went down southeast, maybe thirty feet."

Joe stepped forward. "If the currents get too strong, abort. I mean it. Supplies aren't worth a body."

Manny gave him a firm nod. "Understood."

The three divers moved to the water's edge. A wave crashed against their shins but they held steady. Another hit harder, soaking them to the waist.

Morales crossed her arms. "Go!"

Manny, Guido, and Reed stepped forward together and disappeared beneath the churning surface in a clean, practiced dive.

Ashley watched the bubbles rise, her fingers still unconsciously rubbing the small bolt charm.

Valery looked at Joe. "They'll get it," she said.

Joe hoped she was right. The lake had already taken too much today.

Now it held something they needed—and three men relying on each other to get it back.

The storm broke apart as quickly as it had arrived. The wind faded to a ragged breeze. The heavy sheets of rain thinned to a mist that drifted off the lake in long, pale ribbons. The waves relaxed from violent to restless, still rough but no longer thrashing at the rocks like a thing alive. The sky brightened enough to show patches of gray-blue.

But no one on the ridge relaxed until the divers resurfaced.

Ashley stood at the shoreline, one hand wrapped around her bolt charm, the other shading her eyes. Valery stayed behind the radio, though she kept glancing toward the water. Morales stood like a statue, arms crossed, scanning the waves for bubbles. Joe stayed nearest the edge, boots inches from the froth as he watched the lake with a restless tension he barely hid.

"Movement!" someone shouted.

Three sets of bubbles broke the surface. Then Manny appeared, then Guido, then Reed. All three kicked toward the rocks, fighting the push of the waves.

"They've got it!" Ashley yelled, spotting the crate rising between them. It was bound with two extra lines the divers had wrapped around it underwater, and it bobbed awkwardly behind Manny like a stubborn, half-submerged boulder.

Guido spit out his regulator. "Somebody grab this thing before it decides to sink again!"

Volunteers waded into the shallows, gripping ropes, pulling with steady force. The crate scraped the rocky bottom, then slid up onto the shore with a wet thud. Manny dragged himself up behind it, coughing water out of his lungs.

Joe grabbed his arm and hauled him fully upright. "You good?"

Manny nodded, still catching his breath. "Yep. Found it sitting in a ditch about twenty feet down. Not too bad."

Reed flopped onto the rocks beside them. "Not too bad? A log slammed into my face."

Guido pointed at his forehead. "And I'm pretty sure I bruised a part of me I didn't know could bruise."

Ashley rushed forward, grabbing Manny's arm. "Are you sure you're okay?"

He pushed his wet hair out of his eyes, trying not to look pleased. "Yeah. Promise."

Guido smirked. "Looks like slipping girl's worried. Better hold onto her next time too."

Ashley flushed red. Manny shot Guido a look promising violence later.

Joe ignored the teasing. He crouched beside the crate and tested the lid. "Still sealed."

Valery approached, wiping rain off her face. "Marlin and Kite both confirm they're heading back. No more trouble out there."

Morales exhaled like someone finally releasing a held breath. "Good. Let's get everyone inside. The storm wasn't bad enough to wipe us out, but it was close."

The lake shimmered under the clearing sky, but its calm surface didn't fool Joe. He stared at the horizon with a thoughtful, heavy look.

Valery stepped beside him. "You're thinking about winter."

"Yeah," Joe said quietly. "If this storm hit us this hard and it's only early in the season…"

"We're not ready," Valery finished.

He nodded. "We need more shelter. More stored food. Better fuel reserves. If winter comes early or hits harder than expected, 314 people won't be enough to hold it off."

Morales joined them. "We'll adjust the work schedule. Double the efforts on permanent housing. Expand the garden rows. Reinforce the storage structures."

Joe kept staring at the lake. "And fishing crews need a backup plan. Today showed us how fast things can go wrong."

Ashley knelt near Manny, wringing water from his sleeve. "Maybe I can help with the winter prep. I know some cold-weather training from back on base."

Manny's smile was small but proud. "That'd be great."

Guido flopped onto a crate and sprawled like a soggy cat. "Can we just not have storms for five minutes?"

Reed chuckled. "You chose the wrong camp for that wish."

The volunteers finished hauling the salvaged supplies toward the storage tent. The light thickened into a late-afternoon glow, warm against the wet camp.

Joe glanced at the sky one more time, feeling the air shift toward something colder, sharper.

"We'll make it," Valery said beside him.

"I know," Joe answered. "But we're going to have to fight for it."

The lake lay quiet again.

Deceptively quiet.

And Ridge Camp stood on the shore, realizing the hard truth that survival wasn't a finish line.

It was a beginning.

CHAPTER 27
THE DEEP HOUSE

The morning started with a thin layer of fog rolling across the lake, soft enough to blur the drowned tree line but not enough to hide the broken roofs poking out of the water. The storm had passed, but the lake still felt uneasy—small ripples, shifting colors, a sense of something holding its breath. Joe scanned the surface from the tugboat Marlin's bow, narrowed his eyes, and pointed toward a half-submerged structure about two hundred yards out. "That's the one. Basement level's probably intact under there."

Manny braced his hands on the railing, studying the gray outline of the house. "Looks deep. Strong current."

Calder, one of the volunteer divers with a steady voice and a calm face, checked the tanks lined along the deck. "Depth is manageable. Current's fine as long as we stay anchored." He looked at Joe. "We go in pairs."

Guido adjusted his mask straps and muttered, "Pairs? Great. I'll get stuck with Manny and he'll start giving the house construction advice at the bottom."

Manny didn't look up from the winch he was tightening. "You're insulted because you know I'm right most of the time."

"Most of the time?" Guido scoffed. "Wow. Humble today."

Joe cut them both a look. "Focus. This is a grab-and-go dive. Tools, food cans, sealed containers. No wandering."

Two more divers—Rigg and Owen—finished prepping the tether lines. Rigg tapped his tank. "We marked pressure zones yesterday. The house sits at an angle. The basement might have trapped air pockets."

"Or worse," Guido muttered.

"The storm shifted debris," Calder added. "We'll go slow."

Joe nodded. "Good. Keep comm signals clear. Valery's on the channel from shore. If anything changes, she calls it." He checked his belt once more, pulled on his gloves, then glanced toward the skiff docked behind the Marlin. Ashley stood at the stern, handling rope with quiet precision as she double-checked every knot. She caught Manny's eye briefly. It lasted half a second, but Manny straightened like someone had just evaluated his whole soul and found it mostly acceptable.

Ashley's voice carried across the water. "Make sure you secure your lifelines before you go down. The storm shifted the branches along the bottom. It's messy."

Guido leaned toward Manny and whispered, "She worries about you."

Manny pretended not to hear. He pulled on his mask. "Let's get this done."

Joe climbed onto the platform and waited as the divers lined up. The fog thinned enough to reveal the roofline—shingles peeled back, chimney cracked, a porch roof submerged up to the eaves. A real house once sat here, with a lawn, a driveway, a life. Now it leaned sideways in dark water like a tomb.

"This place went under fast," Calder said. "Windows still intact."

"Means stuff inside might be intact too," Joe replied. "Stay sharp."

The Marlin's engine hummed at idle as Rigg tied off the anchor line. The boat rocked gently, water lapping against metal. Joe ran through one last mental checklist—entry point, signals, extraction, pressure zones—then lowered his mask and clicked the comm test. "Blue leader, check."

"Reading you," Valery's voice answered in his ear. "Visibility looks decent from here. No storm rebound on radar."

"Copy. We're going in."

Joe stepped onto the edge with Manny at his right, Calder and

Guido behind them, Rigg and Owen last. Six figures lined up above the drowned home.

Joe lifted two fingers. "Down."

They dropped together into the cold dark.

The surface closed over their heads with a soft gulp, and the house below rose slowly into view—shattered roof tiles, broken boards, and the eerie stillness of a home frozen in time, waiting for hands brave enough to open it again.

Cold water pressed in from all sides as the team descended. Their flashlights cut narrow beams through the murky green, revealing drifting silt and long strands of drowned grass waving like fingers. The house came into clearer focus—half its roof collapsed inward, but the walls held firm. A front window remained intact, its glass streaked with lake grime but unbroken.

Joe signaled to Calder, pointing toward the lowest entry point along the slanted porch. Calder gave a thumbs-up. The group drifted closer in formation, staying slow and controlled, fins brushing nothing, arms tight to avoid stirring silt. Guido tapped Manny's arm and gestured toward a cracked shutter. "Creepy," his eyes seemed to say behind the mask. Manny only shrugged. They'd seen plenty worse.

Calder reached the porch first and checked structural stability by tapping the wooden beam gently with a metal probe. The sound was dull but solid enough. He motioned forward. Joe reached the window frame and examined it. The latch was rusted, but the glass had stayed sealed, which meant pressure inside might have preserved pockets of air. He drew a small punch tool, braced himself, and tapped the lower corner. The glass cracked in a star pattern. He tapped again. The pane caved inward without exploding, folding like a soft shell into the dark interior.

A cloud of silt belched outward. Everyone backed up until visibility cleared. When it settled, Manny shined his beam through the window. Floating debris filled the room—sofa cushions, toys, paper, a picture frame drifting upside down. "Looks like a living room," Manny said through the comm.

"Entry's doable," Calder replied. "We go slow."

Joe eased himself through the broken window, using the frame to steady his body. Inside, the furniture was shifted sideways from when the house had tilted underwater. But sealed cabinets along the far wall still looked intact. Guido followed next, kicking off the sill. Manny came in after, scanning corners with focus. Rigg and Owen secured the entry point from outside.

Joe motioned toward the kitchen doorway. Manny swept his light across the flooded hallway. The table was still upright, chairs scattered. A calendar drifted near the ceiling. They swam through the narrow hall one at a time, careful to avoid hanging wires and splintered boards.

The kitchen was the jackpot. The pantry door on the right wall was swollen shut but unbroken. Joe tugged it once—no give. Calder approached with a pry bar and wedged it between the frame and door. With two solid pushes, the door cracked open. A rush of trapped air escaped upward in a burst of bubbles.

Inside the pantry, shelves held tight rows of sealed cans, jars, and emergency containers. A few floated loose, but most remained stacked, untouched by decay. Manny grabbed one can, rotated it, and grinned behind the mask. "Beans. Sealed. Label's still readable."

Guido tapped a jar and held it up like treasure. "Peaches. Actual peaches. Oh man, Amy's gonna kiss us on the mouth."

Joe ignored Guido's commentary and began clipping sealed goods into the haul nets they carried. The divers worked efficiently—cans, jars, water packets, even a small metal toolbox tucked behind a shelf. Manny inspected the toolbox quickly before stowing it. "Still sealed. Could be useful tools inside."

Calder pointed his beam deeper into the pantry. At the back, behind a row of jars, sat a tightly sealed vacuum container, the kind meant for long-term storage. The lid showed no rust at all. Joe nodded, grabbed it, and clipped it into the main net.

The pantry wasn't the only find. In the utility corner of the kitchen, a thick metal case sat on a counter. Manny moved closer, wiped algae off the lid, and found the faded label: EMERGENCY SEED KIT. He signaled quickly. Joe swam over, inspected it, then nodded sharply.

"This alone makes the dive worth it," Joe said. "Bag it."

Guido looked around the kitchen, eyes wide. "It's like the family left for a weekend and forgot the apocalypse happened."

Joe lifted a hand. "All right, that's the haul point. Let's take it up."

They turned toward the living room, nets heavy with sealed food and tools—unbelievable finds in a world where everything was scarce.

What none of them noticed yet was the faint patch of darkness shifting along the bedroom hall, too large to be debris, too slow to be current. Something was moving deeper inside the house.

Joe led the way back toward the living room, hauling the first net of sealed food behind him. The kitchen glow faded as the beams of their lights shifted forward. Silt drifted lazily around them, stirred by their movement. Everything felt heavier now—the water, the walls, even the silence. Manny followed behind him, toolbox clipped to his vest, eyes scanning the ceiling for loose beams. Guido stayed near the window exit, humming nervously through the comms like he knew he shouldn't.

Calder moved last as rear guard, sweeping the hallway with a wider, slower arc of his flashlight. "Hold up," he said into the comm. His voice stayed calm, but the tone sharpened. "Something moved."

Joe rotated in place. "Where?"

Calder pointed his light toward the far bedroom corridor. The beam cut across a tangle of floating curtains, bumped a dresser stuck sideways against the ceiling, then hit the open doorway at the end of the hall. Inside the room, something large shifted—slow, deliberate.

Guido muttered, "That better be a waterlogged mattress."

It wasn't.

The shape slid along the back wall, too fluid for furniture, too large to be any normal fish. The light hit what looked like a curved silhouette—dark, broad, thicker than Manny's torso. Then it slipped behind a fallen bookcase, vanishing again.

Rigg's voice broke in from outside through the comm. "We've got movement down here too. Big wake near the porch. Nothing surfacing."

Calder's breathing hitched once. "This isn't debris. Whatever's here is alive."

Joe kept his flashlight pointed down the hallway, steady and controlled. "Everyone stay calm. We back out slow. No sudden kicks, no thrashing."

Manny swallowed hard, but his voice stayed even. "Copy. Moving."

They drifted toward the living room windows, keeping formation. The hallway behind them grew darker as their lights shifted away, but the faint outline moved again—closer this time. A soft thump echoed through the frame of the house, like something heavy brushing the floor—or what used to be the floor.

Guido cleared his throat. "No offense, but if that's a catfish, it's been doing steroids."

"Be quiet," Joe said. "Everyone out the window. One at a time."

Calder held position until the others moved, keeping himself between the team and the dark corridor. Manny slipped through the window frame first, pulling the heavy net through behind him. Rigg reached out from outside and grabbed the load, securing it to the lift line.

Guido exited next, shoving himself through the gap with a frantic determination that he tried and failed to disguise as skill. He swam out and immediately looked under the porch. "I don't see anything out here. But something stirred up the sand."

Joe passed the emergency seed kit to Rigg, then eased himself through the window frame. Calder was last, still watching the corridor. When he finally moved, the shadow slid forward again—as if it had been waiting for the one-on-one moment.

"Calder, move," Joe snapped.

Calder kicked off the doorframe and cleared the window just as a dark shape swept across the hallway behind him. The water shifted in a sudden pressure change, pushing silt outward like a low underwater gust.

All six divers regrouped outside, hovering beneath the slanted porch roof. The house loomed in front of them—quiet, still, but wrong.

Manny whispered over comms, "What the hell was that?"

Joe didn't answer for a moment. He kept his beam on the window. Nothing moved. Nothing followed. But the feeling of being watched didn't go away.

Calder finally said, "We're not alone down here. That's for damn sure."

Joe clenched his jaw. "We're ascending. Slow and steady. Keep eyes on the structure."

They rose in controlled formation. The house grew smaller below them, dark windows staring back like sunken eyes.

Halfway to the surface, something stirred in the shadow under the porch—broad, smooth, and silent.

Guido whispered, "Yeah… we're not diving alone anymore."

None of them disagreed.

The skiff rocked gently in the shallow chop as Ashley stood near the bow, hands gripping the railing tighter than she meant to. She kept her gaze locked on the dive line, searching for the first hint of bubbles. The lake looked calm on the surface, but something in the water felt wrong—like the air itself carried tension. She couldn't shake it. She'd paced so much that one of the volunteers finally muttered, "You'll wear holes in the deck," and wandered away. She didn't care.

Valery stood beside her, headset on, monitoring radio noise. "No panic signals," she said quietly. "They're fine."

Ashley nodded but didn't unclench her jaw. She kept replaying the moment Manny slipped underwater—how he had flashed her a quick, lopsided smile that didn't hide how nervous he actually was. She'd pretended not to notice. She always pretended. But the fear in her stomach hadn't eased since he disappeared beneath the lake.

A small splash made her jolt. Just a drifting piece of bark hitting the hull. She let out a slow breath, shaking her head at herself.

Valery adjusted the volume on her headset. "Signal's moving. They're coming up."

Ashley's heart kicked hard. "How many?"

"All of them," Valery answered. "Net load too."

Ashley's breath left her in a rush she didn't mean to show. The surface stayed smooth for another moment—too long, too still. Then bubbles finally broke through, spreading outward in a soft burst. A second later a head emerged—Joe. He cleared his regulator, lifted his mask, and gave a quick hand signal toward the deck.

Ashley didn't care about Joe. She scanned past him immediately.

Another diver surfaced—Guido—blowing out a loud curse just because he could. Then Calder emerged behind him, calm and steady. Two more divers surfaced next.

Still no Manny.

Ashley leaned forward involuntarily, knuckles white on the railing.

Finally, Manny broke the surface with a splash, hauling a mesh bag of gear. He pushed his mask up, shook water from his hair, and took a steadying breath.

Ashley felt her knees nearly give out.

Valery smirked softly. "You can breathe now."

Ashley swallowed hard. "I was breathing."

"Not well," Valery said.

Manny spotted the skiff and raised a hand when he saw Ashley on deck. The gesture was small, but it sent warmth through her chest. She tried to look casual and absolutely failed.

Joe swam toward the ladder. "Ashley, drop the hook line!"

She snapped back into motion, grabbed the hook line, and tossed it over the side. Joe clipped his gear net to it. Manny reached the ladder next, handing up a sealed toolbox. Ashley leaned down to take it from him, fingers brushing his for a moment longer than necessary.

"You okay?" she asked quietly.

He nodded, but his voice came thin from the cold. "Yeah. Big haul. Weird vibes down there, though."

"What kind of weird?"

Manny hesitated. "Like we weren't alone."

A chill shot through her. "Did something follow you?"

"No," he said, "but something watched us."

Joe climbed aboard and stripped off his regulator. "We'll debrief once everyone's up. For now, get the nets secured. We don't want to lose anything on deck."

Ashley helped Manny over the rail. When his boots hit the deck, he wobbled once, and she grabbed his arm instinctively. He steadied, offered a grateful smile, and whispered, "Thanks."

Her face warmed. "Anytime."

Valery raised an eyebrow from across the deck, but she didn't comment. She simply handed Ashley a towel for Manny and moved to help the others.

As the divers settled, Ashley kept sneaking glances at Manny—his wet hair plastered to his forehead, his hands shaking from cold and adrenaline, the faint grin he couldn't hide.

She exhaled slow, her chest finally loosening.

He was safe.

For now, that was all she needed.

By the time the skiff reached the shoreline, word had already spread across Ridge Camp that the dive team had returned with a major haul. People gathered along the waterline, forming a loose semi-circle as Joe, Manny, Calder, Guido, Rigg, and Owen unloaded the nets. Volunteers rushed forward to help, lifting crates and sealed cans with a mix of disbelief and hunger.

"Are those peaches?" one woman asked, voice cracking with hope.

Guido grinned and held up the jar. "The good stuff. The pre-apocalypse dessert of champions."

Several kids cheered like he'd lifted a treasure chest out of the deep. Even Joe cracked a small smile. The camp needed this—something good, something easy to celebrate.

Manny set down the toolbox and the seed kit. Donnie spotted them immediately and let out a sharp whistle. "You're kidding me. Seeds? Actual viable seeds?"

"Looks that way," Manny said.

Donnie clapped Manny's shoulder hard enough to make him wince. "Kid, you just handed us a future garden."

Ashley hovered nearby, towel around Manny's shoulders even though he'd claimed he didn't need it. She watched him talk to Donnie with a soft smile she quickly hid whenever anyone looked her way.

Valery jogged over with Morales, both breathless from excitement. "Inventory everything," Morales ordered. "We need a full count of what lasts, what needs pressure-testing, and what we can cook today."

Amy arrived next, wiping her hands on her apron. "I'll check the jars. If they're sealed right, we'll ration them for the next few weeks."

People gathered around, touching cans, examining jars, whispering in awe at the emergency seed kit. A few survivors who had barely spoken since Harrow's camp actually laughed—sharp, startled, healing sounds.

The dive had done more than bring supplies. It had lifted spirits.

Joe stepped away from the noisy crowd and looked out over the lake. Clouds rolled low and heavy across the sky, thickening by the hour. The wind carried a colder bite than yesterday. He felt it first as a chill along his neck, then as a knot in his gut.

Valery noticed the shift and moved beside him. "What's wrong?"

He nodded toward the darkening clouds. "Storm season's here. And after that comes winter."

Valery folded her arms, her sling creaking faintly. "We've dealt with storms."

"Not like these," Joe said quietly. "Not with this many people. Not with limited fuel, no greenhouse, and barely enough shelter."

She followed his gaze to the half-built huts and patched tents. The new families were settling in, but nothing felt sturdy yet. A strong storm could take out half the camp. A harsh winter could take lives.

Valery exhaled. "So what do we do?"

Joe didn't answer at first. He watched Donnie showing a group of volunteers how to strengthen tent frames. He watched Morales drawing new diagrams in the dirt with a stick, planning efficient labor crews. He watched Manny cracking open the sealed toolbox while Ashley crouched beside him, leaning close enough that their shoulders brushed.

"We build," Joe said finally. "Faster. Smarter. Before the cold hits."

Valery nodded. "Winter's coming whether we're ready or not."

Joe sighed. "Exactly."

Behind them, the excitement around the haul continued—people passing out jars of peaches for inspection, the new pantry cans, Donnie planning how to plant the seeds once they identified what they were.

The dive had given Ridge Camp a burst of good news. But Joe couldn't ignore the way the wind shifted.

Supplies were great. Morale was great.

But winter didn't care about morale.

And Ridge Camp still looked fragile.

Joe squared his shoulders. "We start tonight. Planning, organizing, assigning. No wasted daylight."

Valery smiled faintly. "There's the version of you Amy loves yelling at."

He huffed a small breath that almost counted as a laugh. "Let her yell. She's usually right."

The wind picked up—colder, sharper. The lake rippled under a darkening sky.

Joe stared at the horizon and made himself a promise.

They would be ready.

Because they had to be.

CHAPTER 28
ICE ON THE WIND

The frost hit before sunrise. Not a light dusting, not a warning shimmer—real frost. Crisp white crust on the ground, thin ice sheets clinging to water barrels, the kind of cold that made breath sting on the way out. Ridge Camp woke to it slowly, confused at first. Then someone shouted, "Hey—look at the ground!" and the panic spread fast.

Joe stepped out of his hut and felt the temperature cut straight through his shirt like a blade. He crouched and pressed a hand to the frozen dirt. It was solid. Hard. Too early for this. Winter had jumped the line.

Morales jogged over from the perimeter path, breath fogging. "This shouldn't be happening yet. We had three mild nights. One cold one. And now this?"

"Patterns don't exist anymore," Joe said. "Not since the impact. Weather can flip overnight."

Valery joined them with a thermal cup against her chest. Her sling peeked from under her jacket. "I checked the temp sensor," she said. "Twenty-nine degrees. It dipped lower before dawn. If this is a trend, winter is coming weeks faster than projected."

People began to gather in clumps, voices rising with nervous urgency. Parents wrapped blankets around kids. Two men tried to unfreeze a water spigot using their breath. Someone cursed loudly when a bucket cracked down the side, split by the sudden cold.

Amy hurried past with a box of gloves. "Hands first," she called. "Frostbite shows up faster than you think. Don't let the kids touch metal."

Joe looked across camp at the thin huts and tent roofs. They were built for wind and rain—not heavy freeze. "We're not ready for this," he said quietly.

Valery nodded. "We have to move fast."

Donnie emerged from the tool shed bundled in two coats, grumbling like a bear. "What the hell is this weather?" he muttered. "It's October. Last week it felt like late summer. Now it feels like January."

"Welcome to the new world," Morales said.

Donnie scrubbed his beard with a gloved hand. "This cold hits again tonight, we'll lose half the water pumps. And pipes. And maybe the garden lines."

Joe's jaw tightened. "Fix what we can today. Prioritize insulation. What about the generator shed?"

"It's fine," Donnie said. "Warm enough from its own heat. But the cabins? The tents?"

Valery exhaled slowly. "We haven't got enough materials to winterize the whole camp yet."

Joe didn't hesitate. "Then we triage. Start with families with infants. Elderly. Anyone with chronic illness. They get the warmest shelters. Everyone else doubles up if needed."

Morales scanned the camp. "We need a whole-camp meeting. People are already spiraling."

They were. Everywhere Joe looked, he saw fear brewing. Not loud fear. Quiet fear. The kind that sinks teeth into the ribs. A man kicked at frozen mud, muttering about his kids freezing in their sleep. A woman paced with her arms around herself, looking at the frost like it was a predator slowly climbing the ridge.

The food anxiety hit next.

A mother tugged Joe's sleeve. "Does this mean winter crops are done? Will the cold hit the storage? Will we lose what the Air Force left us?"

Her voice shook.

Joe placed a steady hand on her shoulder. "The rations are safe. The Air Force left us enough to last a year. But weather like this means we still need to keep gathering food. We can't rely on stores alone. We replenish or we run out, no matter how much we have."

She nodded, inhaled shakily, and hurried back to her family.

Valery watched her go. "People hear 'winter' and think of the old world. But winter after the impact..."

"Is something else entirely," Joe finished.

Guido sprinted over then, breath puffing clouds. "Boss, we've got a problem."

Joe braced. "What now?"

Guido pointed toward the lake. "The tugboats? Kite and Marlin? Their decks are iced. Not bad, but enough to be dangerous. If this keeps up, the lake could start crusting along the edges."

Morales groaned. "We needed another month before that."

Joe didn't flinch. "Then we adapt. Call a camp meeting. Thirty minutes."

Valery met his eyes. "You think they're ready to hear how bad this could get?"

"No," Joe said. "But they need to."

Donnie didn't wait for any more discussion. As soon as Joe called for the camp meeting, he headed straight for the stack of crates they'd brought from Arnold Air Force Base. Guido trotted behind him, blowing into his cupped hands.

"I swear my bones are freezing," Guido said.

"Good," Donnie answered. "Means you're still alive. Grab that crate on the right."

They pried open the nearest box marked HVAC AUXILIARY. Inside were thin aluminum ducts, flex hose, clamps, and brackets—spares that had sat untouched in some Air Force storeroom for years. Donnie grunted in approval.

"Perfect," he said. "We'll use these to build heat ducts from the generator shed out to the huts. Not direct exhaust, just warm air."

Guido eyed the metal skeptically. "You sure this won't turn us into a campfire?"

"Not if we do it right," Donnie said. "Heat exchange boxes only. We pull warmth, not fumes."

He dragged one of the empty ammo crates over and flipped it upside down to use as a worktable. Tools clinked. Metal rang cold. Every piece they handled felt like ice.

"Start cutting sections two arm-lengths long," Donnie said. "Small ducts, lots of outlets. We're not heating palaces, just keeping people above freezing."

Guido pulled on his gloves tighter and started cutting flex hose with a hacksaw. "You know these things were probably meant for some big, fancy base building, right? Offices, hangars, that kind of thing."

"Yeah," Donnie said. "Now they're gonna keep kids in canvas huts from shivering to death. I'd say that's a good reassignment."

Two teenagers from Ridge Camp jogged over. Ben and another boy, Colton, both bundled in borrowed jackets. "Morales said you needed extra hands," Ben said.

Donnie nodded. "Perfect timing. You two sort clamps and brackets by size. Keep the small ones separate—we'll use those for the hut connections."

Colton's teeth chattered as he picked up a handful of clamps. "Is it going to get colder than this?"

Donnie didn't pretend otherwise. "Yeah, kid. It will. But if we work fast, you'll feel it a lot less."

They worked in tight rhythm—Donnie measuring, Guido cutting, Ben stacking, Colton sorting. Breath puffed white over the crate. Lenny hovered nearby, watching each piece like it was a squirrel.

Valery walked over carrying a lantern and a list. "We've got the huts tagged," she said. "Families with infants and elders get first ducts. After that we rotate by how exposed they are to wind."

"How many priority shelters?" Donnie asked.

"Fourteen," she answered. "Another ten we'd like to hit before night if we can."

Donnie winced. "We'll push it. No promises."

Joe approached with Morales on his heels. "Camp meeting in ten,"

Joe told them. "I'll tell everyone what we're doing and what we're not doing. Any surprises I should mention?"

"Yeah," Donnie said. "Tell them not to hang blankets over the vents once we install them. They'll suffocate the airflow and burn up our fuel for nothing."

Morales made a note on her clipboard. "I'll handle that. People like to 'improve' good ideas until they break them."

Joe gave a faint, exhausted half-smile. "Story of my life."

Valery checked her list again. "The good news is the Air Force rations in storage are fine. No temperature sensitivity. They were built for war zones and long-term bunkers."

Joe nodded. "That buys us time. A year's worth of food if we manage it right."

"But not if we treat it like a feast," Morales added. "We still need fresh calories—fish, foraged greens, whatever the divers bring up. If we don't replenish, that 'year' becomes eight months. Or less."

Donnie tightened a clamp with a screwdriver. "I'll make some weld-up heat boxes near the generator shed. We'll tap the warmth off the exhaust housing, run the ducts out from there."

Guido frowned. "And fuel?"

"We've got enough from the base to keep the generator running if we're smart," Donnie said. "Not full tilt, not all-day luxury. Controlled use. Night heating cycles, maybe a couple short blasts during the coldest mornings."

Joe thought about that, already running the math in his head—fuel barrels, generator draw, heater cycles, how many nights of hard winter a year of fuel could really cover. Not enough. Never enough. But better than nothing.

"Do it," he said. "I'll back you up when people start complaining their huts aren't tropical."

"They can complain all they want," Donnie said. "So long as they wake up with all their toes."

Morales glanced toward the center of camp where people were slowly gathering in front of the main fire pit. "Time for your speech, Captain."

Joe exhaled slowly. "Yeah. Time to tell them we're safe enough to live—but not safe enough to relax."

He stepped away from the worktable, the cold air cutting harder as he moved. Behind him, metal clinked, tools scraped, and the first improvised heat system for Ridge Camp began to take shape under numb hands and stubborn hearts.

The first real panic hit midmorning, right after Joe finished speaking to the camp. People dispersed from the meeting with forced calm, but the moment they walked past the storage tent, the tension surged again. A line formed without anyone officially starting one—parents clinging to kids, newcomers whispering, longtime Ridge Camp residents checking the ration lists pinned to the wall.

Manny stood outside the tent, clipboard in hand, trying to keep the flow steady. "One at a time," he said. "We're not running out. We're organizing."

A woman stepped forward clutching a thin jacket around her shoulders. "How do you know we're not running out? How do you know this frost won't ruin everything?"

Manny kept his voice steady. "Because everything in the storage tent is freeze-stable. The military packed it for weather a lot worse than this."

"But what about the fresh food?" someone else asked. "What about the gardens?"

Joe appeared then, calm but firm. "The gardens will take a hit, yes. This frost is early. But the seeds we brought up from Arnold will let us restart in the spring. And the food we have will last if we stay disciplined. No hoarding. No panic."

A man stepped forward, anger cracking through his fear. "We can't just trust food in crates! What if this freeze keeps coming? What if the lake locks? What if the storms keep hitting?"

"Then we plan for all of that," Joe said. "We fish when we can. We forage when we can. We ration when we have to. But right now? None of you are going to starve this winter."

The man didn't back down at first, but his wife gently touched his arm and lowered his hand. "He's right," she whispered. "We're here. We're safe. Let's give them a chance to organize this."

He exhaled sharply and stepped back.

Behind them, Amy walked out of the medical tent with a steaming mug in her hands. "Everyone with small children, come see me first," she called. "We've got vitamins, broth, blankets, and warming stones. Nobody's going cold if I have anything to do with it."

Her voice cut through fear like warm soup cuts through a bad day.

Inside the storage tent, a pair of newcomers stood staring at the stacks—neat rows of sealed crates labeled MRE PACK 24 and WATER RATION UNIT and BULK GRAIN SUPPLY. Morales guided them through.

"These might not look like the meals you used to know," Morales said, "but they'll keep you alive, warm, and strong."

One of the newcomers, a teenage girl with a blanket around her shoulders, traced her fingers across a crate. "Harrow never let us near supplies. He told us we were one bad season away from starving."

Morales softened her tone. "That was a lie meant to control you. Here, we prepare for bad seasons, but we don't let fear dictate decisions."

Joe overheard and stepped inside. "Listen up," he said to the gathered families. "Winter is going to test us. It will test every choice we make. But we are not the same people we were under Harrow or on the ships. We are stronger now."

The teenage girl looked up. "How can you be sure?"

Joe met her eyes. "Because we're facing this together. And that changes everything."

A murmured agreement spread across the tent. Fear didn't vanish, but it loosened.

Behind Joe, crates were opened and volunteers began distributing sealed envelopes of MREs and small tins of nutrient pastes.

Valery ducked inside, rubbing her cold fingers together. "Joe, we have another problem," she said quietly.

Joe braced. "What now?"

"The cold messed with the radio wiring," she said. "Nothing broken, but enough to weaken signal. I'll need help running a new connection before night."

"We'll make it happen," he said.

She hesitated, then added, "People are scared, Joe. Really scared."

"I know."

"Then maybe…" She lowered her voice. "Maybe give them something to focus on. Work, tasks, anything. Fear gets smaller when hands get busy."

Joe nodded. "Good idea. We'll set up work rotations."

Outside, the cold wind pushed across camp again, cutting through coats and jackets. A baby cried. Lenny barked. A man rubbed his hands together, muttering about the lake turning to glass.

Morales stepped out of the storage tent and blew her whistle twice —sharp and commanding. "All volunteers report to the fire pit!" she shouted. "We've got winter work to do!"

People turned toward her voice, drawn to it like a lifeline.

Joe caught Valery's eye. "Let's give them direction."

Valery nodded. "Let's give them hope."

Manny was hauling a bundle of rope across the clearing when Ashley nearly collided with him coming around a corner. She had a crate of kindling in her arms, her breath fogging hard. "Sorry," she said quickly, stepping back before the crate slipped.

"You okay?" Manny asked, grabbing the crate before it tipped.

"Cold," she said. "And trying not to think about how much worse tonight's going to be."

Manny set the crate down beside her hut wall. "We'll get heat ducts up before dark. Donnie's working like a man possessed."

Ashley rubbed her gloved hands together. "Good. The families in the south row are already freezing. One woman said her baby wouldn't stop shaking."

His voice softened. "They'll get the first ducts. Joe and Morales already prioritized them."

Ashley nodded, taking a slow breath. "Okay. Good. That's good."

She looked jittery, the cold waking up old memories that had nothing to do with temperature. Manny recognized the signs—tight shoulders, eyes scanning too quickly, a breath that trembled before she forced it steady.

"Hey," he said gently, "want to help me teach some knot work? Donnie needs rope hitches ready for the duct mounts."

She hesitated. "You want me to teach knots?"

"No," he said. "You and me together. Come on."

He led her to the workbench near the supply tent, where coils of rope sat waiting. Snow-dry frost clung to the fibers. He pulled one end loose and handed it to her. "Start with a hitch. Simple anchor knot. We'll need a hundred of them."

Ashley held the rope like it might bite. "I haven't tied anything since… you know."

"Since before Harrow," Manny finished for her, quiet and matter-of-fact. "And that's fine. Today's a new one."

She exhaled shakily and nodded. "Show me."

He looped the rope slowly, talking her through it with steady patience. "Over… under… around the standing line… pull through. That's it."

Ashley tried to copy the motion but fumbled the twist. "Damn it—my fingers won't work."

"They're cold," Manny said. "Give them a second."

She tried again. This time her hand brushed his—quick, accidental, but enough to send a faint shock up both their arms. Ashley froze. Manny cleared his throat way too fast.

"Sorry," she murmured.

"No, it's—uh—it's fine."

They tried the knot again. Their hands brushed a second time.

Ashley's cheeks flushed pink—not from the cold. She didn't pull away, exactly, but she didn't look up either.

Manny swallowed and pretended to focus on the rope. "Okay, you almost had it. Try again."

She tightened her grip, steadied her breath, and tied the hitch clean on the next attempt.

Manny smiled. "There you go. Perfect."

Ashley looked at the knot as if unsure she'd really done it. "I remember that one. We used to tie down gear racks on the Halcyon with something like this."

"You were always good with hands-on work," Manny said. "I figured it would come back."

Ashley's eyes flicked up—not fully confident, but grateful. "Thanks for… helping me not panic."

He shrugged, but his voice stayed soft. "Anytime."

Lenny trotted by then, tail swishing like a metronome. He stopped at Ashley's foot, sniffed the rope, and barked once like he approved of the knot.

Ashley reached down and scratched behind his ears. "At least someone believes in me."

Manny let out a small laugh. "Lenny believes in anyone who gives him attention."

Ashley's smile grew a little. "Still counts."

Across camp, Morales shouted orders to the volunteers gathering at the fire pit. Frost crept along the tent roofs. Wind pushed through the trees in cold waves.

Ashley picked up another rope coil. "Okay. What's next?"

Manny handed her a second length of cord, his fingers brushing hers again—this time more controlled, more deliberate. "Next," he said, "we get enough knots tied so nobody freezes tonight."

Ashley nodded, determination pushing past her nerves. "Then let's work."

And side by side, hands moving in rhythm despite the cold, they began tying the ropes that would help keep Ridge Camp warm.

The storm clouds rolled in before noon, dark and low, pushing wind across Ridge Camp hard enough to bend the treetops. The temperature dropped again—fast—and people moved with new urgency. Donnie barked orders from the generator shed while Valery coordinated runners between the storage tent and the work crews. Manny and Ashley kept tying hitches, their hands raw but steady.

Joe stood near the ridge edge watching the sky shift into a bruised gray. Winter wasn't approaching—it was sprinting toward them. He pulled his jacket tighter and turned toward Morales as she approached with her clipboard tucked under one arm.

"Wind's shifting north," she said. "If it keeps dropping temperature like this, we'll hit another freeze before nightfall."

Joe looked around camp, noting every weak point—thin tent walls, open garden rows, uninsulated water lines. "We're not ready," he said. "Not even close."

"No," Morales said bluntly. "But we can get as ready as possible."

Down by the water barrels, volunteers worked in frantic rhythm to wrap the tanks in canvas and insulation pads. Two men hauled armfuls of straw and packed it around the barrels to keep them from splitting in the cold. Children helped by carrying small handfuls of kindling and stacking it near the fire pit.

Amy walked between families handing out heat stones she'd warmed by the generator exhaust. "Keep these wrapped," she instructed parents. "Don't let them touch bare skin."

Her calm steadiness grounded the anxious crowd.

Donnie dragged another duct section toward the nearest hut. "Guido! Get me that bracket! No—not that one, the bigger one!"

Guido dug through the clamp pile. "Everything looks big when your hands are frozen!"

Ben and Colton helped hold the duct in place as Donnie secured it. Inside the hut, a baby cried, the sound thin and exhausted.

"Hurry," the mother called from inside. "Please."

"We're hurrying," Donnie answered. "Almost done."

Across the ridge, the tugboats Kite and Marlin bobbed uneasily in the choppy water. Frozen spray crusted their rails. If the lake iced too far out, the boats would be trapped until spring.

Joe made a note to check the boats himself once the ducts were finished. Fishing trips would be harder in winter, but they were still necessary. Even with the Air Force rations, they had to keep replenishing food. One hard season could burn through their stores faster than they wanted to admit.

Valery jogged over, one arm pressed against her sling as she held a coil of wire in her good hand. "Radio's stable again," she said. "But if temperatures drop any further, we might lose signal strength after dark."

"I know," he said. "But we take one step at a time."

A sudden gust of wind blasted through camp then, knocking over a stack of firewood and sending loose leaves swirling into the air. Several children shrieked in surprise. Manny darted forward to steady a tool rack before it toppled.

Ashley grabbed the rope bundles before they slid off the workbench. "This wind's getting worse," she muttered.

"Yeah," Manny said. "Feels like winter's tapping on the door with a sledgehammer."

Ashley stiffened as the gust hit again, instinct pulling her back a step before she caught herself. Manny saw it, softened his voice. "Hey. You good?"

She nodded, though her breath shook. "Just… memories. I hate storms."

"You're safe here," Manny said. "I promise."

Ashley looked at him, grateful and terrified to be grateful. "I know."

Together, they pushed the rope hitches into a crate and covered it with a tarp.

By late afternoon, the first huts on the south row had warm air trickling in through the new ducts. The relief was immediate—mothers cried, older children sighed into blankets, and even the air inside the huts smelled different, warmer, safer.

Morales did a slow lap around the ridge, checking every station. When she returned to Joe, she tapped her pen against her clipboard. "We got through today," she said. "Barely. But we did."

Joe looked across the camp as thin snowflakes began to drift down —slow at first, then steadier. "We survived one frost," he said quietly. "But winter hasn't even started."

Morales nodded. "Then we better get ahead of it before it swallows us whole."

Joe watched the snow gather on the ridge, each flake a quiet warning in the settling cold.

"We'll get ahead of it," he said. "We have to."

CHAPTER 29
THIN HARVEST

The meeting started before sunrise in the main shelter near the fire pit. The air inside was cold enough to see breath, but at least there was no wind. A single lantern hung from a hook, throwing steady light over the table.

Joe stood at the head of the table with a clipboard. Morales, Valery, Amy, Donnie, Guido, and Calder sat or leaned nearby. A stack of supply sheets lay spread out. No one was talking. They were waiting for numbers.

Valery cleared her throat first. "I ran the latest projections," she said. "We have three confirmed food sources. One: the Air Force rations from Arnold. Two: the remaining garden crops that did not die in the frost. Three: whatever we can still catch, forage, or haul up from dives."

Joe nodded. "Start with the rations."

Valery tapped the paper in front of her. "We have enough sealed MREs and bulk supplies to feed three hundred and fourteen people for about one year at normal intake. That assumes moderate work and stable weather."

Morales frowned. "Define normal intake."

"Two thousand calories a day per adult, less for children," Valery said. "Enough to work, think, and stay healthy."

Donnie crossed his arms. "And winter is not normal intake."

"Exactly," Valery said. "In cold conditions, people burn more energy. Heavy work in low temperatures means higher calorie needs. If we try to keep everyone at full strength and do not add other food, that 'one year' drops closer to eight or nine months."

Guido let out a low breath. "That does not get us through winter and early spring."

Calder spoke up from the corner. "So the crates are not a guarantee. They are more like a timer."

"Right now, yes," Valery said. "A very generous timer, but a timer."

Morales leaned over her own sheet. "What about the gardens?"

Amy answered that. "The early frost killed most of the surface greens. We still have root vegetables under the ground, but not many. Carrots, some potatoes, a few rows of beets. Enough to supplement meals once or twice a week for a couple of months, not enough to carry us."

"Or to replant right away," Calder added. "We preserved seed stock from Arnold. That is for spring. We cannot eat those unless we accept starvation later."

"So the garden is a bonus, not a base," Morales said.

"Correct," Valery said.

Joe flipped to the last page. "Fishing and dives."

Calder straightened. "The lake is still open water. The edge has some ice, but the main body is clear. We can run fishing lines from the skiffs and tugboats. With good luck and regular trips, we can add a few hundred pounds of fish a week."

"Best case?" Joe asked.

Calder did not sugarcoat it. "Best case, we add enough to slow the drawdown of rations and keep people feeling full. Worst case, storms hit, the water ices near the shore, and every trip becomes a risk."

"And the dives?" Morales asked.

"Dives are changing," Calder said. "Colder water means shorter dives, more chance of hypothermia. But there are still canned goods, grain, and preserved supplies underwater. The Deep House run proved that."

Amy pressed her hands against the table. "So we have food. We are not empty. But we are on a countdown unless we keep adding to it."

Valery nodded. "Yes. If we treat the Air Force rations as backup only and push fresh food as far as we can, we can stretch that year closer to a year and a half. Maybe more."

"And if we are lazy?" Guido asked.

"Then by next fall we will be in real danger," Valery said.

The room went quiet. The only sound was the crackle of a small stove in the corner.

Joe looked at each of them. "Say it straight," he said. "Can our current supplies, by themselves, carry three hundred and fourteen people through one full winter and into the next harvest without extra fish or salvage?"

Valery met his eyes. "No. Not safely. Not with this many mouths. Not if the cold is worse than expected."

Morales closed her eyes for a second and inhaled. "Then the answer is simple," she said. "We work like it matters. Because it does."

Joe wrote a single note on his sheet: RATIONS ARE A NET, NOT A MEAL PLAN.

He looked up. "All right. Then we plan for hard winter, not easy winter. We move from comfort eating to survival eating. And we start today."

The meeting ended fast after that. No one lingered. The numbers were clear, and the pressure they created settled over the group like a new kind of weather. By the time Joe stepped outside, the morning sun was barely above the ridge. Frost glimmered on the ground even though the hour was early.

Calder walked beside him. "We should run a fishing trip within the next two days," he said. "Before the next cold front moves through. After that, waves will get rough and the winds will shift to the north."

Joe nodded. "We'll take the Marlin. It has the most deck space."

"It also has the highest risk," Calder said. "It sits low in the water. If we get hit by a sudden storm like the last one, we could lose half the gear. Or worse."

Joe stopped and faced him. "Do you think we have another choice?"

Calder didn't answer. He didn't have to.

Manny jogged up from the path that led to the tugboats. His breath fogged in the air. "We checked the engines on both tugs. Marlin is ready if we want to load it today," he said. "Kite still needs one more patch on the starboard seal, but we can finish that in the next hour."

Joe rubbed the back of his neck. "We have food from the Air Force, but not enough to sit still. We've got to stretch every resource."

Manny nodded. "I get it. Fish buys us time. Even one big haul buys us days."

Calder crossed his arms. "It isn't just about time. It's morale. If people see the boats go out and come back full, they'll feel safer. Makes winter seem survivable."

Guido approached then, wiping grease from his hands. "I heard we're heading out again," he said. "Which boat am I getting tossed off of this time?"

"You're on Marlin with us," Joe said.

Guido made a face. "Great. The low one. The one waves slap like they owe it money."

"Bring a rain jacket," Manny said.

"I'll bring four," Guido replied. "And a bucket."

Joe huffed, almost smiling. Then he sobered again. "We head out tomorrow morning at first light. We'll load gear today. Fishing lines, nets, spare tanks, backup fuel, cold weather suits."

"And rifles?" Manny asked quietly.

Joe paused. "Yes."

Calder didn't argue. "There are still animals in those woods. And not all people on the shoreline might be friendly."

They reached the waterline, where the tugs bobbed gently against the floating pier. The early frost made a thin white film along the edges of the planks. Marlin sat closer to shore, faded blue paint scratched by storms and salvage work.

Calder stepped onto the deck and tested the rail. "Still solid."

Manny climbed in after him. "We'll need to check the anchor lines again. Last storm nearly snapped one."

"I'll handle it," Guido said, heading toward the bow.

Joe walked the deck slowly, scanning everything with practiced focus. He checked the fuel drums secured at the stern, the emergency life vests, the coil of heavy line, the stacked plastic tubs waiting for a haul.

They had done this dozens of times on the open sea during the long drift, but doing it here felt different. The lake was quiet. Too quiet. No wind. No waves. Just cold air and a wide stretch of still water hiding drowned houses, old forests, and whatever else the Rain of Fire had left behind.

Morales arrived a minute later, clipboard already out. "I told a handful of volunteers to prepare meals for the crew before dawn. Hot drinks, high-calorie food. It will be a long day out there."

"We'll take it," Joe said.

She looked at him. "Are you sure you need to go personally?"

Joe didn't hesitate. "Yes. The first long-range run of the season? I'm not letting someone else take that risk."

Morales didn't like it, but she didn't argue. "Fine. But keep the radio open every fifteen minutes. If anything seems off, you turn back early."

Joe met her eyes. "We will."

Manny and Guido stood by the rail now, listening.

Calder secured the last rope.

The decision was made.

Tomorrow, before dawn, Marlin would leave Ridge Camp and head into cold, deep water.

The risk was real.

But so was the need.

Valery gathered the leadership group again after midday to present her rationing plan. This time the meeting took place outside near the storage area so newcomers could listen if they wanted. People were already gathering with anxious expressions. Rumors had moved through camp fast.

Valery stood on a crate so everyone could see her. She held a clipboard and several sheets of data. "I'm going to be direct," she said. "We have food. We are not starving. But our current rate of use will not carry us through winter unless we change our habits now."

A few heads lowered. A few crossed their arms.

Joe stepped beside her, not speaking yet, simply making it clear the message was supported.

Valery held up the first sheet. "Here's how it breaks down. If everyone keeps eating the same portions as the last week, we burn through our winter stores by early spring. That gives us no buffer if storms cut us off from fishing or diving. It also endangers people who are older, sick, or doing heavy labor."

A man near the front raised a hand cautiously. "So what do you want us to do?"

"We shift to controlled rations," Valery said. "Not starvation rations. Balanced rations. Everyone gets the same meals. No one is singled out or punished. Workers in extreme cold get more. But portions drop slightly for everyone else."

A murmur ran through the crowd. Some looked relieved. Others frowned.

A newcomer woman with short blond hair stepped forward. "We were starved under Harrow," she said. "I am not going back to that."

Valery shook her head. "This is not punishment. And this is not Harrow. This is math. If we follow this plan, we maintain health and stability until spring."

Another newcomer, a tall man with a long beard, folded his arms. "What if we don't agree?"

Joe stepped forward then, voice level. "Then you talk to us. But the plan stands. We are responsible for three hundred and fourteen lives. Winter is coming fast. Food is not infinite."

A few people nodded. Others hesitated.

The bearded man stepped closer. "We already lost control once. You expect us to just trust a ration plan because someone with a clipboard says so?"

Valery didn't raise her voice. "I expect you to trust the numbers that kept us alive on the water for two years. And I expect you to trust that no one here wants to see a child go hungry because adults refused structure."

The man didn't back down. "Feels like rules again. Feels like control."

Morales cut in firmly from the side. "Control is when someone uses fear to force you into obedience. Structure is when leaders prevent disaster by telling the truth before it arrives."

Amy joined her, wiping her hands on a towel. "You want soft answers or honest ones? Because honest ones are what keep people alive."

The tension shifted. The newcomers looked at one another, weighing the difference between old trauma and present reality.

Valery continued. "Starting tonight, we go to three meals a day with slightly smaller portions. More vegetables where possible. More fish once the Marlin returns. We protect the rations from Arnold as much as possible so they last. Children, elders, and anyone with medical needs get priority."

A woman holding a toddler raised her hand. "What about babies?"

"You come to med," Amy said. "We'll work out a plan with you."

The man with the long beard finally lowered his arms. His voice softened. "If this keeps us alive… fine. I'll go along with it."

Several others nodded. A few sighed in relief.

Valery stepped off the crate. Joe placed a hand on her shoulder briefly, steady and supportive.

The group began to disperse, quieter now, more thoughtful. People whispered to one another, but the panic had eased.

Ridge Camp wasn't starving.

But for the first time, everyone understood they could not waste a single meal.

Winter wasn't approaching.

It was already here.

Ashley carried a crate of rope coils toward the tool rack near the center of camp. Her steps were steady, but her shoulders were tight. The ration meeting had left people tense, and she could feel the left-over anxiety everywhere she walked. Manny was already there, sorting gear before tomorrow's fishing run. He was checking the knots on a heavy towing line, brow furrowed in concentration.

Ashley set the crate down beside him. "These are the last of the rope bundles," she said.

"Good," Manny replied. "We'll need the thicker ones for the nets. The thinner ones can stay up here for camp use."

As he bent to grab another coil, a newcomer approached—one of the younger men rescued from Harrow's group. He had sharp features, restless eyes, and a posture that suggested he didn't know where he fit yet. He stopped a few feet from them.

"You're Manny, right?" the man asked.

Manny straightened carefully. "Yeah. Need something?"

The man shrugged. "People say you're running the gear prep. Thought I'd help, but it looks like you've got plenty of hands." His tone shifted slightly as he eyed Ashley. "Didn't know they let civilians handle equipment too. Guess standards dropped around here."

Ashley's jaw tensed immediately. Manny opened his mouth, likely to defuse it, but she stepped forward before he could speak.

"Excuse me?" she said, voice sharp and clear.

The man faltered but lifted his chin. "Just saying. Back where we were, gear wasn't handled by people who didn't know what they were doing."

Ashley crossed her arms. "Back where you were, equipment was handled by people ordered around by a man who threatened to break their fingers if they made a mistake. Don't confuse that with skill."

Manny tried to edge between them gently. "It's fine, Ash. Let it go."

She didn't budge. "No. He doesn't get to come in here and disrespect you."

The newcomer blinked, thrown off. "I wasn't—"

"Yeah, you were," Ashley cut in. "Manny keeps the boats running. Manny fixes half the things this camp uses. Manny risked his life more times than I can count to get us off the water alive. You don't know him, so don't act like you do."

Manny looked at the ground for a second, embarrassed but also clearly touched. "Ashley—"

She turned to the newcomer again. "You want to help? Then help. You want to complain? Find someone else to listen, because I'm not interested."

The man stared at her, chest rising and falling like he was trying to decide whether to push back or walk away. Finally, he muttered, "Didn't mean anything by it," and stepped away toward the fire pit.

Manny watched him go, then let out a slow breath. "Thanks. But you didn't have to do all that."

Ashley shrugged once, still annoyed. "Yes, I did. You deserve respect. Not sideways comments."

He rubbed the back of his neck. "People are stressed. Winter, food worries, the storm… He probably doesn't know how to talk to anyone without sounding defensive."

"Still not an excuse," she said.

Manny lifted a rope coil and placed it on the rack. "You defending me like that was… nice."

Ashley looked away quickly. "Well. Somebody has to."

Manny smiled a little, soft and warm. "You're good at it."

She felt her cheeks heat. "Stop."

"Can't," he said lightly. "It's true."

Her throat tightened at the sincerity in his voice. She bent down to grab another rope coil just to have something to do. Manny crouched beside her, reaching for the same coil. Their hands brushed.

They both froze.

Neither pulled away immediately.

Ashley swallowed and stood first. "We should finish this," she said quietly.

Manny nodded. "Yeah. We should."

But the moment stayed between them—quiet, unspoken, real.

The sky dimmed earlier than usual, the sun sinking behind the ridge as if winter was already pulling it down by the shoulders. Ridge Camp shifted into evening mode—lanterns lit, fires built, tools put away. The air smelled like pine smoke and cold earth. Tomorrow's fishing run weighed on everyone, even those not boarding the Marlin.

Near the shoreline, the tugboat crews worked by lantern light. The Marlin floated steady at the improvised dock, her hull patched and reinforced after last month's storm. Her deck was loaded with sonar equipment, towing lines, spare fuel drums, and two insulated crates for storing whatever fish they managed to catch. Joe stood beside the ramp, clipboard in hand, double-checking every item Manny and Guido called out.

"Fuel tanks sealed," Guido said. "Checked them twice."

"Sonar calibrated," Manny added. "Perez helped me last shift. It's reading clean."

Joe marked the list. "Good. Anything breaks out there, we can't patch it with driftwood. Make sure you have the cold-weather gear too. The water temp is dropping fast."

Calder, one of the new divers from Harrow's group, stepped forward. He was steady, calm, and had proven himself more than once already. "We packed extra suits. If we have to dive, we'll switch out in rotation. We won't push the limits."

"Good," Joe said. "This isn't about heroics. It's about food."

Across the dock, Valery walked with Morales, reviewing the route on a small laminated chart. The cold air made Valery's breath puff white, but she didn't seem bothered. "The best fishing zones sit along the deeper channels," she said. "If the storm last week shifted debris, you'll see it on sonar before you hit it."

Morales nodded. "We'll go slow. No reason to rush into trouble."

Joe joined them. "The water levels dropped another foot this week. Takes longer for the skiffs to run supplies. If we want stable protein through winter, we need this run to go well."

Valery crossed her arms. "It will. You picked the right crew."

Behind them, Ashley approached with a sealed medical kit. "Here," she said, handing it to Manny. "For the trip. Bandages, antibiotics, cold packs, and heat pads. Just in case."

Manny took the kit carefully. "Thank you."

Ashley nodded once, avoiding his eyes. After their moment earlier, she seemed unsure of where to look. Manny seemed just as unsure of what to say.

Joe pretended not to notice the awkward tension. "Everyone gets some rest tonight," he said. "You're leaving before first light. Check-in every thirty minutes on the radio. If the weather starts to change, turn back immediately."

Guido groaned. "We're going to be fine. We're not planning to attack the water. We're just fishing."

Morales shot him a look. "Disasters usually happen right after someone says something like that."

"Then I'll stop talking," Guido replied.

"Please do," Manny muttered.

The small group cracked tired smiles. It wasn't relief, but it was close.

As the final gear was loaded, the rest of Ridge Camp drifted toward the shoreline. Families stood together, some holding lanterns, some clutching blankets around their children. The newcomers who'd been unsure about rations earlier now watched with quiet respect as the fishing crew did their final checks.

Amy stepped up beside Joe. "Looks like half the camp wants to say goodbye."

Joe nodded. "They know what's at stake."

Valery stepped to his other side. "We're stable on food for now, but winter will hit hard. If the Marlin doesn't come back full, we'll feel it by February."

Ashley finally looked at Manny as he prepared to board the tug. "Be careful," she said quietly.

He paused, hands on the railing. "I will. Promise."

Their eyes met just long enough that both turned away before anyone else noticed.

Morales clapped her hands once. "All right. Crew on board. Everyone else clear the dock."

People stepped back as the crew climbed onto the Marlin and the deck lanterns flickered to life.

Joe raised his voice so the whole shoreline heard him. "Departure at dawn. Get sleep. Tomorrow matters."

The camp murmured in agreement. Families dispersed. Fires dimmed. Guard posts lit their lamps.

The Marlin bobbed gently in the cold dark water, waiting for morning.

And Ridge Camp held its breath, knowing tomorrow would either strengthen their future—or tighten the margin they had left.

CHAPTER 30
RIVER WOLVES

The Marlin pushed upriver just after dawn, its small engine rumbling through the cold morning air. Frost still clung to the rails, and every breath came out white. The river looked different this far inland—narrower, darker, framed by trees bent from years of storms. Guido kept the wheel steady while Manny and Calder checked gear on deck. Joe stood near the bow, watching the current for debris that could tear the hull open.

Ashley was with them too—quiet on the stern platform, arms folded, eyes tracking the shoreline. She'd come along as secondary overwatch and medic backup, though she hadn't said much since they'd left Ridge Camp.

The air smelled like mud and cold metal. The water was low enough that broken road signs and fence posts jutted out at strange angles. Every few minutes, Manny pointed out old rooftops beneath the surface—ghost outlines of drowned towns long abandoned.

Valery stood beside Joe with a map clipped to a board. "We'll follow this bend and push northwest," she said. "Fish cluster around the deeper pockets. If sonar's right, we'll find enough to fill two crates before noon."

"Assuming nothing else finds us first," Guido muttered.

Joe didn't answer. He kept scanning the banks, alert and tight-jawed.

The Marlin carried a small crew—Joe, Valery, Manny, Guido, Calder, Shaun, and Ashley at the stern keeping watch. The plan was simple: go upriver, catch what they could, look for anything worth salvaging, then return by nightfall. Winter was closing in too fast, and Ridge Camp couldn't rely on the Air Force food stores alone.

They needed fresh catches. They needed tools, fuel, anything that could stretch their margins.

Manny checked the spare oxygen tanks. "We'll rotate dives if we need to check nets underwater," he said. "Visibility isn't great."

Calder nodded. "Better than the lake, though. Less debris."

Guido adjusted the throttle. "Let's hope the fish didn't migrate to the next state."

"You talk too much when you're nervous," Manny said.

"I talk too much when I'm awake," Guido shot back.

Valery hid a small smile. Joe didn't. "Focus. Eyes on the banks. This stretch gets tricky."

The river tightened as they moved deeper. Tall pines leaned over the water, their branches forming a canopy that blocked the sunlight. The temperature dropped sharply. Manny rubbed his hands together. "Feels colder than yesterday."

"Storm's shifting north," Valery said. "We're getting the edge of it."

They rounded a final bend—and the scanner beeped twice.

"Fish," Manny said, tapping the screen. "A lot of them. Depth looks good."

"That's our first good news today," Guido said. He slowed the Marlin and dropped speed to a crawl.

Joe leaned over the rails, spotting ripples below. "Let's set nets here."

Calder and Shaun moved fast. They lowered weighted nets into the deeper pocket while Manny checked the tension on the ropes. Joe and Valery monitored from the deck, watching the water churn as the nets sank. Ashley kept watch from the stern, eyes scanning the tree line, but she tracked the net deployment with the edge of her vision.

Minutes later, the ropes jerked hard.

"We got hits!" Manny called. "Multiple. Strong pull."

A surge of energy flowed across deck. This was what they needed

—something that felt like progress. They hauled the first section of net up, and fish thrashed inside, silver bodies flashing in the cold light. Shaun grabbed the bucket crates while Calder worked the line.

"We can fill both crates if this keeps up," Calder said.

Joe exhaled a little. "Good. Keep going."

They reset the nets and harvested again. More fish. Fat ones. Enough to feed Ridge Camp for at least a week if they rationed properly.

Guido checked the horizon. "If the whole run is this good, we'll make it back before dark."

But then he frowned, squinting toward the northwest. "Uh... Joe? You might want to look at this."

Joe joined him at the wheel.

At first it looked like fog. But as the Marlin pushed closer, the shapes became clear—

Boats.

Five of them. Long, narrow, reinforced with welded metal panels.

A blockaded dam towered behind them.

Armed figures stood along the barricade, watching the Marlin approach.

Valery's jaw tightened. "We're not alone."

Joe nodded once. "No. And they don't look friendly."

The River Wolves had found them first.

Guido slowed the Marlin to a crawl as the five unknown boats shifted into a defensive line. The river narrowed so tightly that the newcomers blocked the entire channel. Their boats were patched together with metal sheets, old fencing, and strips of welded rebar. Smoke curled from makeshift chimneys, and rusted harpoons hung along the rails.

Joe raised a hand, signaling the crew to stay calm. "Don't make sudden moves," he said quietly.

Valery folded her map and stepped beside him. "They're guarding the dam," she said. "Look at the barricades. They've turned that thing into a base."

Calder peered over the rail. "You think they built all that?"

"No," Joe said. "They claimed it. Big difference."

The largest of the boats drifted forward. A man stood at its bow—tall, broad-shouldered, wrapped in a patched military jacket with the sleeves torn off. His hair was braided back, and a scar ran across the bridge of his nose. He held a long spear casually at his side, like a challenge.

"Hold your course!" the man shouted. His voice carried across the water, steady and loud. "You're entering protected territory."

Guido muttered, "Protected by who, exactly?"

The man jabbed the spear downward, stopping only a foot from the river. "Protected by the River Wolves. State your purpose."

Joe stepped forward. "We're from Ridge Camp. We're here to fish and salvage upriver. We're not here to take anything that belongs to you."

A woman appeared behind the Wolf leader. Her head was shaved on one side, the other side braided tight. She held a rifle, muzzle down but ready. "Everyone says they're peaceful until they start stealing," she called.

Valery didn't blink. "We haven't stolen anything. And we won't. We're just trying to get enough to feed our people for winter."

The Wolf leader leaned his elbow on the rail. "Winter comes for everyone. Doesn't mean we let strangers walk through our lines. Show us your weapons."

Joe hesitated. He didn't want to look weak, but he also didn't want to escalate. "Guido. Drop speed. Let's hold position."

Guido eased the throttle back. The Marlin rocked in the current, halfway between advancing and retreating.

Joe lifted both hands, palms open. "We're armed, but we're not raising anything. I'll show you."

Slowly, carefully, he unbuckled his sidearm and held it out for them to see. Manny followed, placing his rifle on the deck. Calder and Shaun did the same. Valery didn't reach for her weapon, but she stepped back so they could see she carried nothing in her hands. Ashley stayed at the stern platform, keeping her silhouette visible and her hands empty—close enough to react, but not threatening.

The Wolf leader studied each movement with calm suspicion.

"You've got discipline," he said. "Most boats that come through here aim barrels first."

Joe nodded. "We don't want conflict. We just want to pass through, fish, and leave."

The woman with the rifle stepped closer to her leader and murmured something. He listened but didn't shift his expression.

After a long moment, he straightened. "We don't allow free passage. The dam is ours. The water past this point is ours. If you want to fish our section of river, you either leave a trade or turn around."

Guido muttered, "This guy thinks he owns a whole river."

Valery ignored Guido and stepped up beside Joe. "What are you asking for?"

The leader's eyes sharpened. They weren't dealing with a fool—he recognized who was in command on the Marlin.

"Tools," he said. "Metal. Anything we can use to reinforce the dam. And fuel, if you have it."

Joe felt Manny freeze behind him.

Fuel was precious. More precious than fish.

Valery didn't flinch. "We don't have fuel to spare."

"Then you turn back," the leader said.

One of his boats drifted forward, blocking any chance of slipping by.

Joe met Valery's eyes. They were trapped in a narrow channel with no room to maneuver. If they tried to force their way through, someone was going into the water—or worse.

Valery exhaled slowly. "We need to talk," she told the leader.

He nodded once. "Talk. But don't take long."

Behind him, the River Wolves watched in absolute silence.

Joe looked at Valery.

This was about to get complicated.

Joe pulled Valery, Manny, Calder, Shaun, Guido, and Ashley into a tight circle on the Marlin's deck. The River Wolves watched from their boats, weapons lowered but not relaxed. The current pushed gently on the hull, reminding everyone how narrow the river was here and how fast things could go wrong.

Valery kept her voice low. "We can't give them fuel. That's non-negotiable."

"Agreed," Joe said. "We can't risk the Savannah's reserve dropping any lower. Not with winter coming."

Calder rubbed the back of his neck. "What about tools? Donnie's got duplicates of half the equipment we salvaged from the rigs. We can spare some."

Guido frowned. "Tools for river rights? That sounds like paying a toll to pirates."

Valery gave him a steady look. "If they wanted to be pirates, they'd already be shooting. They're protecting their land. Or what's left of it."

Manny looked out over the water. The River Wolves' boats weren't clean or polished, but they were organized—rails reinforced, hulls patched with technique, not guesswork. These people knew what they were doing. "They're not amateurs," Manny said. "Look at how they're positioned. They've done this before."

Joe nodded. "Which means we negotiate smart. We don't show weakness, but we don't escalate either."

Shaun pointed toward the dam. "That structure is old. If they're reinforcing it, it's probably failing. If it collapses, the whole river system changes. They're desperate."

Valery looked between the men. "Okay. Here's what we offer. A small crate of tools—hand tools only. No fuel. No engines. Nothing that gives them power over us. And in return, they let us pass upriver and fish on our way back."

Joe met her eyes. "You think they'll take that?"

"They'll take something," she said. "Because their dam won't survive winter without repairs. And they know it."

Joe agreed with a small nod. "All right. Let's do it."

Before they moved, Guido muttered, "Someone should tell them not to point that spear at me. It's making me itchy."

Valery ignored him and stepped back to the rail. She lifted one hand toward the River Wolves' leader, the same way she'd do with an armed patrol during recon. "We're coming forward to talk."

The leader tapped the rail with two fingers. No sudden moves."

Joe stayed by her side as they eased the Marlin ahead. The other

Wolves shifted to let them approach, but the gap they left was barely wide enough for a skiff. Every eye stayed locked on them.

Valery stood tall at the bow, feet steady on the wet deck. "We're willing to trade," she said. "Hand tools only. In exchange, we get safe passage upriver and safe return. No harassment, no tolls on the way back, no targeting our boat."

The leader studied her. "Tools won't fix a failing dam."

"No," she said. "But they'll buy you time."

The woman with the rifle scoffed. "We need more than time."

Valery's voice didn't rise. It didn't shake. "We're not giving fuel. We're not giving engines. And you know why. If you want to shoot, then shoot. But we both lose. And winter wins."

The leader's eyes flicked with the smallest hint of respect. He tilted his head. "You speak like someone who's not used to being ignored."

"I'm not," Valery said.

A long pause stretched across the water.

Joe kept his hands visible, body angled to block any sudden threat toward her. Manny watched from behind him, knuckles tight around a rope, breath held. Shaun stood ready near the rail. Ashley watched the Wolves' flanks, making sure no one tried a side approach.

Finally, the leader nodded once. "One crate. Tools only. You get your passage. You fish where you want. No interference."

The tension on deck loosened a notch, though no one breathed yet.

Valery didn't smile. "Agreed. We'll bring the crate over now."

The leader lifted his chin. "Then we have a deal. But understand this—if you break it, the river closes behind you."

Valery nodded. "We won't break it."

He stepped back, signaling his boats to move aside. The path upriver opened slowly, like a gate creaking open.

The standoff was over—at least for now.

Guido and Calder lowered the crate of tools onto the deck and tied it down. The River Wolves' boats moved just enough to make a narrow corridor through the water. Everyone on the Marlin stayed tense, ready for anything.

Joe watched the Wolves take defensive positions along the channel.

"Slow and steady," he said. Guido adjusted the throttle and kept the boat centered. The river churned harder the closer they got to the dam.

Calder braced himself on the rail. "Feels like the current wants to twist us sideways."

Joe nodded once. "Eyes open. Nobody leans over the rails unless I say so."

Manny went to adjust the ropes around the crate. The deck was slick with spray, and the rope coil had loosened. "If this shifts, it'll block the walkway. I'm tightening it," he said.

Guido shook his head. "Do it after we pass the current."

Manny crouched anyway. "Takes two seconds."

The Marlin hit a sharp cross-current. The hull jerked. Manny's feet shot out from under him.

Ashley saw it first. She'd been standing on the stern platform, arms crossed, watching the Wolves. Her eyes snapped toward Manny the second he slipped.

"Manny!"

He slid across the wet deck, one leg going over the rail. His hand grabbed at empty air.

Ashley lunged without hesitation. She caught his vest with both hands and yanked him back with more force than she meant to use. Manny slammed into her full-body, knocking them both off balance.

Joe turned instantly. "What the Hell!"

Ashley held onto Manny until he found his footing again. "He's fine," she said, voice tight. "But he almost went straight into that current."

Manny's face flushed. "I—I almost had it."

Ashley glared at him, still breathing fast. "You almost had nothing. Sit still when people tell you to wait."

Manny rubbed the back of his neck. "Yeah… okay. You're right."

Guido called from the helm. "Save the romantic drama for after we're not in a death trap!"

Ashley spun around. "It is not romantic—"

Manny jumped in fast. "It was just an accident!"

Ashley crossed her arms again, chin high. "Exactly. A stupid accident." But she didn't step away from him. Not immediately.

Joe let out a slow breath. "Everyone stay centered. We're not losing anybody on this run."

The corridor tightened as the Marlin pushed forward. The River Wolves watched every move, weapons down but hands ready.

Calder checked the crate one more time. "It's holding, but if the pull gets stronger, we might need to cut the line."

Joe nodded. "If the current gets worse, we turn around. No discussion."

Ashley stayed near Manny, still visibly rattled even though she tried to hide it. Manny kept glancing at her like he wanted to apologize again but wasn't sure how.

Ashley noticed. "Stop looking at me like that," she whispered.

"Like what?" he whispered back.

"Like you scared me."

"You were scared?"

She looked away quickly. "I said don't make it weird."

Manny lowered his voice. "Okay. But… thanks."

Ashley's cheeks warmed. "You would've done the same thing."

"Yeah," Manny admitted, softer. "I would have."

Ahead of them, the last River Wolf boat began drifting aside, creating the final gap in the passage. The crew on those boats watched with cautious approval, like they respected anyone willing to fight both the current and their rules.

Guido guided the Marlin through the final stretch. Shaun leaned on the rail and scanned behind them to ensure no one was following.

Only when they cleared the last boat did Joe finally say, "All right. We're through."

The tension eased—but just barely.

The river widened as the Marlin pushed past the last of the River Wolves' boats. The water smoothed out, the current weakening where the channel returned to its natural shape. Guido eased off the throttle and let the boat settle into a steady pace. Everyone onboard took a collective breath now that the immediate danger was behind them.

Valery stepped up beside Joe. "That was closer than I liked."

Joe nodded. "They wanted to see if we'd back down. We didn't."

Calder checked the crate one more time, securing the last knot.

"Tools are intact. This should help Donnie build out the new workshop."

Guido rubbed his hands together. "Trade worked out better than I expected. Tools for river rights isn't a bad deal."

Manny looked out over the water, still rattled. "Those guys weren't bluffing. If the wrong person had said one more wrong word…"

"Yeah," Shaun agreed. "We'd be on the bottom right now."

Joe looked at them both. "You held your ground. That's what matters."

Valery took the mic and logged the update into the comms. "Ridge Camp, this is Marlin. No hostilities. Everyone accounted for."

Morales' voice replied after a short burst of static. "Copy that, Marlin. Dinner's going to taste better knowing we're still alive. Come home."

Ashley stayed on the stern, arms crossed again, but Manny hovered a few feet away, still replaying what almost happened.

"You sure you're okay?" he asked quietly.

"I'm not the one who went halfway over the rail," she said. Her tone was sharp, but her expression wasn't.

Manny nudged the toe of his boot against the deck. "I wasn't thinking."

"No kidding," she said. "You scared me."

Manny blinked. "Really?"

Ashley rolled her eyes, irritated at herself more than him. "Don't make a big deal out of it."

He hesitated, then stepped a little closer. "I'm glad you grabbed me."

She let out a small breath. "Yeah. Well. Just don't fall again."

"I'll try," he said.

"You better."

Guido shouted back from the helm, "If you two are done flirting, grab the side ropes. Current's picking up again."

Ashley flushed instantly. "We're not flirting!"

"Sure," Guido called. "And I'm the Queen of England."

Manny laughed under his breath. Ashley gave him a shove—not enough to move him, just enough to hide how flustered she was.

As they approached the bend in the river, the sun dipped behind a long bank of clouds. The water darkened, turning glassy and reflective. Shaun pointed toward a half-submerged billboard. "End of the old interstate. Ridge Camp's two hours out."

Joe leaned on the rail, watching the far tree line. "We'll need to plan for winter fast. River Wolves were desperate. If their supplies are low, others will be too."

Valery agreed. "Trade only works if we have enough left to trade. And we barely do."

Calder looked over the ration sheets again. "This run helped, but one good storm could wipe out everything we gained."

Joe's jaw tightened. "We'll make it work. We always do. But I'm not pretending it's enough."

The Marlin kept moving, pushing through pockets of debris—old logs, coils of roots, bits of man-made scrap thrown around by rising water.

Ashley glanced toward the horizon. "Think we'll run into more groups like them?"

Joe answered honestly. "Probably. The world's not empty. It's waking up."

Manny stood beside her, "Then we stay ready."

Ashley nodded. "Ready."

By the time Ridge Camp came into view—lanterns glowing, smoke rising from the cook tents—the crew's exhaustion had settled into something steadier.

They'd survived the Wolves.

Made a trade.

Brought home fish they desperately needed.

As the Marlin drifted toward the floating pier, Joe looked over his tired crew. "Good work," he said quietly.

Ashley and Manny exchanged a quick look—something warm, something unspoken.

Valery stood at Joe's side, eyes scanning the camp lights. "Home."

Joe nodded. "Yeah. Home."

CHAPTER 31
COLD IRON (PART ONE)

The morning started colder than anyone expected. Frost clung to the edges of the supply crates and turned the dirt around the fire pits into stiff gray patches. Joe was halfway down the ridge path, checking the new storage tarps, when he spotted Dave waving both arms from the floating pier like someone trying to signal a plane. It wasn't the usual "come help me move a crate" wave. It was sharp, tense, and urgent.

Joe picked up his pace. "What's wrong?"

Dave didn't waste time. "Reactor numbers spiked again last night." His breath formed tight clouds in front of him. "I stayed on the Savannah until sunrise to monitor it. We're closer to the limit than we thought."

Joe's shoulders stiffened. "How close?"

"Too close." Dave rubbed his forehead. "We've been running that reactor on minimal output for months. It's old, Joe. It wasn't meant to sit half-idle this long. And now the temperature swings from this frost aren't helping."

Joe followed him down the ramp. The Savannah loomed in the shallow bay, silent except for the groan of metal shifting in the cold. Perez stood at the stern, arms crossed tight, jaw hard enough to crack.

"You saw it too?" Joe asked.

Perez nodded. "Yeah. The pressure fluctuation hit red twice in the last fourteen hours. We stabilized it, but that's a bandage. Not a fix."

Dave braced his hands on the rail. "We've been buying time since the day we reached Tennessee. But we're out of time now."

Joe let out a long breath. "Give it to me straight."

Dave met his eyes. "If we don't shut the reactor down soon, something is going to fail. And if it fails the wrong way… we lose the ship and everything near it. Maybe more."

A cold weight settled in Joe's chest. "And shutting it down means no more emergency power."

"No more top-deck lights," Perez added. "No more water pumps. No more reserve heating coils. Nothing that the fleet relied on before we came to land."

Joe stared out across Ridge Camp. Lanterns flickered faintly between huts. Smoke rose from breakfast fires. People moved quietly in the cold like they could feel winter pressing down faster every day.

He straightened. "All right. Tell me what needs to happen."

Dave pulled a folded log sheet from his pocket. "Perez and I can handle most of the shutdown. It has to be a controlled cold shutdown. Slow, steady, precise. No rushing it. Once we start, it'll take at least forty-eight hours."

Perez looked up the ridge toward camp. "Everyone needs to know what we're losing when that reactor goes quiet. People rely on backup power more than they realize."

Joe nodded. "I'll call a meeting."

Dave hesitated. "Joe… this isn't just about safety. When that reactor goes dark, the Savannah is done. She won't run again."

Joe knew it. But hearing it out loud tightened something in his throat. The Savannah had carried them through fire, storms, starvation, loss. She had been more than a ship. She was the spine of their survival.

"We do it anyway," Joe said quietly. "We end it before it ends itself."

Perez clapped him on the shoulder. "Good call."

Joe turned back toward the ridge. "Give me one hour. Then meet me in the command tent with everything you need."

Dave nodded. "We'll be ready."

Joe climbed the path, boots crunching on frozen ground. The wind picked up, carrying the sharp tang of cold metal from the bay. Ridge Camp looked peaceful—lanterns, cooking fires, kids carrying buckets of water—but underneath it all sat a truth he hadn't wanted to face.

Their safety depended on a machine that had already given them more than it should.

Now he had to ask the camp to let it go.

Winter was coming fast. The world was changing again.

And today, the old world had delivered one final warning.

The command tent felt colder than the air outside. Joe stood at the head of the long table while Perez and Dave spread out their notes, diagrams, and reactor logs. A lantern swung lightly from the support pole above them, casting slow circles of light. Joe could tell both men were exhausted, but they were steady. This part they knew how to do.

Perez tapped the main page with his pen. "Here. This is the heat curve from last night. You can see where the pressure jumped. It only dipped back into the safe range because Dave adjusted the coolant flow manually."

Joe leaned in. "Could it jump like that again today?"

"Easily," Dave said. "Temperature outside is dropping. Metal contracts, seals strain, lines groan. All it takes is one crack in the wrong place."

Joe folded his arms. "So we shut it down."

Perez nodded. "Yeah. But it's not a switch we can flip. This is a big job. We need forty-eight hours, uninterrupted. And once we start, there's no going back."

Dave slid over a checklist. "We'll drain the auxiliary loops first. That'll take six hours. Then we step down the reactor core in stages. Even one mistake can warp the rods or damage the coolant chambers."

Joe scanned the list. "Who else do you need besides the four watchmen?"

"Just me and Perez," Dave said. "Too many hands slows us down. The rest of the team can help with external systems."

Perez pointed at another line. "But we need the camp ready. No emergency power. No backup heat. No top-deck lights. The Savannah will be dark."

Joe straightened. "We'll adapt."

Dave hesitated. "There's something else. Once we finish the shutdown, the Savannah's reactor can't restart. Not in this lifetime. She becomes a shell."

Joe took that in quietly. The Savannah wasn't just a ship—they'd lived on her. She'd saved them more than once. He let himself feel the weight of that for a moment, then he nodded. "Then we honor her by ending her safely."

Perez exhaled, relieved. "We'll start when you say."

Joe shook his head. "Not tonight. The camp needs a full briefing first. They need a day to adjust."

Dave nodded. "So tomorrow?"

Joe said, "Tomorrow. Cold shutdown starts in twenty-four hours."

Perez marked the time on his log. "We'll make it happen."

Joe rolled up the diagrams and set them aside. "Now we tell everyone."

Dave let out a slow breath. "They won't like it."

"Doesn't matter," Joe said. "They deserve the truth. And they'll handle it."

Perez chuckled faintly. "You have too much faith in people."

"No," Joe said. "I have the right amount."

He pushed open the tent flap. Cold air swept in, carrying a sharp metal scent from the bay. "Let's get the briefing set up."

Dave and Perez followed him out. The morning sun was pale and weak, struggling through a thin layer of cloud. Smoke from the cook tents drifted sideways in the wind. People were starting their tasks—hauling lumber, checking water barrels, preparing fishing gear—but their movements looked stiff, like everyone felt the frost creeping in.

Joe walked to the center of camp, scanning the ridge, the tents, the children gathering around the new fire pit. This was their home now. And part of its foundation—the ship that carried them here—was about to fall silent forever.

He turned to Perez. "Get the word out. Full camp briefing in two hours."

Perez saluted lightly. "On it."

Dave headed for the shoreline. "I'll start prepping the core room. We'll need everything perfect."

Joe watched them go, then looked out at the Savannah resting quietly in the bay.

In twenty-four hours, her heart would begin to slow.

He hoped the camp was ready.

He hoped he was ready.

By midday, the camp had gathered near the central fire pit. People formed loose rows—families in front, workers behind them, patrol members on the outer edges. The cold made everyone stand closer together. Breath fogged in the air. Even the kids were quiet, sensing something serious was coming.

Joe stepped onto a flat stump that had become the unofficial meeting platform. Morales stood on one side of him, Valery on the other. Perez and Dave waited nearby with a folded map of the Savannah's systems. The lantern behind them flickered in the wind, casting sharp shadows across the crowd.

Joe raised his voice. "All right, everyone. Listen up. This is important."

The murmuring died off immediately.

He continued. "Dave and Perez have been monitoring the Savannah's reactor for months. Last night, the pressure levels moved into the danger zone. Not for long—just a spike. But it was enough to show us the system is wearing out."

A few people gasped softly. Someone in the back whispered, "Is it going to explode?"

Joe shook his head. "No. Not if we take action now. But we can't keep running it. The reactor is old, stressed, and colder weather is making the metal contract. If we wait too long, we risk a failure we can't control."

Valery took one step forward. "A reactor failure would mean losing the ship, injuring anyone on watch, and possibly starting a fire we can't stop. Joe isn't exaggerating. We caught it early, but we can't take chances."

Dave held up his log sheet. "We're recommending a full cold shutdown. That means we power it down completely in a controlled sequence. It also means once the process starts, the reactor can never restart again. The Savannah will be offline permanently."

There were ripples of shock through the crowd. The Savannah had been home, refuge, lifeline. Losing her felt like losing a person.

Perez added, "A shutdown takes forty-eight hours. During that time, emergency power from the ship will go offline. External lights on the deck, backup heat coils, and the water pumps connected to the pier will all stop."

One man lifted his hand. "So what does that mean for us? Are we losing power here on camp?"

Morales answered. "Our camp runs on solar and manual pumps. We'll be fine. But we won't be able to rely on the Savannah for backup if something fails. No extra heat on freezing nights. No emergency lights if we get hit with a storm. We'll need to conserve and stay coordinated."

A woman from the rescued families stepped forward, holding a toddler. "Do the kids lose anything? The heat tents? The med tent lights?"

Amy shook her head from the med area. "We'll manage. We have enough solar to keep the vital systems running. No one will freeze. No one will be left in the dark during care."

Joe lifted his voice again. "This is not a crisis. It's a decision. The Savannah has served us longer than anyone expected. She carried us here. She kept us alive. Now we return the favor by shutting her down safely."

Toby raised his hand. "When do we do it?"

"Tomorrow," Joe said. "We need time to prepare. Dave and Perez need to reorganize the core room, check every valve, log every sequence. Everyone else needs to get used to the idea that the ship's lights won't be there to guide us at night."

A low hum spread through the group—fear, sadness, understanding.

Joe didn't soften his tone. "We'll get through it. We always do. But I want every single person in this camp to know what's happening and why."

No one argued. No one pushed back.

The Savannah had earned their respect.

Joe stepped down from the stump. "If you have questions, now's the time. If not, get ready. In twenty-four hours, we begin the shutdown."

Silence held for several seconds.

Then someone in the crowd started a slow nod.

Then another.

And bit by bit, Ridge Camp accepted what was coming.

The crowd broke apart slowly as people returned to work, but the tension lingered like a cold draft. Joe stepped down from the stump, rubbing his hands together as if the briefing had drained the heat from him. Before he could take two steps, a loud clang rang out across the eastern side of camp. It wasn't dangerous—just a metal panel slipping from a storage rack—but the sound was sharp, sudden, and violent.

Ashley flinched hard enough that she stumbled. Her breath caught. Her shoulders locked. The metal echo froze her in place. For a split second, she wasn't in Ridge Camp. She was back on the Halcyon. Back under Harrow's command. Back hearing metal slams that meant punishment, not accidents.

Her eyes widened. The color drained from her face. Her lungs forgot how to breathe. Her hands shook even though she tried to hide them. She backed up until she hit the side of a crate. She stayed there, frozen, staring at nothing.

Manny saw her immediately. He had been talking with Guido about repairing the pier lights when the sound hit. When he turned and saw her pressed against the crate, something in him snapped to attention. He didn't yell. He didn't rush. He just moved to her quietly, steady as a heartbeat.

He stopped a foot away, giving her space. "Ashley," he said softly. "Hey. Look at me."

She tried, but her focus kept drifting. Her chest rose and fell in short, sharp breaths that didn't pull in enough air.

"It's okay," Manny said, his voice calm and even. "You're here. You're safe. Nothing's wrong."

Her hands clenched at her sides. She tried to say something, but only a shaky breath came out.

Manny stepped just a little closer. "You're not on the Halcyon," he murmured. "No one's yelling. No one's coming for you. It was just a dropped panel. That's all."

Her breathing stuttered again.

"Can you match my breathing?" Manny asked. "Just copy me."

He inhaled slow. Exhaled slow.

She tried to follow him. The first attempt broke halfway through. The second made it to the exhale.

"Good," he said quietly. "Keep going. You're doing great."

She finally dragged her eyes up to his. She looked embarrassed, angry at herself, and shaken all at once. "Sorry," she whispered.

"Don't apologize," Manny said. "Your brain got scared before you did. Happens to everyone."

"It's stupid," she muttered. "It was just a noise."

"It wasn't stupid. It was loud metal. And you've had enough of that for a lifetime." Manny shifted slightly so his body blocked the view of the storage racks. "You okay to sit?"

Ashley nodded a small, jerky nod. Manny guided her to a crate and sat beside her without crowding her. She stared at the ground, still working through leftover tremors.

"I thought I was past this," she said quietly.

"Doesn't work like that," Manny replied. "It hits when it wants to. You didn't do anything wrong."

She swallowed. "Thanks. For… catching it."

He shrugged, but his eyes softened. "I see you."

Ashley's lips parted for a second, like she was about to say something important, but she closed them again and shook her head. "You always show up at the right time."

Manny tried to play it cool, but the faint blush climbing his neck betrayed him. "Yeah, well… someone's gotta keep you upright."

She nudged him weakly with her elbow. "I was upright before the metal tried to kill me."

He smiled. "Barely."

Some color returned to her cheeks. The trembling eased.

After a long moment, she stood slowly. "I'm good."

Manny rose with her. "Tell me if you need a break."

"I will," she said. Then, quieter, "I mean it."

He nodded, and for a second, neither moved.

Then Morales shouted across camp, "Manny! Ashley! Need you both on water system checks!"

Ashley straightened. "Back to work."

Manny grinned. "Lead the way."

By late afternoon, the camp settled back into a steady rhythm. People returned to their tasks, though nervous glances still drifted toward the waterline where the Savannah sat rusted and silent, waiting for its final command. Joe crossed the ridge path with long, determined strides, heading toward the temporary operations tent. Perez waited inside with a clipboard tucked under his arm and a map spread across the table.

Perez looked up. "We need to finalize everything for the shutdown. Exact timing. Exact prep."

Joe rubbed the back of his neck. "Dave says one more day before temperatures drop too far."

"One day," Perez confirmed. "We'll pull every tool we need tonight, run final diagnostics tomorrow, and have containment staged. Once shutdown starts, we commit and finish."

Joe tapped the map with one finger. "We'll station guards at the pier. No one goes near the ship except authorized personnel. Last thing we need is someone panicking and trying to board once the lights go dark."

Perez snorted. "Wouldn't blame them. End of the last big reactor left in their world? People don't love that kind of symbolism."

Joe gave him a look. "We're doing the right thing."

Perez didn't argue. He simply said, "Yeah. We are."

They walked outside together. The sky had shifted into a dull, winter-colored blue. The air felt thinner. The wind carried a cold bite that promised harder days ahead. Joe scanned the camp—families sorting supplies, kids stacking firewood, volunteers checking tarp lines. Everything was alive. Everything depended on the next few choices.

Morales approached from the north path. "We good?" she asked.

"Shutdown in a day," Joe said.

She inhaled slowly. "People will panic."

"I know."

Morales adjusted the strap of her rifle. "They took the news better than I expected, but they'll wobble once the lights actually go out. No rumors. No surprises. We keep talking to them."

Joe nodded. "We'll brief the squad leads after dinner, make sure everyone knows what to expect during shutdown."

Joe nodded. "We'll do it right after dinner."

She gave him a short nod and walked off, barking orders before she was even out of earshot.

A quiet voice broke through the wind. "Joe."

Valery approached, her sling still tight, her expression unreadable. "Perez told me. One day."

"Yeah," Joe said. "Has to be done."

She studied his face. "You feel it, don't you? The weight of it."

"Everyone will," he answered. "It's the last piece of the old world still humming."

Valery stepped closer. "You're not ending anything. You're choosing safety. The reactor gave us time. Now we give it rest."

Joe gave a faint, tired laugh. "You make it sound poetic."

"It's not poetic," she said. "It's practical."

Perez called from the tent, "Joe! Got one more thing for you!"

Joe nodded at Valery. "I'll see you at the briefing."

She touched his forearm lightly. "Don't carry this alone."

He didn't answer, but he didn't pull away either.

Down the ridge, the camp moved through the cold like they could feel the shift coming, even if they didn't know its shape yet. Lanterns flickered. Smoke drifted low. Every breath felt like the world holding itself still.

Tomorrow, the Savannah would go dark.

And Ridge Camp would have to light the way without her.

CHAPTER 32
COLD IRON (PART TWO)

The shutdown began at dawn.

The sky was only a pale stripe over the ridge when Joe walked the frosted path toward the floating pier with Perez, Dave, and the last four Navy techs. Every rope and rail glistened with ice. The air tasted metallic, as if the world itself sensed the finality of what was about to happen. The Savannah sat darker than usual, conserving her last heartbeat.

Dave checked his watch. "All right. Time to begin."

His voice held steady, but his shoulders told the truth. He'd lived in that engine room more than anywhere else since civilization cracked apart.

Joe stepped aside as the team boarded. No jokes. No chatter. Only the silence of men entering a place they'd once prayed inside.

Down in the core chamber, the temperature bit colder. Their footsteps echoed off the steel. Dave opened the first panel, eyes scanning readouts like scripture.

"Auxiliary loops stable. Begin drain sequence."

Perez flipped back the protective housing on the main control panel. Soft flickers rolled across the lights.

"Engaging drain valves."

The ship answered with a low groan that rolled through the walls like a long, exhausted exhale. Joe stayed back, letting them work. The weight of the moment pressed on him—this wasn't maintenance. This was a goodbye.

Dave studied the gauges. "Flow rate looks good. Hold it steady."

Steam hissed faintly somewhere in the pipes. The engineers moved with surgeon precision—valves, logs, sequences—hands steady, brows set. It wasn't dramatic. It wasn't loud. It was the kind of meticulous work that had kept hundreds alive.

Twenty minutes later, Dave wiped his hands on a rag.

"Stage one done. Loops will take six hours to drain. After that, we start stepping down the core. That's when the ship goes dark."

Joe nodded. "Warn me before the first dim. Camp needs a heads-up."

"You'll have it."

While the team split off to check exterior coolant lines, diagnostics, and shutoff locks not touched in a year, Perez approached Joe.

"We're good. Better than good."

Joe followed him back to the deck, watching Ridge Camp stir in the cold—smoke rising from breakfast fires, kids hauling water, volunteers bracing for another hard day. Ready or not, change was coming.

Dave joined them. "We'll stay on schedule. No shortcuts."

"Start it right," Joe said. "Finish it right."

Dave returned to the core room.

The Savannah had begun her final shutdown. No going back now.

By late afternoon, the ship's hum had dropped to a shaky, uneven vibration—an old familiar sound turned thin and tired. Joe stood on the pier beside Morales, eyes on the dimming hull. The sky hung low and gray, matching the mood settling across Ridge Camp.

Perez's voice crackled over the radio. "Beginning core step-down. First dim in two minutes."

Joe pressed the mic. "We're ready."

People drifted toward the shoreline, quiet and bundled in blankets. Parents lifted children so they could see. No one joked. No one filled the silence. The Savannah had been their home, their shelter, their world.

A faint flicker washed over the ship's exterior lights.

Morales let out a breath. "There it is."

The lights steadied, then dimmed two degrees. It wasn't dramatic. But it hit like a slow punch. Joe felt it in his chest.

"Primary step-down complete," Perez radioed. "Secondary starting."

Silence settled over the crowd—tight, reverent.

Valery approached, gaze fixed on the ship. "People aren't ready for how this will feel."

"No," Joe said quietly. "They're not."

A deeper vibration rolled up through the pier. Then, one by one, the deck lights faded—upper deck first, then midline, then the lowest. Children moved closer to their parents. Adults swallowed emotion they didn't want seen.

Someone whispered, "She's shutting down."

Another: "She got us here."

Perez's next transmission came softer. "Core at twenty percent. Final step beginning."

The ship seemed to sigh—a gentle shudder, like old lungs finally resting.

Ten percent.

Five.

Three.

Then silence.

"Core offline," Perez said. "Shutdown complete."

The Savannah was dead.

No cheering. No applause. Just quiet breathing, shared grief, and the knowledge that something ancient and loyal had taken its final bow.

Morales wiped her eyes, pretending she hadn't. Valery stood tall beside Joe, jaw tight. The Savannah had carried them farther than they ever imagined.

Now Ridge Camp would stand alone.

As the lights died, Ridge Camp shifted. People turned from the bay toward one another instead of the ship. For the first time since the firestorm began, their lifeline rested on land, not water.

Joe stepped down from the pier. Morales followed.

"Feels like we cut our last anchor," she said.

"We did."

Valery caught up, moving carefully with her sling. "Let people grieve. They need it."

Families gathered around the fire pit—old fleet survivors, newcomers. Children whispered stories about cramped bunks and metal halls. Amy ladled hot drinks with practiced calm, giving warmth to anyone who looked close to unraveling.

Manny and Guido secured the pier, hammering supports, tying down ropes. The work was quiet, respectful.

Guido shook his head. "Feels like putting away a ghost."

"Better than leaving her half-alive," Manny muttered.

Up the ridge, kids climbed a boulder to watch the ship fade into the dusk.

Morales checked the crowd, making sure no one spiraled into panic. She stopped beside an older woman staring at the bay.

"You okay?" Morales asked.

The woman nodded, though tears glossed her eyes. "I lived on that ship."

"You're not losing anything," Morales said gently. "You're gaining ground that won't sink beneath you."

The woman breathed out. "I know. Still feels strange."

"It should."

Joe raised his voice. "Everyone—just a minute."

People formed a loose semicircle. The fire cracked behind them.

Joe spoke simply. "We said goodbye to the ship that kept us alive. The Savannah carried us through the worst months anyone's lived through. She did her job. Now we do ours."

Heads lifted. Shoulders eased.

Valery stepped beside him. "We don't lose anything by letting go. We gain stability. We gain safety. And we gain each other."

Guido called, "And maybe Manny finally fixes that damn heater!"

A few laughs cracked the tension. Manny groaned.

Amy shouted, "Warm stew and real bread tonight!"

Kids cheered. Adults exhaled.

"Go eat," Joe said. "Tomorrow we build."

The crowd dispersed, lighter but still grieving.

Ridge Camp was standing on its own for the first time.

Evening deepened. People gathered near the overlook, drawn by instinct more than invitation. A quiet ceremony formed under the fading sky.

Joe and Valery stood at the front. Ashley and Manny arrived last—Manny holding two cups of tea, Ashley empty-handed. Her fingers still trembled from the earlier trigger, but she hid it well. Manny stayed close, steady as bedrock.

The Savannah blurred into the dusk.

Joe stepped forward.

"She's quiet," he said. "But not gone. She carried us from one world to the next. Tonight we honor that. Tomorrow we move forward."

No applause. Just soft breaths.

Ashley shifted uneasily as the crowd pressed close. Shadows moved. Someone dropped a tin cup behind her. The metal clatter hit her nerves like a lash.

Her breath hitched.

Manny didn't touch her—just leaned close enough she could feel him.

"You're okay," he whispered. "I'm right here."

Something inside her eased. Without thinking, she leaned into him—shoulder to shoulder, grounding herself. Manny froze for half a second, then steadied himself so she could lean fully.

He didn't move.

She didn't move.

Guido turned to joke, saw them, and spun back around like he'd witnessed a classified event. Valery caught it with the corner of her eye and smirked.

Joe kept talking, but the ceremony belonged to the moment between Manny and Ashley—quiet, fragile, real.

She stayed leaned into him the whole time.

And he stood solid, as if he'd been waiting for her weight without knowing it.

When the ceremony faded, people drifted down the ridge. The cold sharpened. Dave and the techs lingered above the bay, staring at the silent ship.

Perez approached Joe. "Sequences complete. She's stable."

"How's Dave?"

Perez exhaled. "Barely holding it together."

Joe approached. Dave didn't turn until Joe was beside him.

"She's quiet," Dave said. "Quieter than I thought."

"You did it right," Joe said. "Better than right."

Dave nodded sharply. "We're packing up tonight. Moving to camp at sunrise."

"You've got a place waiting."

The techs arrived with half-packed duffels—Kepler, Medina, Hughes, Santos. Medina breathed out.

"No more alarms," she said. "No more ladder climbs. No more praying nothing blows."

Kepler tried to joke. "Guess we're civilians now."

It came out thin, but no one corrected him.

"Come on," Joe said. "Warm tents. Hot food."

Morales met them near the north huts. "Bunks are ready. Close to the water lines so you don't freeze."

Dave swallowed. "Thank you."

"You're Ridge Camp now," Morales said. "Welcome home."

The words hit deeper than she realized.

Ashley and Manny approached from the fire pit.

"Need help carrying anything?" Manny asked.

Hughes smiled. "A night without reactor ladders sounds perfect."

Ashley gestured toward the path. "Come on. We'll walk with you."

The group moved together—Manny and Ashley side by side, techs talking softly about real ground under their feet, Morales pointing out tomorrow's task lines.

Valery handed Joe a mug of broth. "Mom's bringing blankets. Techs are settling in."

Joe wrapped cold fingers around the cup. "Feels like a turning point."

"It is," Valery said. "The last echo of the old world is gone. Everything from here on out—we build it."

Joe glanced toward the bay. The Savannah rested dark and peaceful for the first time since the firestorm.

"She did her job," he murmured.

"And now we do ours."

Ashley called, "Joe! Manny says we need more lanterns for the north huts!"

Joe lifted the mug. "On my way."

He walked toward the noise, the firelight, the work waiting for him—toward the future being shaped one frozen night at a time.

Behind him, the Savannah rested.

Ahead of him, Ridge Camp breathed as one.

A new chapter—not a new era—began.

CHAPTER 33
ASH & WATER

The water looked darker than usual, almost black under the gray sky. Joe stood on the deck of the Marlin, watching the small swells slap against the tug's hull. The cold bit through his jacket and into his bones, but he kept his focus on the gear spread out at his feet. Two sets of dive tanks. Extra lines. Mesh bags. Pry bars. A portable sled rigged from a broken pallet and rope so they could haul heavier finds without dragging them by hand.

Manny checked the valves on a tank, listening for leaks. "Air seals are good," he said. "You'll get a full bottom time if you don't go sightseeing."

Guido snorted and adjusted his own straps. "You say that like you don't go sightseeing every time you see a shadow and ask, 'You think that's a shark?'"

"There are no sharks here," Manny said.

"Yeah," Guido replied. "You say that every time too."

Joe cut in before they drifted. "Remember the plan. We're not out here for scrap nails. We're looking for metal stock we can melt down, intact gas cylinders, and anything from the medical list Mom marked."

He tapped the waterproof clipboard on the rail. Mom's handwriting covered the page—IV lines, sealed bandage packs, anything marked sterile, any crates with a red cross or hospital stamp.

Beyond the tug, the broken tops of drowned buildings poked out of the water at odd angles. This section of the old town sat deeper than the "deep house" they'd explored before. A former industrial strip, according to old maps they'd pieced together. Warehouses. Loading docks. Maybe a small clinic near the riverfront.

Ashley glanced at the buildings and then at Joe. "Current's stronger out here. You sure we shouldn't triple-line them?"

"We're set," Joe said. "Primary line from the bow, secondary safety from the mid-cleat. You and I are topside. We watch the lines every second."

Manny rolled his shoulders, loosening muscles. "Depth?"

"Deeper than the house, but still safe," Joe said. "You follow the main line down to the entry point. Stick together. Stay in visual contact. Nobody wanders. If visibility goes bad, you back out."

Guido nodded once. "Got it."

They moved through the last checks—regulators, harness clips, knife sheaths, wrist lights. Joe watched Manny's hands as he went over every buckle on Guido's gear and then his own. He didn't rush or skip anything.

"Reg test," Manny said.

He bit down on his regulator and pulled a few deep breaths, then gave a thumbs-up. Guido did the same.

Joe looked each diver in the eye. "If anything feels wrong, you come up. We're not heroes today. We're smart."

"Since when?" Guido muttered, but the grin didn't reach his eyes. He was nervous, and Joe saw it.

Ashley caught Guido's shoulder lightly. "You get in trouble, you yank that line twice. Hard. We'll feel it."

Guido gave a short nod. "Yeah. Okay."

Joe stepped back. "All right. Let's go."

They moved to the edge. The bow line ran out from the Marlin and disappeared into the dark water, weighted at intervals to keep it steady in the current. They'd dropped a marker float over the spot where the old street dipped toward what used to be a service road and, hopefully, a loading zone.

"On my mark," Joe said. "Manny first, then Guido."

Manny lowered his mask, slipped the regulator into his mouth, and eased off the edge in a controlled backward roll. Water swallowed him with a clean splash. A moment later, his hand broke the surface, giving the agreed okay signal.

Guido went next. His entry was less smooth, but he recovered quickly. Two hands up. Okay.

Ashley watched the line. "They're following it."

Joe watched too, tracking the faint movement of the rope as the divers pulled themselves along its length. The float near the bow bobbed, then steadied.

"Depth should level out soon," Joe said. "They'll hit street level, then follow the map."

Ashley adjusted her grip on the secondary line. "How deep are we here again?"

"About thirty, maybe thirty-five feet to the road," Joe said. "Structures rise from there. We scoped it last week with the cam. Remember?"

Ashley nodded. "Yeah. Just thinking about pressure. And cold."

"They're layered up," Joe said. "They'll be fine."

Under the water, Manny and Guido moved slowly through the green-black murk, lights cutting short tunnels through the drifting silt. Broken rooflines and leaning walls appeared at the edges of their beams. Old signs, half-rotted and covered in algae, hung at angles that made no sense anymore. They followed the main line down to the sunken street and then along a cracked curb toward where the maps said an equipment warehouse once stood.

Up on deck, Ashley held the secondary line in both hands. She watched the rope with tight focus, her shoulders tense. Joe worked beside her, testing the slack, checking the knots, keeping his eyes on the bow line for any irregular movement.

"Feels steady," Joe said.

Ashley didn't look away from the rope. "They're deeper than last time. Current's rougher."

He gave her a sideways look. "You worried about them?"

She shot him a sharp, embarrassed, warm look. "Both. Shut up."

Joe snorted quietly and went back to the line, and she tried not to smile.

Down below, Manny's light swept across a twisted doorway, half-blocked by debris. He pointed, then glanced at Guido. Guido nodded. They moved inside the warehouse, slipping into the drowned shadows.

Manny steadied himself on a broken metal beam as he pushed deeper. The water swallowed their light quickly, turning the place into a shifting tunnel of silhouettes and debris. Guido flashed his wrist light toward the ceiling—what was left of it—and then toward the floor where scattered equipment made the descent uneven. They stayed close, breathing slow, checking each other's bubbles and distance.

Manny pointed left. An overturned shelving unit lay wedged against a far wall, trapping whatever had been stored behind it. Guido motioned to clear the silt first. They worked methodically. Manny lifted one end of a broken crate while Guido used his pry bar to loosen tangled wires and torn plastic covers. The contents spilled out—rusted cans, scraps of fabric, shattered glass. Useless. They pushed deeper.

A few minutes later, Guido's beam stopped on something rectangular and solid. He tapped his light twice—check this. Manny swam over and brushed away muck. A sealed canister lay half-buried in grime. Metal intact. Weight promising. He clipped it to his mesh bag. One good find wasn't enough, but it was a start.

They moved on, entering a smaller room with toppled storage lockers scattered like fallen dominoes. Some were rusted through. Others were sealed. They headed for the nearest one—but before they reached it, Manny's light struck something pale resting against the wall.

He froze.

Guido's beam followed—and stopped.

A human arm.

The body floated slightly above the floor, pinned by debris. The uniform was dark and faded. A cracked flight helmet hung loosely from the skull. The water had taken the face long ago, but the patch on the sleeve remained clear beneath the algae.

A blue circle with a stylized wing.

An Air Force flight crew patch.

Manny's pulse kicked hard. He drifted closer, careful not to disturb the remains. Guido hovered beside him, staring at the patch. They exchanged a wide-eyed look—shock, disbelief, and recognition all in one.

Manny's beam caught on a metal tag tangled in the harness strap. He reached carefully into the debris, closed his fingers around the tag, and gently pulled it free. Dog tags.

He turned them toward his light.

HOLT SQUADRON – 119th RESCUE.

Guido inhaled sharply through his regulator. A set of dog tags under a flooded warehouse was never a good sign. Whoever this man was, he hadn't died in the lake. He'd come from somewhere else—somewhere dangerous enough that he never made it out.

Manny tucked the dog tags into the zip pocket of his suit. He signaled: We move. Stay focused. Keep working.

Guido steadied himself and nodded.

They turned back to the lockers. Guido pried one open. Metal cylinders drifted out. He caught one, tested the seal—good. Manny grabbed another. They filled the second mesh bag with oxygen, acetylene, and one marked MED AIR.

Then Guido's beam caught a box wedged beneath the lockers. Rectangular. Riveted. Sealed. Manny swam down, braced his boot against the wall, and pulled it free. The faded stencil across the side read:

ARNOLD AFB – EMERGENCY STOCKPILE.

Manny stared at the words. Guido stared at him.

They grabbed it together.

Manny clipped the crate to the haul rope, securing it with two knots and a reinforced clasp. The weight pulled the rope taut. It held.

He signaled upward: Time to go.

Guido nodded—but tapped Manny's arm and pointed toward the far wall. A shape hovered at the edge of their beams. Large. Slow. Watching.

Neither of them breathed for a second.

A long, dark form glided just beyond the reach of their lights—too big to be a loose beam, too smooth to be a cluster of debris. It hung there in that unnatural, waiting way that made every hair on Manny's neck rise beneath his hood.

He tightened his grip on the haul rope.

Guido reached for his dive knife.

The shadow drifted deeper into the dark and disappeared.

Manny mouthed through his regulator: Go.

Guido didn't argue. They started their ascent fast, hauling the crate and canisters behind them. More than once, Manny felt the rope jerk like something brushed past the sled, but he didn't look back. Not today. Not down here.

Up on deck, Ashley tightened her grip on the line. "They're coming up fast," she said. "Way faster than usual."

Joe stepped closer. "Trouble?"

Ashley watched the rope jerk again. "Feels like it."

They set their feet and waited for the first breach. The water heaved once, then again.

The water broke as Guido surfaced first. He shoved his regulator aside. "Get the sled up!" he yelled.

Manny surfaced right after him, dragging the haul rope with one hand. "Pull! Now!"

Joe and Ashley grabbed the rope together, muscles straining as they dragged the weight toward the platform. The crate rose, heavy and dripping, bumping against the hull. As soon as it cleared the surface, something struck the underside of the pier—a dull, heavy thud that vibrated through the boards under their feet.

Ashley startled at the sound and grabbed Joe's arm without thinking. "What was that?"

Joe steadied himself and her both. "Something hit the pier. Just stay back."

Manny yanked off his mask, breath fast. "Something big in the warehouse. Could've been a beam settling. Could've been… something else."

Guido collapsed onto his back on the deck. "Whatever it was, it didn't want company."

Joe nodded toward the crate. "Get this secured. Crate goes to the main deck—now."

They muscled it onto the central platform. Ashley unclipped the remaining canisters from the divers' gear and set them on the drying rack. She glanced at Manny and Guido, scanning for bleeding or torn suits. "Either of you hurt?"

"Mostly not," Guido said, still panting. "Something down there had teeth."

Ashley's eyes dropped to his hand. A thin, curved slice ran along the base of his thumb—clean, but already bleeding through the glove. "Guido," she said sharply. "Glove off."

He winced as he peeled it back. "Locker hinge. Or demon building. Hard to tell."

Ashley blew out a breath. "You idiots always act surprised when the water tries to kill you. Sit."

Guido sat without argument. Ashley snapped her med kit open and worked quickly—antiseptic, sterile cloth, clean wrap—her movements brisk but careful.

"Flex it," she said.

He flexed. Winced. "Still attached."

"Barely," she said. She taped the bandage down. "Don't rip it open again."

Manny snorted. "That's literally his job."

Ashley shot him a look. "Then tell him to poke things with the other hand."

Guido groaned. "Yes, ma'am.""

Manny just smiled.

Guido jerked his chin toward the crate. "Whatever's in that better be worth almost getting flattened."

Manny nodded. "It is."

Joe stepped in. "Show me."

They crowded around the crate. Joe and Manny unfastened the latch, and the metal hinges squealed as they swung the lid open.

Inside lay neatly wrapped bundles, protected by waterproof foam. Manny unpacked the first layer—medical supplies: bandages, antibiotics, saline bags, sterile packs.

Guido lifted out a set of labeled gas cylinders. "All intact. These are gold."

Then Manny brushed his hand over the last bundle—a heavy pack with a printed label in block black letters:

AFB SECURITY—PRIORITY RESCUE PACK.

Joe's expression shifted—surprised, tense. "That's Air Force command-level gear."

Manny exchanged a look with Guido. "Matches the patch on the body."

Joe's gaze sharpened. "Body?"

Manny swallowed. "Air Force flight suit. Helmet. Patch. And the tags."

He dug into his suit pocket, pulled out the dog tags, and laid them in Joe's palm.

HOLT SQUADRON – 119th RESCUE.

Ashley's eyes widened.

Joe's jaw tightened as he read the dog tags. "Holt Squadron," he said quietly. "Air Force rescue unit… that's all I know."

He looked at the crate stencil: ARNOLD AFB – EMERGENCY STOCKPILE.

"Arnold Air Force Base," Joe murmured. "I've heard the name, but I don't know how they operated. Morales will. She's worked under enough captains to recognize this kind of gear."

Manny and Guido exchanged a look.

Ashley frowned. "So none of us knows what Holt was doing out here?"

Joe shook his head. "No. We bring Morales in before we make assumptions."

Ashley frowned. "Then why would one of them be under an old warehouse in the middle of nowhere?"

Manny asked quietly, "So what were they doing way out here?"

Joe didn't answer right away. He lifted the heavy pack from the crate and unzipped it. Inside lay coded maps, emergency beacons, ID tags, and a black hard case stamped with Holt Squadron's insignia.

He held the case like it might bite.

"This isn't random," Joe said. "Someone left this for a reason. And that reason died down there."

The wind picked up, sharp and cold across the water. Everyone fell silent.

Joe closed the black case, jaw tight. "Get everything logged and secured. We'll go through the maps tonight."

Guido dragged in a slow breath. "Boss… there's more down there. That body wasn't alone."

Joe nodded once. "We'll deal with that tomorrow."

He turned toward the water—toward the drowned city stretching under the gray waves.

And toward the truth rising to the surface much faster than anyone wanted.

Back on deck, Ashley checked Guido's hand one more time and gave his wrist a quick squeeze before she let go. "You're lucky that didn't go deeper," she said.

Guido flexed his fingers again. "Hurts, but it works."

"Good," Ashley said. "Because Manny's not doing your half of the work."

Manny groaned. "Why am I suddenly responsible?"

"Because you're the one who keeps volunteering to go deeper," she shot back.

Her tone was sharp, but her eyes kept flicking to Manny as if checking he really was okay. The cut hadn't been his, but the adrenaline was fading fast, and all of them were starting to feel the cold and the weight of what they'd found.

Joe glanced over. "Warm up. We'll unload the rest once we're tied off."

"I can still work," Manny said.

Ashley's eyes narrowed. "You can still sit down like a sane person. If you fall over, I'm blaming you, not the ocean."

Joe clapped his hands once, bringing everyone back to focus. "All right. We need to document this crate properly. Manny, help me with the inventory board. Guido, log the canisters."

Ashley straightened. "What about the locker contents?"

"We'll sort those after the crate," Joe said.

He crouched beside the opened lid again, eyes tracing the edges of the foam and the way everything inside had been packed like someone expected it to matter. "These seals haven't been cracked. Whatever's in here—it's been waiting a long time."

Manny set the inventory board down beside him. "Those Holt maps... they looked different."

Joe nodded. "They are."

Ashley frowned. "Different how?"

"Not search grids," Joe said. "Targets."

Guido's eyes widened. "Targets for what?"

Joe didn't answer. Not yet.

The wind pushed across the deck, rattling the loose rigging. Somewhere near the bow, a rope slapped against metal. Everyone tensed for a moment—still shaken from whatever had chased the divers inside the warehouse.

Ashley stepped a little closer to Manny without seeming to notice she was doing it. Manny noticed, but he didn't say a word.

Joe stood over the crate again, studying the Holt insignia burned into the black case. There was weight here—not just physical, but historical. Something the Air Force hadn't meant to leave behind.

"We'll go through everything tonight," he said. "Not here. Too much noise, too much movement."

Ashley nodded. "Back at camp?"

"Yeah," Joe said. "In the operations tent. After dinner."

Guido pointed at the dog tags. "And the body?"

"We'll go back for him," Joe said quietly. "But not until we know what this crate is telling us."

Manny tightened his jaw. "Whatever Holt was doing out here... it wasn't good."

"No," Joe said. "It wasn't."

He lifted the black case, weighing it.

"And it didn't end down there."

They secured the crate, lashed the canisters, and turned the Marlin back toward shore. The Savannah sat dark and quiet in the distance, a dead giant resting in the shallows. Ridge Camp rose on the ridge line beyond—lanterns like scattered stars between tents and rough-built huts.

By the time they tied off at the dock, the light was already fading.

Cold bit harder. People waited at the pier, ready to help. Word of "Air Force gear" traveled faster than Joe would've liked.

He shut it down quick. "Supplies first," he said. "Questions later."

They carried the crate up the path toward camp. Manny and Guido hauled the worst of the weight. Ashley walked beside them, keeping an eye on Guido's bandaged hand. Joe walked ahead, mind already spinning.

Joe didn't go to dinner. He tried—walked halfway up the ridge toward the big fire pit where Mom and Morales were serving stew—but the noise pulled at something unsettled inside him. Instead of forcing it, he turned off the main path and headed toward the quieter side of camp where the lanterns were spaced farther apart and the wind drowned out most of the voices.

The crate sat on a work table outside the operations tent, already dried off and neatly organized. Manny had stacked the medical packs in rows. Guido had logged every cylinder and hauled the gas to storage. The only thing left untouched was the black Holt Squadron case sitting dead center, like an accusation.

Joe stepped up to it and rested his hands on the table. The lantern beside him flickered, throwing shaky shadows across the stamped insignia. Holt Squadron—just the name carried weight.

"What were you doing out here?" he murmured.

He flipped open the case. Coded maps. Sealed orders. Rescue gear. He lifted one of the maps toward the lantern. It was marked with three red circles. One near Nashville. One south of it.

One nearly on top of Ridge Camp.

He lowered the map slowly. "This wasn't random."

The wind cut colder. Somewhere down by the main fire, kids laughed. Lenny barked once. Someone clanged a pot. Camp sounded like a small, stubborn town trying very hard to pretend the world hadn't ended.

Joe folded the map and closed the case. His fingers lingered on the emblem. Holt Squadron hadn't been sightseeing. They'd been on a mission.

And at least one of them had died trying to get out of that warehouse.

He picked up the case and carried it across the ridge toward his tent. The air bit at his cheeks. Lanterns cast long, thin shadows. Inside most of the shelters, people were eating, arguing softly, telling stories. Ridge Camp breathing. Alive.

His tent sat a little apart from the main cluster—close enough to reach in seconds if something went wrong, far enough to give the illusion of privacy. He ducked inside, set the case on the small table, and lit the tin lantern hanging from the center pole.

Canvas walls glowed gold. Their bedrolls lay side by side, blankets piled thick against the coming winter. A pair of Valery's spare boots waited near the flap. Her sling was folded on the crate they used as a trunk. Home, such as it was.

He sat on the edge of the bed and stared at the case. He didn't open it again. He didn't need to. The three red circles were already burned into his mind.

He wished Morales were here. She'd know the meaning of the order bands, the markings, the color codes. Joe only recognized that they weren't standard rescue packets.

Boot steps crunched outside. The flap lifted.

Valery stepped in, cheeks pink from the cold, hair loose around her face. She closed the flap behind her and let out a breath. "Morales says if you skip dinner again, she's going to write you up for insubordination."

Joe huffed a tired sound that might have been a laugh. "She can add it to the list."

Valery's gaze went straight to the black case. "You brought it in here."

"Didn't feel right leaving it out there," he said.

She crossed the tent, set a mug on the table beside the case, and then stepped between his knees, close enough that he could smell smoke and cold and the faint citrus soap she liked. "You're doing the thing again."

"What thing?"

"The thing where you stare at a problem until it either solves itself or catches fire."

He let his forehead rest lightly against her sternum for a second.

Her good hand slid into his hair, fingers warm against his scalp. The contact grounded him more than he wanted to admit.

"You okay?" she asked softly.

"No," he said. "But I'm working on it."

She eased back just enough to see his face and then sat beside him, shoulders touching, their knees pressed together. "Talk to me."

He nodded toward the case. "Holt Squadron. Maps with three red circles. One of them is here. We found one of their bodies under that warehouse. He died reaching for the exit, not hiding. That's someone running from something."

Valery listened without interrupting, thumb tracing idle circles on the back of his hand.

"This wasn't a survivor sweep," Joe said. "Those orders in there—band color's wrong for standard rescue. That's high-clearance stuff. Sensitive mission. Classified objective."

"You think they were tracking something," she said.

"Yes."

"And you think whatever it was… it might still be out there."

He didn't answer. He didn't need to. She could read it on his face.

She sighed, leaning her head against his shoulder. "Ridge Camp is finally starting to feel safe. People are sleeping at night. Kids are planning stupid games for tomorrow. You don't want to tear that open."

"That's exactly why we can't ignore it," he said. "If the Air Force was chasing something that led them here—and if it's dangerous—we need to know before winter locks us in."

She was quiet for a moment, then nodded. "Then tomorrow, we start digging. Together."

"You should be resting," he said. "Your shoulder—"

She cut him off with a look. "You're not doing this alone. So either we both rest tonight, or we both stay up and obsess. Pick one."

He let out a breath that was almost a laugh. "You're bossy."

"Today?" she said, echoing Guido's earlier mutter.

He turned his hand, laced his fingers with hers. "Thank you."

"You're welcome," she said, softer now. She leaned in and kissed him—slow, familiar, a promise instead of a distraction. When she pulled back, her forehead stayed against his. "We'll deal with Holt tomorrow. Tonight, we sleep."

He glanced at the case one last time. "Tomorrow morning, we open it with everyone there. Morales, Mom, Perez, Manny, Guido. No secrets."

"Good," she said. "Secrets are how everything went to hell the first time."

Joe stood long enough to hang the lantern lower and dim the wick. The tent sank into a softer gold. Outside, the ridge creaked and sighed under the cold. Inside, blankets rustled as they lay down together, shoulders and hips aligned in a way that felt as natural as breathing.

"Whatever the Air Force didn't want dying with them," she murmured, already half asleep, "we'll make sure it doesn't."

Joe stared at the faint outline of the Holt case on the table until his eyes finally blurred. Three red circles. One dead airman. One drowned warehouse full of questions.

Tomorrow, they'd start pulling at the threads.

Tonight, Ridge Camp slept. And Joe, for the first time since he'd seen that Air Force patch in the dark, let himself sleep too.

CHAPTER 34
THE COUNCIL

The next morning came in thin, gray light and breath you could see. A dusting of frost clung to the tent ropes and turned the ridge path into a crunching line of white. Joe woke to the familiar weight of Valery's arm across his chest and the soft sound of her breathing against his shoulder. For a second, before the world caught up, it felt like a normal cold morning on any ship, any shore. Just the two of them and the day ahead.

Then he saw the black Holt Squadron case on the small table and remembered.

Valery shifted, eyes blinking open. "You're staring at it again," she murmured.

"At you," he said. "Better view."

She huffed a quiet laugh and nudged him with her forehead. "Liar."

He kissed her once, brief and warm, then sat up. The tent was cold enough that his breath fogged. "We need the council together," he said. "All of them."

"The Seven," she corrected softly, using the camp's nickname. "Even though there are eight of us now and it makes absolutely no mathematical sense."

"Camp branded it early," he said. "Nobody wants to redo the legend."

"You're still the chair," she said. "You call it, they'll come."

"We need Morales, Donnie, you, Mom, Manny, Guido, and Dr. Elise in the council tent," he said. "And Perez for the tech part, even if he's not one of the Seven-That-Is-Eight."

She pushed herself up slowly, cradling her injured arm. "I'll meet you there. Mom's already up. I heard her threatening someone's life over burnt porridge."

"That tracks," he said.

They dressed fast, layering sweaters and jackets, pulling on boots still stiff with yesterday's cold. Outside, the sky had gone from gray to a washed-out white. Tiny flecks drifted down—not rain, not quite snow, just the first stubborn hint of flurries. The air had that heavy, waiting feel. Winter wasn't knocking anymore. It was halfway through the door.

Joe picked up the Holt case. It felt heavier today. He carried it down the ridge toward the big tent they'd reinforced for meetings. Valery walked beside him, close enough that their sleeves brushed. People moved out of their way without being asked. News had already spread: something about the Air Force, something about maps. Eyes followed the black case. No one asked out loud.

The council tent sat near the center of camp, close to the fire pit and the water lines. Inside, the big table was already set with chipped mugs and a lantern at each end. Morales stood with her back to the entrance, talking quietly with Donnie over a stack of reports. Mom was there too, hands wrapped around a mug, her expression serious and tired but steady. Manny and Guido sat side by side near the end of the table, both freshly scrubbed but still carrying the hollow look of divers who'd seen something they couldn't easily file away. Dr. Elise sat near Mom, notebook open, pen ready, her dark hair pulled back in a tight knot, eyes already tracking everything. Perez hovered at the far corner with a battered tool kit and a coil of wires, clearly the invited specialist, not a council seat.

Joe ducked inside with Valery and set the Holt case down in the center of the table. Everything else in the tent seemed to fall away.

Morales glanced at the case, then at Joe. "All right, Chair," she said.

"Everyone's here. Let's do this before the weather decides to punch us in the teeth."

Joe looked around at them—Mom, Morales, Donnie, Valery, Manny, Guido, Dr. Elise. Seven council members plus him as the chair. The camp still called them "the Seven" out of habit, but everyone in this tent knew how many hands actually steered Ridge Camp.

"Okay," he said. His voice came out rougher than he meant. "This is council business. But it's also camp business. The Seven are here, plus the idiot who chairs you. What we decide ripples out to everyone."

Mom nodded once. "Then say it straight, Joey."

He opened the case and laid the dog tags on the table first. The metal clinked softly, louder than it should have been.

HOLT SQUADRON – 119th RESCUE.

Manny went still. Guido swallowed. Perez leaned forward. Morales' jaw flexed. Dr. Elise's eyes sharpened, taking in the name as if it were a symptom.

Joe set the folded map beside them, the three red circles already burned into his mind. "We found a dead airman under a drowned warehouse," he said. "Holt Squadron. Arnold Air Force Base. And this," he tapped the map, "says they were here on purpose."

Something cold moved through the tent that had nothing to do with the air outside.

Perez reached for the tags with careful fingers, as if they might vanish if he grabbed too fast. "This is real," he said quietly. "Not some leftover junk somebody traded through a black market."

"Real enough," Manny said. His voice was flat. "We took them off his harness ourselves."

Guido nodded, eyes unfocused for a second, back in that warehouse. "Helmet, flight suit, the whole thing. Patch still on his arm. The water ate his face, but the patch was clear."

Mom tapped the edge of the table. "If there's more down there, we'll want pictures—evidence we can study without risking divers."

Morales nodded. "Good thing we still have the Polaroid from the Savannah's inventory. Low-tech, no batteries, no corruption. It's the only reason we can document anything in this cold."

Guido huffed. "That old brick? I thought Manny broke it."

"I fixed it," Manny muttered. "Don't look at me like that."

"I only knew the name," Joe said. He looked at Morales. "You've worked under enough captains. What were they really?"

Morales' expression tightened. "Long-range rescue. Recon. They were the ones who went where nobody else would. They flew out of Arnold AFB before the impacts."

Joe tapped the crate manifest he'd brought from storage. "Same base stamped on the emergency stockpile box we pulled from the warehouse."

Donnie straightened. "So they were here with supplies on purpose. Not just crashing wherever the weather dumped them."

"Exactly," Joe said.

Morales studied the map. "Walk me through the circles."

Joe unfolded it completely and weighed the corners down with mugs. "Here—" he pointed near the top "—Nashville area. Here—south of that—looks like a corridor along the old interstate." He tapped the last circle and felt the tent tighten around him. "And this one is us. Ridge Camp. Our ridge. Our bay."

Guido let out a low whistle. "Well, that's not creepy at all."

Manny frowned. "You sure it's not just some generic grid?"

Joe shook his head. "Search grids are broad. Sweeps. These are tight and deliberate. Somebody selected three key points and told Holt Squadron to hit them. This wasn't a fishing expedition. It was a mission."

Perez flipped the map, searching the margins. "No legend. No ops code. Just the circles and route numbers."

"That's where the rest comes in," Joe said. He opened the padded pack from the crate and slid out sealed plastic envelopes stamped with Air Force codes. "These are the mission orders. Or part of them," Joe said. "Band color's wrong for rescue."

Morales' eyes narrowed. "Rescue orders use a blue band," she said. "This?" She tapped the seal. "This is red. Command-level, high-risk. You don't slap red-band orders on a rescue squadron unless it's serious and classified."

Dr. Elise leaned in, gaze narrow. "Anything in there about biological or chemical containment?" she asked. "Key phrases. Quarantine. Hot zone. Transmission vector."

"Not that I can see from the outside," Joe said. "But Perez hasn't opened them yet."

"Yet," Perez echoed. "I can soften the seal with steam and maybe alcohol. If they used standard cipher format, I can at least get headings and intent without ruining the ink."

Mom's eyes narrowed. "Classified how, Joey?"

"Could be anything," he said. "Containment. Asset retrieval. Targeted evac. But whatever it was, they thought it mattered enough to send their best into a collapsing map."

Morales crossed her arms. "And at least one of them died under a warehouse chasing it."

"Yeah," Manny said. "He died with his hand on the exit rail. Not curled up in a corner. That's someone running."

Dr. Elise tapped the table lightly with her pen. "Running from an environmental hazard? Biological? Hostile humans? We need to know what kind of threat profile we're talking about."

"The orders might spell it out," Perez said. "At least in code."

The tent went quiet for a beat. Outside, the wind hissed at the canvas, and a few stray flakes scratched against the roof.

Mom broke the silence. "What do you think they were doing here, Joe? Not what scares you. What do you actually think."

He didn't sugarcoat it. "I think the Air Force wasn't just looking for survivors," he said. "I think they were looking for something—or trying to keep something from spreading. And I think whatever it was, they didn't finish."

Donnie shifted, mind already working. "So we're sitting on top of whatever they tagged as important. Or dangerous."

"Feels that way," Joe said.

Morales' mouth pressed thin. "And now the Seven—" she jerked her chin toward the table, including him in that whether he wanted it or not "—are supposed to decide if we poke it or pretend it isn't there."

"Pretending hasn't worked great for humanity," Dr. Elise said dryly. "From a medical standpoint, I'd rather identify a risk than be surprised by it mid-winter."

"Of course you would," Morales muttered. "You like labels on your nightmares."

Elise didn't flinch. "Labels save lives."

The argument sat there, sharp and unavoidable.

Flurries thickened outside, tapping a soft, steady rhythm on the canvas. The lantern flames flickered as a draft shouldered through the seams. Inside, the table might as well have been a live wire.

"We are not chasing ghost missions from a dead Air Force squadron," Morales said. "Winter is here. Look at that." She jerked her chin toward the wall, meaning the wind, the flurries, the creaking ridge. "We have tents to reinforce, roofs to brace, food to ration. I'm not diverting half our strength to go dig up whatever got those people killed."

"We're not talking about half our strength," Joe said. "We're talking about a small recon team and Perez working on paper in a warm tent."

"It's not the size, it's the risk," Morales snapped. "You start talking about secret missions and classified objectives, the camp gets spooked. Spooked people make stupid choices."

Mom didn't raise her voice, but everyone heard her. "Enough. Nobody here is wrong. We just need to be smart." She looked at Joe. "Can we afford to ignore it?"

He forced himself to picture it honestly—winter locked in, snow piling up, some unknown Air Force problem buried under their feet like a landmine. "No," he said finally. "If Holt Squadron came here on purpose and died trying to get out, we can't pretend that has nothing to do with us."

Donnie spoke up, steady as a keel in crosswind. "I've got roofs to fix and stoves to keep running. But I've also spent my life being real careful around machines I didn't understand. You don't leave a reactor running when it's about to crack, and you don't sleep on a mystery somebody died trying to solve."

Morales frowned. "So you're voting to chase it."

"I'm voting we understand it before it bites us," Donnie said. "Joe's not talking about taking a caravan to Nashville in the snow. He's talking about answers under our own waterline."

Dr. Elise nodded slightly. "From a medical perspective, an unidentified 'critical asset' with containment language around it is a red flag. If it's biological, we need protocols. If it's technological, we need to know what people might try to turn on."

Mom turned to her. "You think we move forward."

"I think ignorance won't keep anyone safe," Elise said. "Cautiously, yes. But forward."

Mom nodded slowly. "I'm with Donnie and Joe and Elise—carefully. Perez, can you work on those orders here without risking the camp?"

Perez nodded. "I'll steam the seal, not tear it. If they used standard cipher formatting, I can at least get headings and intent. If it's more complex, I'll need time, but time indoors I can do."

"And in the meantime?" Morales asked.

Joe pointed to the map. "We already dove one circle—warehouse sector. That's where we found the body and the crate. We mark it as hot, no random dives without council approval. Next step is controlled recon. A second dive to bring up the rest of the bodies, if there are more, and whatever else they were guarding down there."

Manny's jaw tightened. "We go back in? Same spot?"

"Not tomorrow," Joe said. "Not until we know more. But yes, we'll have to. You and Guido are the ones who know the layout."

Guido didn't flinch. "I'll go. Just don't send me back down there blind."

"You won't be," Joe said. "We go in with a tighter plan, better lines, maybe a third diver if we can spare one."

Dr. Elise tapped her notebook. "If you go back, I want full decontamination when you come up. Gear rinse, suit check, eye and throat exam. Just in case."

Manny gave a short nod. "Yes, ma'am."

Mom looked around the table. "So the question is simple: do we move forward at all, or do we lock this case in a box and hope it never matters?"

Morales' mouth tightened, but she didn't interrupt.

Mom raised her hand. "I vote we move forward—cautiously. Perez on the codes. Joey and his people on a limited recon once we have more intel. No mass panic. No rumors."

Donnie lifted his hand. "With strict safety limits? Yeah. I'm in."

Dr. Elise raised hers. "In. With medical oversight."

Manny and Guido exchanged a look, then lifted theirs. "In," Manny said.

All eyes went to Morales.

She scowled at the map, then at the dog tags. Then she raised her hand too, sharp and annoyed. "Fine. In. But the second this threatens camp stability, I pull us back."

Joe nodded. "Deal."

Valery was the last. "You already know my vote," she said. She raised her hand. "In. We face it before it finds us."

Seven council hands up. Joe let out a breath he hadn't realized he was holding.

"So it's decided," Mom said. "The Seven move forward."

Outside, the wind whistled through the guy lines. Inside, the decision settled on all of them like fresh snow.

By afternoon, the flurries had turned the ridge into a thin, patchy quilt of white. Not deep enough to slow work, just enough to remind everyone that the real storms were coming. People moved through camp with armloads of scrap lumber and canvas, reinforcing tent lines, packing moss and cloth into hut cracks, checking stoves and chimneys. The air smelled like woodsmoke, wet canvas, and the sharp bite of coming winter.

Joe spent the next few hours helping Donnie check roof braces and stove vents, mind only half on the work. Morales stalked the ridge with a checklist, making sure patrol routes were clear even in low visibility. Manny and Guido hauled their dive gear to a covered rack, inspecting every strap and seal, already thinking ahead to the next time they'd have to go under. Ashley rotated between med tent and water lines, eyes sharp, checking fingers and toes for early frostbite and making sure nobody ignored a cough "because it's just cold air."

Dr. Elise worked beside her, triaging complaints, quietly putting together a mental list of what would kill people fastest when the temperature really dropped.

By the time the last nail was hammered and the last flapping tarp tied down, Joe's shoulders ached and his thoughts had gone in too many circles to count.

He ducked into the operations tent once to check on Perez. The engineer sat hunched over a notebook covered in tight, neat writing. The Air Force packet lay open beside him, the seal mostly intact but softened enough that he'd been able to slide the documents out.

"Tell me you got something," Joe said.

Perez rubbed his eyes. "Not everything. But enough. The orders are red-band. Designation: Operation Storm Glass."

Joe didn't like the sound of it. "What was the objective?"

"Language is vague on purpose," Perez said. "But the gist is this: Holt Squadron was tasked with reaching three inland points and either securing or destroying 'critical assets' tied to pre-impact contingency planning. There's mention of 'containment risk' and 'unauthorized appropriation.'"

"Meaning?"

"Meaning somebody in a bunker somewhere was afraid their toys were about to end up in the wrong hands," Perez said. "Could be tech. Could be weapons. Could be black-site research. Arnold AFB was one of the nodes coordinating it."

Joe's stomach sank. "And Ridge Camp is sitting on one of those 'assets.'"

"Looks that way," Perez said. "The warehouse you dove looks like it was another. The third circle—up near Nashville—no details yet."

Joe closed his eyes for a second. "All right. Council in an hour. Same crew as this morning. Elise too."

"You sure you don't want to sleep on it?" Perez asked.

"No," Joe said. "I've done enough staring for one lifetime."

Dusk settled early, the sky bruising into purple and then black. Lanterns bloomed across the camp. Voices faded from work tones to evening ones—lower, closer, full of tired jokes and small comforts.

Joe finally made it back to his tent, hands numb, face stung raw from cold. When he pushed the flap aside, Valery was already inside, sitting cross-legged on the bedroll with a blanket over her lap and two mugs steaming on the crate.

"You look like someone beat you with a snowman," she said.

"Feels about right," he answered, dropping onto the blanket beside her. "Donnie had me crawling on roofs like I'm twenty."

"You're not," she said. "In case you forgot."

He leaned his shoulder against hers. "Rude."

She handed him a mug. "Drink. You'll live."

The tea was hot and faintly citrus. Miami in a cup if he squinted hard enough. He let the heat burn his tongue on purpose.

For a while, they sat in comfortable silence, listening to the muffled sounds of camp outside—the clank of a pot, a kid laughing, Lenny barking twice and then being shushed by Mom.

"Remember when cold meant sixty-eight degrees and we complained about it?" Valery said eventually.

He laughed once. "Miami cold. Yeah."

Her eyes went distant in that way that meant she'd left the tent and gone back years. "Valentina's deck used to burn my feet by ten in the morning," she said. "You'd hose it down and swear the water sizzled."

"It did," he said. "We should've bottled that heat."

She smiled, small and real. "Key Biscayne in May. Remember? We'd take her out past the sandbar and kill the engine. Just us, the gulls, and that stupid Bluetooth speaker that only played three good songs and twelve terrible ones."

"You picked the songs," he said.

"You danced to them," she shot back.

"Lies."

"Absolute truth. You, on the aft deck, doing that thing with your shoulders—"

She mimed it with her good arm and made him snort tea up his nose. He wiped his face with the back of his hand, laughing harder than the joke deserved. It felt good. Wrong and good.

"We'd anchor for the night sometimes," he said once he caught his breath. "Remember that storm off Biscayne Bay? Lightning all the way up the coast, and you still refused to go inside because you wanted to watch it hit the water."

"It was beautiful," she said. "Like the sky was trying to write something on the ocean."

"Marco thought we were insane," he said.

"Marco thought anyone who wasn't on a ship's clock was insane," she replied.

He let the memories roll: hot fiberglass under bare feet, diesel and salt, the way the Valentina rocked gentle in a good swell and tried to throw you across the cabin in a bad one. Nights where the only light was Miami's glow on the horizon and a single masthead star.

"You know what I miss most?" Valery asked.

"Showers that lasted longer than two minutes?"

"Okay, second most." She gave him a look. "That feeling when we'd bring her back through the channel and the air would change. You could smell mangroves and sunscreen and fried fish from the docks. Everyone salty and loud and alive. Nobody looking over their shoulder for the sky to fall."

He went quiet. That part hit too close.

She reached over and laced her fingers through his. "We had a whole life before the water came up," she said. "Sometimes I think we forget that. We weren't just survivors. We were idiots on a fifty-foot boat burning gas we didn't need to burn and arguing over whose turn it was to wash the galley."

"It was always my turn," he said.

"You were just slower at ducking out of the room when Tina yelled."

He smiled, then sobered. "Feels like another world."

"It was," she said. "But we're the same people. The guy who kept the Valentina's engines from exploding is the same guy who kept the Savannah moving. The woman who used to plot weekend trips to Bimini is the same one reminding you not to jump headfirst into classified Air Force messes without a plan."

He squeezed her hand. "I like that woman."

"She likes you too," Valery said. She leaned in and kissed him, slow and sure, the way you kiss someone you've chosen a hundred times already. When they separated, her forehead rested against his. "We'll figure out Holt. We'll figure out winter. We always have."

"It's getting harder not to be scared," he admitted.

"Good," she said. "Scared people make careful plans. It's the confident idiots who get killed."

He huffed a laugh. "Inspiring."

"Shut up and drink your tea," she said. "You have a council to wrangle."

Morning came with a pale sun and a thin layer of snow clinging to the tent tops like someone had dusted the camp with flour. It wouldn't last, but it was a warning shot. Breath smoked in front of every face. Kids skidded on frozen patches until Mom barked at them to stop before they broke their necks.

Ridge Camp moved like a hive. Donnie and a crew checked every roof beam. Morales ran through a checklist with patrols, making sure rifles were dry and paths were marked for bad-weather visibility. Manny and Guido hauled their dive gear, inspecting again, already planning lines and fallback signals. Ashley and Dr. Elise worked side by side in the med area, quietly putting together what they'd need when the cold really bit: extra blankets, hot drinks, frostbite checks, decon kits ready in case the mystery turned ugly.

In the operations tent, Perez looked like he hadn't slept much. Joe stepped in and found him hunched over a notebook covered in tight, neat writing. The Air Force packet lay open beside him.

"The orders are red-band," Perez said. "Designation: Operation Storm Glass. Holt was tasked with reaching three inland points and either securing or destroying 'critical assets' tied to pre-impact contingency planning. There's language about containment risk and unauthorized appropriation. Someone didn't want this stuff in the wild."

Joe's jaw tightened. "And our ridge is one of their points."

"Yeah," Perez said. "You're sitting on something they cared about."

When the council gathered again, the tent felt even colder than before, despite the extra lantern. Snowflakes peppered the canvas overhead, each impact a soft tap. Joe laid out what Perez had found, keeping it clean and simple—no extra drama, no speculation he couldn't back up.

"Operation Storm Glass," Morales repeated. "God, they really never missed a chance to make things sound ominous."

"Point is," Perez said, "they were worried enough to send Holt after whatever this was. And they didn't finish."

Mom looked at Joe. "Details. No generalities."

He didn't hesitate. "Step one: Perez finishes decoding everything he can. He works with Morales and Elise so we have technical, security, and medical eyes on it. Step two: limited recon dive back at the warehouse. Manny and Guido only, one extra line handler topside just outside the main structure, no heroics. They bring up what they safely can and confirm how many bodies are down there. Step three: landside survey." He tapped the circle on the map that overlapped Ridge Camp. "We grid this ridge properly. I want to know what Storm Glass thought was so important under our feet. If there's a sealed bunker, a depot, anything—that's a bigger risk buried under our latrines than it is open and understood."

Morales considered it, jaw tight. "You're going to spook people if you start talking about buried bunkers."

"Then we don't talk about bunkers," Mom said. "We talk about checking the ground before winter sets in. Safety sweep. The council doesn't lie, but we don't feed nightmares either."

Manny rolled his bandaged hand, thinking. "When?"

"Forty-eight hours," Joe said. "Tomorrow we finish winter prep and gear checks. Day after, if weather holds, you two dive again. Same day, a land team starts a careful survey of the circle zone. We do it fast, controlled, and we don't push past what we can handle."

Valery nodded. "I'll coordinate the land team with Morales and Donnie. No abandoned shafts, no crawling into unknown holes without rope and backup."

Ashley, hovering near Elise, looked pale but steady. "Med will be ready," Elise said. "For cold exposure and… anything else."

"This is what we do," Mom said, looking around at the eight of them. "We don't hide from hard things. We look at them straight, together. Holt Squadron didn't get to finish their job. That doesn't mean we walk away from what they were trying to contain."

Outside, the snow had thickened into real flurries, dusting the ridge, the tents, the hull of the dead Savannah down in the bay. Inside, the council—The Seven, plus the chair who never quite counted himself—moved as one.

Later, Joe stood alone in the operations tent, looking at the map pinned to the board. Three red circles. Operation Storm Glass. A dead airman in a drowned warehouse. A ridge camp full of people who were finally starting to believe in tomorrow.

He glanced at the Holt case on the table, a fine powder of snow clinging to its top.

"Storm Glass," he muttered. "Let's see what you were really trying to stop."

Outside, Ridge Camp tightened lines, checked fires, and got ready. Shaken, yes. But unified.

And somewhere under ice-cold water and frozen ground, the Air Force's unfinished business waited for them.

CHAPTER 35
GHOST FIRES

The smoke showed up on a day that should've been ordinary. Cold, but ordinary.

Joe was halfway up the ridge path, hauling a crate of cut firewood toward the main storage tent, when the wind shifted and brought him a smell that didn't belong—burnt something, but not like their cooking fires. Sharper. Drier. Old wood and metal.

He stopped, set the crate down, and squinted out over the drowned valley. The sky was a flat slab of gray. Flurries drifted in thin, lazy spirals, not enough to stick, just enough to remind everyone winter was edging closer hour by hour. Ridge Camp's smoke curled straight up from the central fire pits, familiar gray threads in the cold air.

This other smoke came from farther out. Much farther.

"There," Joe muttered.

From a rise on the ridge, a thin column of darker smoke clawed up from the far side of a distant hill, beyond the edge of the flooded plain. It wasn't big enough to be a forest fire, but it was too tall and straight to be anything small. It didn't match the direction of their fires or the Savannah's old stacks. It belonged to something else.

Someone else.

Boots crunched behind him. "You break something?" Donnie asked.

Joe lifted a hand. "Look."

Donnie stepped up beside him, breath huffing white in the air. He followed Joe's line of sight and went still. "That's not us."

"No," Joe said. "It's not."

They stood in silence for a few seconds, watching the column rise. The distance made it hard to judge scale, but Joe guessed it was at least a few miles out, maybe more. Too far for a stray ember from Ridge Camp. Too straight to be an accident.

Flurries thickened for a moment, briefly veiling the smoke. When the air cleared again, the column still burned strong and narrow, like whoever lit it knew exactly what they were doing.

"Signal fire?" Donnie asked.

"Could be." Joe's jaw tightened. "Could also be someone burning something they don't want found."

Donnie grunted. "Comforting."

Voices rose from below as people worked—kids hauling kindling, adults checking lashings on tent frames, Manny yelling for someone to stop overloading a sled. Camp life. Normal, or as close to normal as they got now. None of them had seen the smoke yet. None of them knew something had changed out on the horizon.

Joe's mind jumped straight back to the black Holt case sitting in the ops tent, to the maps with three red circles. One of those circles sat almost exactly where that smoke was now.

Of course it does, he thought grimly.

"You going to tell Morales?" Donnie asked.

"In a minute," Joe said. "I want to make sure this isn't a trick of the wind first."

As if on cue, the wind shifted again, colder now, blowing straight into their faces. The smoke line wavered but didn't vanish. It held steady, rooted to the same spot beyond the far hill.

"Not a trick," Donnie said.

"Yeah," Joe murmured. "That's what I was afraid of."

Behind them, someone called his name. "Joe!"

He turned. Manny jogged up the path, gloves shoved in his pocket, cheeks flushed from work. "We just finished reinforcing the north huts," Manny said, then followed his gaze out toward the valley. His expression changed. "Please tell me that's just some old junkyard burning itself out."

"No junkyards left," Donnie said. "Just ghosts."

Manny frowned. "You think it's from Holt's people?"

Joe didn't answer right away. He watched the smoke, narrowing his eyes, tracing the line of the distant hill and the way the land dipped beyond it. The maps in his head, the circles, the coordinates—all of it stacked in the same place.

"It's too close to their mark to be an accident," he said finally. "Whoever lit that is either military-trained or copying someone who was."

Manny's shoulders tightened. "Friendly or not friendly?"

"Unknown," Joe said. "And I don't like unknown sitting that close to my people."

The wind picked up again, sharper now, a promise of the real winter that hadn't quite sunk its teeth in yet. Snowflakes spun sideways, dusting Joe's jacket and Donnie's beard, clinging briefly to Manny's hair before melting. The smoke in the distance kept rising, thin and deliberate.

A few kids down by the fire pit finally noticed. One pointed. Another shouted.

"Joe?" Manny asked quietly. "We say anything yet?"

"Not to the whole camp," Joe said. "Not until we know more than 'there's smoke.' I'll call the council. The seven—" He corrected himself automatically. "The eight of us will look at the maps."

"Scouting run?" Donnie asked.

"Soon," Joe said. "Carefully."

He picked up the crate of firewood and started toward the operations tent. The weight bit into his arms, a familiar, grounding ache. Behind him, Manny and Donnie fell into step. Below them, Ridge Camp went on working, tightening tarps and stacking wood, getting ready for the cold.

Above them, the sky lowered.

Out there, beyond the drowned streets and broken roofs, a single thin column of smoke drew a straight line from the ground to the gray.

Like a finger pointing.

Like a warning.

The council meeting didn't wait until nightfall. Joe called it the minute he'd finished splitting the firewood and the minute Manny stopped hovering like he expected the smoke to walk into camp and introduce itself. Word traveled fast. By the time Joe reached the operations tent, the eight were already gathering—Morales, Mom, Donnie, Manny, Guido, Valery, Dr. Elise—and Joe himself as chair. Only seven chairs fit around the crate-turned-table, so Guido stood, leaning on the tent pole like it was part of him.

Joe didn't sit right away. He pulled down the Air Force map he'd pinned beside the Holt case two nights earlier and slapped it flat on the table. The paper crackled under his hands.

"There's smoke," he said. "North ridge, about three miles out."

Manny crossed his arms. "Straight column. Controlled burn."

Morales' brows drew tight. "Military-style?"

"Very possible," Joe said.

Mom let out a long breath, one of those tired, measured ones that meant danger was circling again. "And we still have the Air Force crate sitting in this tent. Holt Squadron maps. Holt Squadron dead. Holt Squadron mission circles. Now we have smoke in one of those circles."

Valery leaned forward, her good arm braced on the table, her bad one resting in her sling. "Same direction?"

"Same mark," Joe confirmed. "Almost exactly."

Dr. Elise adjusted her glasses. "That could mean survivors. It could also mean hostile remnants." Her tone stayed even, but her fingers tapped the table. Nervous.

Guido swallowed. "We not assuming friendly this time, right?"

"No," Joe said. "We assume nothing."

Donnie looked around the tent. "You're thinking of sending a scout team."

"Yeah." Joe pointed at the map. "We don't approach the smoke directly. We go wide, cut across the flooded plain, use the old highway ridge as cover. If it's a threat, we come back unseen. If it's survivors, we approach on our terms."

Morales nodded. "You want Manny and Guido?"

"Yeah," Joe said. "They're used to moving quiet. They know the terrain. Ashley stays here—she's not ready for another scare after yesterday."

Manny lifted his chin. "Fine by me." He paused, catching Joe's look. "We're not engaging, right? Stealth only."

"Stealth only," Joe echoed.

Valery squeezed his forearm gently. "You're not going alone?"

"No," he said. "You and Mom run the camp while I'm gone. Manny and Guido come with me. Dad stays here—your job is to reinforce the shelters while we're out."

Donnie nodded without argument.

Mom folded her hands. "If this is Holt-related, we need proof. Tracks. Gear. Anything identifiable."

"We'll look," Joe said.

Morales looked from face to face. "We vote. Standard. Eight voices."

The room went quiet.

Mom: "Scout."

Morales: "Scout."

Donnie: "Scout."

Manny: "Scout."

Guido: "Scout."

Valery: "Scout—but only if you stay off the direct path."

Dr. Elise: "Scout."

All eyes fell to Joe.

He didn't hesitate. "Scout."

Decision unanimous.

Wind rattled the tent walls as if punctuating it.

Valery leaned close enough that only Joe felt the brush of her sleeve against his arm. "This feels connected," she murmured. "Holt, the maps, now this."

"It is," Joe said quietly.

Then he stepped back and addressed the entire council.

"We leave in ten minutes."

They broke apart, each moving fast—Morales toward the storage tent, Mom toward the medical bins, Manny and Guido grabbing packs and winter layers. Valery stayed behind just long enough for Joe to notice.

"You good?" he asked.

"No," she said honestly. "But I trust you."

He touched her cheek once, gentle, grounding. "We'll be back before dark."

She caught his wrist, held it a second longer than necessary. "Joe… if this is someone signaling, be careful. And if it's someone hiding… be more careful."

He gave a faint smile. "You always liked giving me impossible instructions."

"And you always liked pretending you don't need them," she said.

They shared the briefest kiss—soft, familiar, not distraction but reassurance. Then Joe stepped into the cold.

Outside, the air had shifted again. Flurries blew sideways. The ridge wore a thin dusting of white now, just enough to crunch under Joe's boots. Ridge Camp was busy—lashings being tightened, kids ushered into the bigger tents, every head turning toward that distant column of smoke.

Manny jogged up beside him. "Got the packs, boss."

Guido followed, carrying a rifle in a sling. "And the fun stick."

"No shooting unless we have no choice," Joe said.

"Understood," Guido said, though he didn't look thrilled.

Donnie met them at the path. "Watch the flooded bend. Ice is starting to form on the shallow edges."

Joe nodded. "We'll avoid the low spots."

Mom intercepted them long enough to shove a bulky canvas pouch into Joe's hands. "If you find tracks, I want to see them without freezing my butt off," she said.

He unzipped it just enough to check the contents—Ridge Camp's battered Polaroid and a small stack of film. Low-tech, no batteries, no excuses. He slid it into his pack. "We'll bring back pictures," he said.

Valery stood behind them with Mom, both watching like hawks refusing to blink.

"Let's move," Joe said.

The three men headed down the ridge, boots crunching on the frost-crusted dirt, breath fogging in front of them. Behind them, Ridge Camp stayed small, lanterns flickering in the gray.

Ahead of them, the smoke rose steady and thin.

Ghost fire.

A signal.

A warning.

Or the start of something none of them were ready for.

Snow flurries stung their faces as they crossed the lower ridge. The last of the afternoon light sat low and weak behind a blanket of gray clouds, turning the world into a half-lit smear of white, dead grass, and flooded pockets of black water. Manny pulled his scarf higher over his nose, muttering something about how Florida never prepared a man for this, and Guido nodded wordlessly because his beard was already icing over.

Joe kept the pace steady but not rushed. Stealth meant patience, not panic. The distant column of smoke rose in thin pulses across the sky—controlled, deliberate, not the kind of smoke that came from an accidental fire or collapsed shelter. Someone was tending it.

They dropped into the flooded ravine next, stepping onto the makeshift plank route that Ridge Camp had built weeks back. The water was only knee-deep at the widest point, but the current was unpredictable. Ice formed at the edges. One wrong step and a man could end up in a freezing channel he wouldn't climb out of.

Guido tested each plank before putting his weight on it. "Feels slick," he murmured.

"Just go slow," Joe said.

Manny stepped lightly behind them, eyes scanning the banks. "Tracks on the far side. Could be deer."

Joe glanced. "No deer walk in a straight tactical line."

Manny nodded. "Figured you saw it too."

They reached the far bank without incident, boots sloshing through mud that sucked and held like it didn't want to let go.

Guido wiped his gloves on his pants. "All right, boss. Which route?"

Joe pointed left. "Upstream along the old access road. It'll take us behind the hill. Smoke's on the other side."

They climbed the slope in slow, controlled steps. Snow clung to the dying brush and the twisted remains of guardrails, turning the path into a narrow cut of white and rust. Manny found a half-rotted mile marker sticking out of the dirt.

He tapped it. "This was the old service road to the industrial strip."

Joe nodded. "Which means Holt might've used the same route."

Guido let out a breath that fogged instantly. "Feels like we're walking in their footprints."

"Maybe we are," Joe said.

They reached the top of the ridge, crouched low, and peered down the other side.

The campsite below wasn't recent, but it wasn't old either. The snow lay unevenly across messy patches of churned mud and footprints. A ring of stones marked where a fire pit had been built. A few charred sticks sat within, burned low but controlled. Someone had lit this place intentionally.

Manny whispered, "That's military formation."

Joe nodded once. "Definitely not civilian."

They slid down the ridge in a crouch, spreading apart the way Manny and Guido had trained back on the Savannah when they used to drill with the Navy techs. Quiet steps. Low silhouettes. Slow breaths.

They reached the edge of the camp.

Joe raised a fist—stop.

Guido and Manny froze.

Joe crouched beside the fire pit. Fresh enough to smell faintly of smoke the snow hadn't completely smothered. He brushed away a thin layer of frost and exposed a circle of blackened stones that were still warm.

Manny mouthed, "Hours old."

Joe nodded.

Guido pointed at the ground. "Boss. That's… that's an MRE tear tab."

Joe knelt, brushing snow aside. A half-torn military ration packet lay in the mud, caked with ice but unmistakable.

ARMED FORCES—MEAL KIT

LOT: AFB ARNOLD

Manny blew out a slow breath. "That's Air Force."

Guido crouched beside Joe. "Holt again?"

"Maybe," Joe said. "Or someone who knew where to find their supplies."

Manny stood and walked toward the tree line. "Tracks," he whispered. "Four people. Maybe five. Mixed shoe sizes."

Joe joined him. The footprints led east, deeper into the woods, then vanished where the ground grew rocky.

"They were moving fast," Manny said.

"Retreating fast," Joe corrected.

Guido's eyes darted to the fire pit. "You think they saw our smoke? Or the camp lights last night?"

"Maybe," Joe said. "Or maybe they were already watching us."

The three men exchanged a look that meant the same thing:

Something was out here.

Something organized.

Something military.

Manny kicked lightly at the edge of the pit. "Why leave a fire burning? That's sloppy."

"Or intentional," Joe murmured. "Could be a signal. Could be bait. Could be a warning."

Guido swallowed hard. "Boss, I don't like any of those options."

Joe stood and scanned the ridge again. The cold wind cut across his jacket, carrying the smell of distant pine and the faintest trace of wood smoke.

"We collect everything," he said. "Food wrappers. Tracks. Ash samples. Anything with a label. Morales will want to examine it."

Manny nodded and began gathering debris into a small evidence pouch. Guido photographed the boot tracks with the camp's battered

Polaroid, the one Mom had dragged off the Savannah's inventory and declared too useful to ever trade away.

Joe walked a slow circle around the camp perimeter. His steps crunched on the frozen ground. His breath came out in plumes. A familiar stillness settled around him—the same stillness he'd felt in the bunker after the meteor impact; the same stillness on the Savannah the night they'd lost Raines; the same stillness in the drowned warehouse when he saw the Holt insignia.

The world wasn't done with them.

Not by a long shot.

He turned back to the others.

"We head home," Joe said. "Council needs to see this. We've got proof—someone military passed through here within twenty-four hours."

Manny tightened his pack. "And they're not looking for a fight."

Guido added, "But they're not avoiding one either."

Joe adjusted his gloves and looked once more toward the trees, where the footprints vanished like ghosts slipping into the dark.

"Let's move," he said.

Snow fell harder as they began the long walk back to Ridge Camp.

They didn't look back. Not one of them wanted to see what might be watching from the trees.

Snow thickened by the time the team was halfway back toward Ridge Camp. It came down in fast little bursts, more flurry than storm, but heavy enough to coat their jackets and fog their glasses. The wind cut sideways across the ridge, turning each breath into a small fight. Manny tightened his scarf and stomped across the frozen brush, boots cracking through thin sheets of ice with every step.

Guido kept his eyes on the path. "Feels like we're walking through freezer burn."

"Better than Florida humidity," Manny said.

Guido shot him a look. "Shut up."

They kept moving.

Joe stayed a few feet ahead, scanning the tree line instinctively, listening for anything that didn't belong—too many footsteps, the wrong kind of quiet, echoes that moved when wind shouldn't. The Air

Force tracks still tugged at the back of his mind. He didn't like mysteries that breathed.

By the time they reached the frozen creek crossing, daylight was fading fast. The sky had that early-winter color—flat steel with a hint of blue, like evening wanted to come early. Joe stepped onto the first log plank to test it. It held.

"Single file," he said. "Ice is new."

Guido went next, arms slightly out for balance. Manny followed behind him. The planks creaked under each step but held steady. They crossed without incident, climbed the opposite bank, and stepped onto the snow-packed trail that fed toward Ridge Camp.

They'd gone maybe twenty yards when the near miss happened.

A loud CRACK split the air—sharp, mean, violent.

Joe dropped instantly, hand up. "Down!"

Manny grabbed Guido by the back of his jacket and yanked him sideways as a massive branch snapped free overhead. It sheared off the trunk with a groan, plummeted straight down, and slammed into the ground exactly where Guido had been standing a heartbeat earlier. Snow burst upward in a white cloud.

Guido hit the ground hard. "Shit!"

Manny stared at the branch, chest pounding. "That was close. Too damn close."

Joe scanned the trees. "Wind did that. Wood's frozen. Let's move."

They pushed off the ground, hearts still hammering, and hurried down the trail—none of them eager to linger under trees heavy with snow and old storm scars.

Back at camp, Ashley had been pacing near the rope line, watching the ridge for the first sight of them. When she saw shapes move through the flurries, she jogged forward automatically—then stopped when she spotted Manny's expression.

"What happened?" she demanded, breath fogging.

Manny shook snow out of his hair. "Branch fell. Missed Guido by inches."

Ashley's eyes widened. She stepped closer and grabbed Manny's arms, scanning him for injuries even before checking Guido. "You weren't under it?"

"No," Manny said. "But I was close enough that I smelled the bark."

Ashley's hands trembled—just slightly—but Manny noticed. She swallowed and tried to hide it, but her breathing tightened the same way it did during thunder or sudden metal impacts.

Joe stepped away to deliver the evidence pouch to Morales. Guido headed to the fire pit, muttering about cheating death again. That left Ashley and Manny alone in the thin swirl of flurries.

She ran her fingers once more over his sleeves. "You scared the hell out of me."

"I'm fine," he said gently.

"You always say that," she shot back.

"And I'm always fine."

"One day you won't be," she whispered.

Manny's expression softened. "Not today."

She tried to steady her breathing. He noticed. He always noticed.

"You're doing the tight breathing again," he murmured.

"I know," she snapped—and then immediately softened. "I know. I just… when you came over that hill, you didn't look okay."

"I'm okay now." Manny reached out and lightly tapped two fingers against her wrist. "Breath with me."

"I'm not—"

"Just one," he said quietly.

She hesitated, then nodded.

He breathed in slow. She followed.

He breathed out. She did too.

Her shoulders loosened. The panic faded.

When their breaths synced, Manny smiled—small and warm. "See? Not dying today."

She tried to glare, but it came out as a shaky laugh. "You're an idiot."

"Yeah," he said. "But I'm your idiot."

Ashley's cheeks flushed. She looked away fast. "Shut up."

Manny didn't push. He didn't tease. He just walked beside her toward the fire pit, close enough for his elbow to brush hers, steady and quiet the whole way.

Behind them, snow kept falling.

And Joe watched the two of them with a mix of worry and something like hope.

Because fear meant care. And care meant risk. And risk meant there was something worth fighting for beyond ration counts and winter prep.

Something human. Something fragile.

Something that made the cold feel a little less sharp.

By the time the team made it back into the heart of Ridge Camp, the flurries had thickened into a steady curtain of white. Lanterns flickered along the walk paths, casting warm halos through the falling snow. People drifted toward the main fire pit to hear what the scouting party had found, but Morales waved them back with a sharp hand.

"Go eat," she ordered. "If it's important, you'll hear it at the meeting."

That settled the crowd.

Joe, Manny, and Guido headed straight for the operations tent where heat drifted from a barrel stove. Mom was already inside, sorting blankets with Dr. Elise. Donnie stood near the stove, rubbing warmth back into his hands. When he saw Joe's face, he straightened.

"What'd you find?"

"Better if I show you," Joe said.

Morales arrived behind them, boots heavy with snow. She closed the tent flap and faced Joe with her clipboard tucked under one arm. "Talk."

Joe didn't talk first. He set the evidence pouch on the center table, poured out the contents, and stepped back.

Everyone leaned in.

Burned wood scrap. Dirt samples. A half-torn MRE wrapper marked AFB ARNOLD. A boot print photo. A melted piece of para cord. And the kicker—a still-warm ash fragment wrapped in cloth.

Manny crossed his arms. "That campsite wasn't old. Whoever used it was there this morning."

Guido nodded. "Fire pit was still warm. And they ate military rations. Not Navy. Air Force."

Mom frowned. "Air Force? Again?"

Joe placed the ration wrapper where everyone could see it. "This wasn't dropped by accident. They were staying there. Watching something. Or someone."

Dr. Elise leaned closer, tapping the wrapper. "There is no way civilians would have this. East Tennessee didn't have much military presence pre-impact."

Morales turned the evidence pouch over and examined the boot track photo carefully. "This is size eleven. Military tread. Deep heel cut. Standard issue."

Joe pointed at the next photo. "And that one's a size seven. They were mixed. Four to five people total."

Donnie whistled softly. "That's a squad."

Guido swallowed. "Boss… you think it's Holt again?"

"It's somebody with access to their supplies," Joe said carefully. "Could be remnants of Holt. Could be another unit."

Morales straightened sharply. "Another unit? What unit? We haven't seen a living uniform outside of our own in almost two years."

"Exactly," Joe said. "Which means whatever unit it was—they're hiding."

The tent went quiet.

Mom was the one who finally broke the silence. "You think they saw us?"

Joe nodded. "They were moving fast. Tracks showed they left in a hurry."

Manny shifted uneasily. "Retreating fast."

"Or repositioning," Morales added.

Donnie rubbed his jaw. "Well, that doesn't sound good."

Morales gathered the evidence into a cleaner pile, then looked at Joe. "We need the council."

"They're coming," Joe said. "They're already on call for winter prep. This moves to the top of the list."

Dr. Elise closed her notebook. "Medical implications aside, any unknown group is a risk. Disease, conflict, desperation… we cannot assume friendly intentions."

Mom crossed her arms. "We also can't assume they're a threat. Someone might just be lost."

Joe shook his head. "This wasn't a lost group. They followed tactical formation. They cleaned their prints where they could. They kept the fire controlled. They're trained."

Morales fixed her eyes on him. "What's your gut say?"

Joe hesitated—not because he didn't know the answer, but because knowing didn't make it easier.

"My gut says Air Force," he finally said. "Maybe Holt. Maybe survivors of another squadron. Someone with orders."

"Orders to do what?" Dr. Elise asked.

Joe didn't answer.

Because he didn't know.

And because the last group of Air Force survivors they found had died under a drowned warehouse.

A soft knock hit the tent flap.

Ashley poked her head in. Snow scattered from her hood. "Everyone asked me to tell you—dinner's ready for the council whenever you're done."

Mom touched her arm. "You okay, sweetheart?"

Ashley nodded. "Just cold. And… glad they're back."

Her eyes flickered to Manny for half a second. He looked away, rubbing his bandaged hand.

Morales clapped once. "All right. This moves to priority. We meet in ten minutes. Bring the evidence."

Everyone began gathering coats and notes.

Before leaving, Joe stepped back and looked at the items on the table one more time—the ration wrapper, the burned ash, the para cord, the photos.

Then he looked at Morales.

"We're not alone out there," he said quietly.

Morales nodded. "No. And whoever they are—they're close."

Snow hit the tent roof in steady tapping bursts.

Winter was settling.

And now they had ghosts on the ridge.

CHAPTER 36
THE FROZEN TUNNELS

The cold came in layers, like the world was shutting down one switch at a time. First the ground went hard. The paths that had been mud and slush turned to ridged, frozen ruts that caught the toes of boots and twisted ankles if you didn't watch your step. Then the air went thin and bitter. Breath burned going in, scraped going out. The last layer was the worst—the silence. Birds stopped calling. The creek's steady whisper turned to a muffled grind under new ice. Even the trees sounded different when the wind hit them, branches cracking instead of swaying. Ridge Camp hunkered down and tried not to break.

By late afternoon, they knew it wasn't just "a cold snap." The deep freeze hit like a dropped hammer. Water barrels along the south line went solid in hours, not overnight. The thin tarps some families had stretched as extra windbreaks stiffened and snapped along the edges. The sky was clear, a hard glass blue, but there was no warmth in it. The sun looked like something painted on the world instead of fueling it.

Joe walked the ridge with his scarf up over his nose, making notes as he went. Frost rimed every tent line. The smaller huts—especially the ones rushed together in the first weeks—were already showing strain. A seam here. A sagging roofline there. The kind of flaws that didn't matter in fall but turned deadly when the wind cut down from the north at twenty miles an hour.

He stopped at the north edge, where the worst of the wind knifed through the gap between two hills. A cluster of smaller family tents huddled behind makeshift pallets and scrap-wood fences. Breath fogged thick around the people moving there—mostly moms hauling extra blankets, older kids pounding in stakes with numb hands.

Mom stood beside one of the thinner tents, her hair wrapped in a wool hat someone had crocheted from unraveled sweaters. She was arguing softly with a man Joe recognized from the newest arrivals—a dad with that hollow, always-counting expression of someone who'd lost too much and was terrified of losing the last pieces.

"You can take the cot," Mom was saying. "The kids and your wife go into the hut with the others. We'll figure out the rest later."

He shook his head. "We can't crowd them like that. Someone's already coughing in there. My youngest—she..."

"She'll be worse if she sleeps in this ice box," Mom cut in, voice firm but kind. She looked up as Joe approached. "We need more insulation on this side. Wind's coming straight through the gaps."

Joe already had it written on his slate. "We're reinforcing all north-facing walls before dark," he said. "We'll move a heater closer to the hut door. Morales is reallocating fuel."

Mom gave him that look that mixed pride and worry in equal parts. "We're short on everything," she said quietly. "You know that."

He knew. The numbers in his head never stopped moving anymore. How many sacks of root vegetables. How many jars of dried beans. How many barrels of potable water. How many people with fevers. How many days until they had to choose between keeping people warm and keeping food thawed.

"We'll stretch it," he said. "We always do."

She snorted softly. "That's not a plan. That's a prayer."

"Sometimes that's all we've got," he answered.

By dusk, Ridge Camp looked like a siege scene. Every spare tarp had been doubled over and nailed down. Fire pits were ringed with salvaged bricks and scrap metal to trap as much heat as possible. People moved slower than usual, conserving energy without even thinking about it. Every task took twice as long because gloves were thick and fingers were numb.

Food lines formed earlier than normal. People clutched bowls like lifelines, holding them close to their chests to catch the rising steam. Tonight's meal was thinner than last week's—stew stretched with more water, less meat, extra root vegetables boiled until they almost dissolved. They'd been cutting portions in tiny, careful increments for days. Most people suspected. No one said it out loud.

Sickness moved quietly through camp in the cold. Coughs that had been mild got sharper. A few kids developed fevers that came back as soon as the medicine wore off. Mom, Dr. Elise, and the volunteer med crew moved from tent to tent, wrapped in layers, eyes tired, hands steady. A handwritten board outside the med hut listed names and temperatures and breathing rates. Joe knew every number by heart.

He stopped there as night settled in, reading the latest updates. Elise stepped out, cheeks bright red from the cold, breath fogging around her face. She rubbed at her eyes with the back of her wrist.

"Any worse?" Joe asked.

"Not yet," she said. "I'd like a few more blankets and a miracle, but I'll settle for blankets."

"You'll have them," he said. "We're raiding the least-used storage tarps. Nobody sleeps with bare canvas tonight."

She nodded. "Good. Because if we lose power to the med hut heaters, hypothermia becomes our main enemy."

He didn't flinch. He'd already thought it. "Fuel ration schedule stays as-is," he said. "No one touches the med heaters unless you say so."

"Good," Elise repeated. She studied him for a second. "You look like hell."

"Busy day," he said.

"It's going to be a busy night," she replied. "We'll need runners if any of the kids take a turn."

"You'll have them," he said again. "Morales already assigned shifts."

He left the med area and walked toward the central fire, boots crunching on snow that hadn't been there three days ago. People clustered close, not just for warmth, but for company. Families who'd once guarded their spaces were sharing benches and hanging pots together.

A woman from Harrow's group stirred a pot while an older Savannah mechanic sat beside her, fixing a broken lantern handle. Their kids—once wary strangers—shared a blanket and argued in whispers over a scavenged deck of cards.

Exhaustion hung over everything like frost—on shoulders, on faces, in the way conversations tapered off into heavy silence. The deep freeze wasn't the dramatic disaster people imagined when they thought of the world ending. It was slower, meaner. It just kept taking. Heat. Food. Health. Sleep.

Joe stood at the edge of the firelight, taking it all in. Every crack in a roof. Every cough. Every empty space where food stores used to be higher. The weight of it pressed between his shoulder blades, made his breath feel heavier in his chest.

The deep freeze had turned Ridge Camp into a test.

And somewhere under all of it—the cold, the hunger, the fear—he knew he didn't just have to keep them alive. He had to keep them believing that "alive" meant more than just not dead.

The fires burned low that night—lower than anyone wanted, lower than anyone would ever admit. Fuel had to last until spring, and spring felt like a rumor whispered by someone who'd never lived through a Tennessee winter. Families huddled close, wrapped in blankets that smelled like smoke and wet wool. Kids sat on laps, two or three per parent, little shoulders tucked under coats to steal warmth. People talked in softer voices, the kind used in waiting rooms and disaster zones. The cold had stolen their confidence, and the food shortage had taken the rest.

Joe walked through camp again, slower this time, watching rather than fixing. Sometimes a leader needed to see the things that couldn't fit on a supply chart. The lines at dinner had been tense. People kept glancing at the stew pot like it owed them more. That subtle panic—that quiet calculation of how many meals before the bottom of the barrel—he recognized it. He'd seen it in bunkers. On ships. In flooded cities. And now in Ridge Camp.

A few families had pulled their tents together and tied the flaps open so one fire could do the work of three. Two mothers from different arrival waves—women who'd barely spoken a month ago—

were stitching blankets together to make a bigger one for the kids to share. A cluster of men took turns boiling water over a single flame, waiting for the moment it reached steaming so they could sip it like tea and pretend it was something comforting. Even Guido, who never stopped talking, was quiet tonight, sitting near the edge of the circle stacking scrap wood for the morning.

Joe crouched beside the biggest fire pit, letting the heat hit his face. It wasn't much. Not enough to reach his back. Flames snapped against the cold like they were losing the fight one spark at a time. A teenager added another handful of kindling, careful not to use more than he should. Joe nodded at him. The kid nodded back, face pale but proud.

Morale was slipping. He could feel it in the air, like camp itself was exhaling in short, scared breaths.

A low cough echoed from the med hut. A woman shushed her child in a nearby tent. A man stirred the thin stew again even though it was already as mixed as it was going to get. The night crawled with small, worried motions.

Joe stood and turned toward the south line. That's where he saw her—Ashley, sitting on an overturned bucket near a smaller fire, elbows on her knees, staring into the flames like they'd betrayed her. Manny was talking to someone a few feet away, but his eyes kept flicking toward her.

Joe walked quietly, boots crunching. Ashley didn't look up. Her breath shook once, barely perceptible, but enough for him to recognize the signs. She was frayed. So was everyone else. But she was young, still learning how to hold fear without letting it swallow her.

When Manny finally finished talking and stepped toward her, she wiped her eyes fast, too fast, the way someone does when they don't want to draw attention. Joe paused, letting the moment belong to them.

"Ash," Manny said softly, kneeling down. "You okay?"

She nodded automatically. Reflex, not truth. She looked away, jaw clenched, shoulders small. "Just cold."

Manny let out a slow breath. "Yeah. But it's not just that."

Ashley's gaze flashed to him, defensive. Then the crack formed—

subtle, but real. She looked down. "I'm tired. Everyone's tired. And sick kids… I don't know if we're enough."

Manny reached out, stopped, waited for her to move first. When she gave the tiniest nod, he touched her arm. "Hey. You're doing more than enough. You're here. That counts."

Her breath trembled. "What if winter gets worse?"

"It will," he said honestly. "But so will we."

She let out a wet laugh, small but real. Manny shifted closer, slow and careful, letting her choose the pace. She leaned in just an inch. Then another. Then finally rested her forehead against his shoulder—quick, almost embarrassed—but it was permission. Manny wrapped an arm around her, pulling her in tight but gentle.

Joe stepped back into the shadows and let them have the moment. He'd needed moments like that once himself.

Valery found him minutes later. She appeared beside him with the quiet way she always moved in winter, the cold making her steps softer. "Watching them?" she asked.

"Yep."

She followed his gaze. Ashley was tucked against Manny, chin pressed to her coat, eyes closed. Manny held her like she might break, or like he might if he let go.

Valery exhaled. "We need joy soon," she said. "This camp… it's losing its color."

Joe didn't argue. He felt it too—the way the cold was sucking the life out of everything they'd built. The way people were surviving, not living.

"We do," he said. "And I think I know what it has to be."

Valery's brows lifted. "Oh?"

Joe guided her away from the fires, just far enough that the noise softened into a distant heartbeat. Snow glittered like ground glass, and the cold wrapped around them but couldn't quite touch the warmth between them. Valery sat on a crate, resting her sling, her breath fogging in faint silver clouds.

He took one long breath. Then another. Then he crouched in front of her, not rushed, not shaky — just full of something he'd been holding for too long.

"Val..." His voice was low, steady. "Do you remember the Valentina?"

Her eyebrows lifted, surprised. "Of course I do."

"You remember us in the water? Biscayne Bay. Warm. Quiet. Before... everything." His smile tugged crooked, soft. "You said something then."

She tilted her head. "I said a lot of things."

"You said..." He huffed out a laugh, gentle and full of memory. "You said, 'When we get married.' And I shot back, 'We're getting married?' acting all shocked like an idiot."

Valery's laugh cracked the cold, bright and breathless.

"And you told me," Joe continued, "that when we got married, you wanted it somewhere beautiful. Blue water. Flowers everywhere. Something like Hawaii. And you said you wanted me in white. All white."

She covered her face, laughing harder now, cheeks pink from more than the freeze. "Oh God."

"And I said I'd look terrible in white," Joe murmured.

"You would," she teased, eyes shining.

He let that warmth settle between them. Then he reached into his pocket.

"I've been thinking about that day," he said, voice suddenly soft. "About how you weren't joking. You were dreaming. And I was too scared then to let myself dream with you."

He opened his hand, and the small ring gleamed in his palm — simple, silver, shaped by care more than perfection.

"I had Dad make this," Joe said quietly. "He melted down an emblem we salvaged from the Savannah. Something that kept us alive. Something that survived the wreck. And I wanted... I wanted our future to come from something that survived too."

Valery's breath hitched, sharp and full.

Joe shifted, lowering himself onto one knee in the snow. The cold bit through him, but he didn't feel it — not with her looking at him like that.

"Valery," he said, steady as a heartbeat, "I love you. I have loved you through storms and fire and loss. And I want whatever life we get next — whatever pieces of a world we build — to be with you."

Her eyes shimmered.

He took her good hand gently in his. "Will you marry me?"

Her exhale broke — a laugh, a sob, a breath of pure light — and she nodded before the word even formed.

"Yes," she whispered. "Joe… yes."

He slipped the ring onto her finger. It fit like it belonged there. She pulled him up, pulled him into her, pressing her forehead to his as her breath trembled and steadied against his skin.

"You look perfect," she whispered.

"In white?" he joked softly.

"In everything," she murmured.

The freeze pressed in around them, the camp silent except for the distant wind. But in that small circle of firelight and snow, the world warmed — just enough.

The cold deepened until it felt like the air itself had teeth. Ridge Camp dimmed into a labyrinth of flickering lanterns and low fires, each one guarded like it was the last warmth on Earth. The wind rolled down the ridge in sharp, knifing waves that made the tents tremble and the cooking tarps snap like warning flags. People curled closer to the flames. Kids stopped asking questions. Even the noise of the camp took on a hollow, brittle quality, like the winter had pressed a hand over everyone's mouths.

Ashley tried to help Mom distribute blankets, but her fingers were stiff, clumsy. She dropped a stack twice. Mom didn't scold her —Mom just put a warm hand on her cheek, the kind a mother gives a kid who's trying too hard. That alone almost broke Ashley. She muttered something about needing air and stepped out before her breath could betray her. But outside wasn't any better. The cold carved straight through her coat, and the darkness pressed in like it wanted something from her. The fires cast long, lonely shadows across the snow-crusted dirt, and every tent she passed held someone coughing, shivering, or crying quietly so they wouldn't wake children.

She stopped near the south fire pit, hugging her arms around herself. Her breath came too fast, too shallow. She couldn't slow it. Couldn't ground herself. Not tonight. Not after the dive. Not after the tracks they found. Not after the ration shortage meeting. Everything was stacking. Too many sounds, too many responsibilities, too many things she couldn't fix. The cold didn't help. It felt like her old life—those nights after the meteors when she'd been alone, hiding in a tool shed, listening to the world break.

She sat on the edge of a crate, elbows on her knees, staring at the fire as if looking away from it would let the dark swallow her whole. Her throat tightened painfully. She blinked hard, desperate to stop the tears. She didn't want to cry in camp. She didn't want to be the girl who kept breaking. But the weight was too much tonight. The wind howled across the ridge and hammered her last defenses flat. A soft sound escaped her—barely a breath—but it was enough. Manny, who'd been stacking scrap wood nearby, turned instantly. He knew that sound. Everyone else might've missed it under the cracking fire, but Manny didn't miss anything when it came to her.

He approached slowly, like she was a wild thing that would run if he stepped too hard. "Ash?" he murmured.

She didn't look at him. Couldn't. Her jaw was locked too tight. Her throat felt raw. She shook her head once, quick, embarrassed. "I'm fine," she whispered.

"No," Manny said gently. "You're freezing and you're shaking."

She hadn't even noticed the shaking until he pointed it out. Her hands trembled in her lap, small and sharp, like she was vibrating from inside her bones. Manny crouched in front of her, hands hovering, waiting for permission he wasn't sure she'd give. "Can I…?" he asked quietly.

The permission was tiny—just a slow inhale and the faintest tilt toward him—but Manny took it like it was sacred. He touched her wrist first, warm fingers wrapping gently around cold skin. When she didn't pull away, he moved closer, guiding her hands into his. His palms were rough, warm, steady.

"Look at me," he said softly.

It took her a moment, but she lifted her eyes. And Manny's expression—God, it almost undid her. He wasn't pitying her. He wasn't annoyed. He wasn't overwhelmed. He just looked... present. Solid. Like a wall she could lean against without it collapsing.

"You're not alone," he whispered.

Her breath shivered, escaping in a small gasp. The dam cracked. A few tears slipped down her cheeks, too fast for her to stop. She inhaled sharply and turned her head away, ashamed. "I'm sorry," she said. "I don't... I don't know what's wrong with me."

"Nothing is wrong with you," Manny said. He moved to sit beside her on the crate, shoulders touching. Heat radiated from him instantly. "You've been carrying too much. You're tired. You're cold. And you're allowed to fall apart for a minute."

A choked laugh escaped her. "I don't fall apart."

"You do," he said softly. "And you get back up. That's what matters."

She pressed her face into her hands. Manny slipped an arm around her waist—slowly, giving her every chance to pull away. She didn't. She leaned into him, head resting against his shoulder. Manny let out a breath he'd been holding for hours and pulled her closer, his cheek brushing the top of her hair.

"Just breathe with me," he murmured.

Her breaths were jagged, uneven. Manny took slow ones—deep, steady—letting her follow the rhythm. Their chests rose and fell together, gradually syncing. She clung to his jacket with cold fingers. Her shaking eased.

A long moment passed in soft, shared silence.

"You're okay," Manny whispered. "I've got you."

Ashley didn't thank him. Didn't apologize. She just breathed against him until the world stopped spinning.

The wind eased for a moment, just long enough for Joe to hear the small sounds of camp settling for the night—canvas shifting, boots thudding softly, a child murmuring in sleep. People drifted toward their shelters, carrying what little warmth they'd gathered from the fires. The ridge felt worn down, but not broken. Not tonight.

Joe drew a slow breath and let it out, watching it curl into the cold. They were bruised, hungry, exhausted—but they were here. Still standing. Still trying. The freeze would take another run at them come morning, but tomorrow wasn't just another day of survival. Tomorrow held something good. Something bright. Something their world hadn't seen in far too long.

And for the first time in weeks, the night didn't feel like an ending.

It felt like the pause before something beautiful.

CHAPTER 37
COMMANDER HOLT, THE LOST WING

Commander Nathaniel Holt had been living on borrowed time for six months. Ever since the joint NASA–UN broadcast confirmed the Rain of Fire would strike the Earth, he'd known this day was coming. He'd sat in secure rooms while officials tried to sugarcoat the numbers. He'd watched classified models spill red across continents. He'd read the projections showing how the atmosphere itself might ignite if the wrong meteor mass hit at the wrong angle. None of it surprised him anymore. But the alarm that ripped through Arnold Air Force Base at 0420 that morning wasn't the usual test or readiness drill. This alarm had a pitch he'd only heard in simulations marked NO SURVIVORS EXPECTED.

He was on his feet before consciousness fully returned. His boots hit concrete, his hand grabbed his jacket, and he stepped into the hallway just as Captain Reeves rounded the corner. She'd clearly run straight from her barracks room. Her face was pale, her ponytail uneven, and she clutched her tablet like it held the last truth on Earth.

"It's time," she said. No preamble. No greeting. Just the truth between two people who had been preparing for the same nightmare.

"Show me," Holt said.

They walked fast through the corridor toward the command center. Reeves pulled up the live satellite feed, her thumb shaking once before she forced it steady. The screen was a blur of color-coded heat maps and atmospheric readings. Red bands were beginning to bloom over the eastern seaboard—not fires yet, but precursors. The first ignition zones.

"NASA confirmed the first atmospheric ruptures an hour ago," Reeves said. "The models flipped from projected to active. The Rain of Fire is officially underway."

Holt didn't answer. He didn't blink. He had known this moment was coming ever since the initial six-month warning. What he felt now wasn't surprise—it was the heavy settling of inevitability.

They entered the command center. It was no longer the calm, humming nerve hub he'd known for years. Officers rushed across the floor. Techs barked updates. Alarms pulsed. The air smelled faintly of hot electronics and fear.

Lieutenant Avery turned toward Holt. "Sir, ignition indicators rising across the East Coast. New York, D.C., and Baltimore grids are fluctuating. Thermal spikes detected. The first debris impacts likely began offshore."

"What about the Midwest corridor?" Holt asked.

"It's holding. For now."

Holt stepped to the main display. The United States glowed with pockets of shifting color. Behind the hot zones were projections of future atmospheric destabilization—swaths of red that would sweep inland over the next hours and days. The Rain of Fire wasn't a meteor strike anymore. It was a rolling atmospheric weapon.

Reeves stood beside him. "Sir, this matches the worst-case models they briefed us on. The ones they said probably wouldn't happen."

"They lied," Holt said. "Or they hoped."

Reeves swallowed. "Charlie-Nine?"

He gave a short nod. "Activate it."

Charlie-Nine. The emergency survivor protocol they'd been preparing quietly for months. It wasn't designed for the military. It was designed for the civilians who would come after—the ones who might crawl out of bunkers or climb over ruins searching for anything that hadn't burned. Under Charlie-Nine, Holt's base became a last-chance lifeboat. They'd hidden caches in hardened storage wings and underground annexes, filled with rations, medical kits, solar arrays, water purification units, and fuel drums. They had pre-recorded scripts, pre-staged supplies, pre-wired automated loops. All they needed today was the final trigger.

Reeves hurried away shouting orders. Avery shouted coordinates to the drone teams. Holt sank into the command chair and grabbed the edge of the console as another tremor gave the building the faintest rock—distant, but real.

"Status of our bunkers?" Holt asked.

Avery checked his panel. "Prepped and powered. Civilians can shelter for months on scrubbers."

"And our evac route?"

"Open. The northern ascent road is clear. Weather stable."

Holt stood again. "Begin full lockdown sequence. I want caches sealed within the hour."

Outside the command center windows, dawn crawled across the sky. But the sunrise looked wrong—orange where it should have been gold, darker at the edges like someone had set fire to the horizon.

Reeves returned, breathing hard. "We're at thirty percent completion on cache sealing. Teams are moving fast."

Holt didn't slow. "Good. Prepare the recording room. The beacon goes live today."

Reeves nodded. "You always knew this day would come."

"And we prepared for it," Holt said. "Now we execute."

He turned toward the corridor as the first faint boom rolled through the air—far away, but unmistakable.

Impact.

The beginning of the end had arrived.

The boom faded into a low, grinding rumble that seemed to ride the bones of the earth. Conversations in the command center faltered for a heartbeat, then surged louder. Everyone knew what that sound meant, even if they'd never heard it before. Somewhere east, a piece of the sky had hit the ground.

Avery looked up from his console. "Seismic network confirms impact off the Atlantic seaboard," he said. "Secondary tremors inland. Tsunami models are… bad."

Holt didn't ask for numbers. Numbers wouldn't change anything.

"Focus on our sector," he said. "We're inland, we're elevated, and we're sitting on top of one of the most stable logistics hubs left in the region. That means we don't get to panic."

Reeves tapped her tablet, eyes flicking across status bars. "Caches one and two sealed. Three and four at eighty percent. Medical annex is loading final pallets now. Fuel depot ninety percent. We won't get every drum locked, but we'll get most of them."

"Seal what you can," Holt said. "Anything that's not locked down by the time the ignition band reaches us is forfeited."

Reeves paused. "Forfeited to who?"

"Whoever's luckier than we are," Holt said. "Or nobody at all."

He stepped away from the main screen long enough to glance at the smaller monitor showing national status. Most of the country was a patchwork of flickering power grids and communication outages. The rolling news feed at the bottom showed a familiar face—Lester Holt, voice calm as he narrated the end of the world to anyone with a functioning screen.

Six months earlier, Nathaniel Holt had sat in a secure conference room in D.C. while that same broadcast went wide to the public. He hadn't watched it on a living room TV. He'd seen the unedited version. The one that didn't trim casualty curves or soften language. He'd listened as NASA and international agencies laid out impact trajectories and best-guess projections—tsunamis chewing coastlines, firestorms turning cities into furnaces, "atmospheric disturbance" standing in for the parts they couldn't model cleanly.

He remembered the second briefing even more clearly: the one branded under the quiet code name Order Phoenix. Nuclear drawdown. Nationwide cold shutdown. The screen had filled with four simple lines:

COLUMBIA GENERATING STATION – WA

INDIAN POINT FACILITY – NY

CALLOWAY NUCLEAR PLANT – MO

TURKEY POINT NUCLEAR GENERATING STATION – FL

Those were the last four beating hearts of the American grid. Skeleton crews. Lifeline status. Direct federal control. He'd read the language that never made it to the evening news—the phrases like "probable long-term grid loss," "regional government collapse," "continuity-of-survival in lieu of continuity-of-government."

He'd understood, immediately, what that meant.

Now the models had rolled from theory to reality.

"Sir," Avery said, dragging Holt back into the present. "We're receiving a secure flash from Strategic Command. Priority one."

"Put it through," Holt said.

The audio feed cracked and hissed, the long-distance line already degraded. A clipped voice came on—flat, official, the kind of tone that belonged to a man reading humanity's last to-do list.

"Station Arnold, this is StratCom. Be advised: Rain of Fire impact phase has entered full onset. Eastern coalition forces are in disarray. Western grids compromised. Continuity-of-government protocols are suspended. I repeat, suspended. All strategic nodes shift to continuity-of-survival posture. Surviving bases will transition to autonomous operation within twenty-four hours."

Holt's jaw tightened. "This is Arnold Actual. Confirm Charlie-Nine authority."

"Confirmed," the voice replied. "You are authorized to deploy survivor caches and automated emergency beacons per pre-approved schema. Evacuation orders: all mobile personnel to consolidate at predesignated mountain shelter complexes. Cheyenne priority one. Raven Rock, Mount Weather, and related facilities secondary. Use best judgment. Communication beyond this transmission not guaranteed."

"Understood," Holt said. "Any word on the lifeline reactors?"

"Last confirmed: Columbia at sixty percent, Calloway at fifty, Indian Point and Turkey Point operating under skeleton crews. Radiation levels nominal. But impact modeling suggests all coastal and low-elevation facilities are at extreme risk. Expect loss of contact."

The line crackled. Someone on the other end muttered something that didn't make it into the transcript. Then: "Final directive, Commander. Leave something behind. The rest is up to whoever outlives us."

The transmission cut.

The room went quiet except for the hum of machines.

Reeves blew out a breath. "They're not even pretending anymore."

"They haven't been for a while," Holt said. "We're just catching up."

He turned to her. "You heard the order. We're officially off the leash. Once we roll north, there's no more backstop. No more higher authority. It'll just be what we carry and what we can build."

Reeves nodded once. "Then we make sure we don't roll empty."

A klaxon blared near the far wall. Sergeant Malik called from the perimeter station. "Commander, this is North Gate. We've got incoming vehicles. Civilian. Dozens of them. Maybe more behind."

"Of course we do," Holt muttered. Word traveled, even when networks died. People would have seen the news, the sky, the early impacts. They would remember every film where the base meant safety.

He hit the comm. "Malik, hold them at the outer line. No weapons drawn unless somebody makes you. We're starting triage."

"Yes, sir."

Reeves glanced at him. "We always knew they'd come."

"Yeah," Holt said. "The models didn't show faces. That part's on us."

He straightened his jacket and squared his shoulders. "You have the caches. Avery has comms. I'll handle the gate."

Reeves frowned. "You going to tell them the truth?"

"I'm going to tell them what they can use," Holt said. "Truth's too big for today."

He left the command center and moved down the hallway toward the open air. The base vibrated faintly under his boots, a distant echo of impacts he couldn't see. As he stepped outside, the sky over the eastern horizon glowed a sick, pulsing orange. Smoke streaked the clouds. Ash already dusted the wind.

Holt walked toward the gathering crowd at the gate, knowing exactly what he could give and exactly what he could not.

The crowd outside the North Gate stretched down the access road —cars pulled halfway into ditches, trucks left running with their doors open, families clustered together under a sky that looked sick at the edges. People pressed forward as Holt approached, hope and terror mixing into something volatile. A mother held a toddler so tight the child's face was buried in her shoulder. A man in a business suit clutched a duffel bag against his chest like it held his life inside. An elderly couple stood arm in arm, their faces set with resigned determination. They all watched Holt climb onto the concrete barrier.

He didn't shout for attention. He didn't need to. Silence rolled outward as people realized the man standing above them wore a commander's insignia.

"I'm Commander Nathaniel Holt," he said, his voice carrying without strain. "This base is under emergency operations. We will shelter who we can. But we cannot take everyone."

A ripple went through the crowd—fear, anger, desperation—but they stayed quiet. That alone impressed him.

A woman near the front raised her hand. "Please," she said, her voice cracking. "My daughter can't breathe. The air—she started coughing hours ago."

Holt signaled to the medics. "Children, medical emergencies, mobility-impaired—priority one. Bring them forward." Then he addressed the crowd again. "We have two bunkers prepared with air scrubbers capable of sustaining limited populations. Those spaces will go to the most vulnerable first."

A man stepped forward, face pale. "Is it true? The Rain of Fire… it's already falling?"

Holt didn't soften it. "Yes. Impacts have begun on the East Coast. Atmospheric ignition is spreading. We have hours—maybe less—before Tennessee feels the effects."

People murmured, clutching their families tighter. But they didn't break. Not yet.

"Listen to me," Holt continued. "We have prepared for this. This base has spent months loading emergency caches: food, water, medical supplies, fuel, communication gear. We don't have space to shelter everyone here—but we are leaving resources behind for those who continue north on foot, by car, however you can. You will not be empty-handed."

"Where do we go?" someone shouted. "What direction?"

Holt shook his head. "Not here. Not now. If I give you a single destination, you'll trample each other to get there. What you need is simple: get to higher ground, stay mobile, and follow the caches we've placed north of this location. You'll get maps. You'll get instructions."

A father stepped forward, his voice trembling. "Can you protect us? Army? Air Force? Anyone coming?"

"No reinforcements," Holt said. The truth landed like a brick. "No outside cavalry. The United States military is fractured, and command networks are failing. What we can do is buy you time. Supplies. Clean air inside the bunkers for the vulnerable. After that, survival will depend on movement."

The crowd didn't scream or riot. They sagged inward, absorbing the blow like a single wounded organism.

Reeves jogged up beside him. "Sir, priority list is ready. Medics are moving twenty-four civilians to Bunker One. Thirty more undergoing assessment. We'll hit limits fast."

"Acceptable," Holt said. "Move them."

A small girl tugged her mother's sleeve and pointed at Holt. "Is he the boss?" she whispered. The mother tried to hush her, but Holt crouched slightly so she could see him.

"For today," he said gently, "I am. And I'm doing everything I can."

She nodded, eyes too old for her age, then buried her face in her mother's coat.

Holt stood again and surveyed the line. "You will get instructions. You will get coordinates. You will get supplies. But some of you will not get shelter here. That is reality. We're preparing you as best as we can."

Someone at the back shouted angrily, but the crowd hushed him. Even fear had its limits.

Holt nodded once—respect, acknowledgment—and stepped down from the barrier.

Reeves walked with him toward the gatehouse. "Sir, I've got soldiers asking for clarity on evac timeline. If the ignition band reaches the valley, they want to know when we roll."

"When the last bunker seals," Holt said. "And not before."

"Understood."

They reached the gatehouse as Sergeant Malik jogged over, out of breath, sweat streaking soot across his forehead. "Sir, we've got heat signatures rising just east of the county line. It's climbing faster than projections."

"Atmospheric ignition?" Holt asked.

"Yes, sir."

He looked toward the horizon. The glow had deepened into a violent orange streak across the clouds, pulsing with every fresh ignition ripple. Distant booms rolled through the air like a heartbeat in the sky.

Reeves lowered her voice. "If we don't move soon, we'll be trapped between the ignition front and the lower valley."

"We'll move," Holt said. "But not empty."

They started back toward the command center when the first ash flakes drifted down—gray specks swirling in the hot wind. Not wood ash. Not smoke. The kind of particulate matter that meant the atmosphere itself had begun to burn.

Avery's voice crackled over Holt's comm. "Commander, caches three, four, and five sealed. Medical annex locked. Fuel depot at ninety-five percent."

"Seal it," Holt said. "Lock everything."

"Yes, sir."

Reeves exhaled. "That's it, then. Charlie-Nine is live."

"No," Holt said softly. "It's the beginning of it."

He glanced once more toward the civilians, the bunkers, the horizon, the sky burning in slow motion. Then he turned toward the base interior with grim purpose.

"Prep the recording room," he said. "It's time to light the beacon."

Holt moved quickly across the base, the heat in the air rising with every minute. The sky above Arnold AFB had shifted into a bruised palette of orange and black, colors twisting together like the atmosphere couldn't remember how to behave. The distant booms came faster now—irregular, overlapping, like artillery fired by a drunk god. He didn't let himself look long. He already knew what it meant: more impacts, more ignition bursts, more heat rolling toward them.

Reeves kept pace beside him. "Sir, bunker capacity will be reached in ten minutes. After that, we have no more sealed haven space."

"Lock them as soon as they're full," Holt said. "If the ignition band hits while those doors are open, everyone inside dies."

Reeves nodded tightly. "Yes, sir."

They entered the recording wing, a narrow corridor reinforced with thick steel ribs. Two techs were waiting by the door to the broadcast booth, both pale, both trying to act like they weren't. Holt pushed the door open and stepped inside.

The booth swallowed sound. The walls were padded, the air still, the red standby light blinking with mechanical patience. He took his position before the microphone.

Reeves stood behind the soundproof glass, tablet ready. She lifted a hand. "Commander, channel is clear. Begin on your mark."

Holt lowered his head for half a second. Not in prayer. In memory. He saw the first meeting six months earlier, when the words *Rain of Fire* were spoken out loud in that secure DC room. He remembered the silent horror of knowing the numbers were real. He remembered returning to Arnold AFB with orders to prepare quietly, without sparking panic among the junior ranks. And he remembered looking at the faces of his airmen, young and eager, some barely twenty, and thinking: *I will be the one who decides who lives and who dies.*

He opened his eyes.

He pressed the button.

"This is Commander Nathaniel Holt, United States Air Force, Station Arnold, Tennessee," he began. His voice filled the silent booth. Steady, grounded, unshakable. "Emergency Protocol Charlie-Nine is now active. If you are receiving this broadcast, you have survived the initial Rain of Fire."

He paused, ensuring clarity.

"Emergency supply sectors one through five contain food, water, medical equipment, and fuel reserves. These caches are available to any survivors who reach this location. This is not a rescue beacon. It is a survivor beacon."

Reeves' face behind the glass softened. Even now, Holt kept his tone calm, never raising his voice, never letting the fear outside bleed through.

"Use what you need," Holt continued. "Take what keeps you alive.

These supplies exist because we planned for this moment. You are not alone. You are not abandoned. This message will continue until power loss or structural collapse."

He stopped the recording.

Reeves pressed her earpiece. "Copy. Audio clean. Running it on loop."

Holt stepped out of the booth, the door clicking shut behind him. Outside in the corridor, the hum of the base hit him again—boots pounding, doors slamming, alarms flickering.

Avery appeared at a run. "Sir, the ignition front is accelerating. The thermal wall's hitting the outskirts of Coffee County now. We've got readings that are—" He hesitated, then said it plainly. "Sir, it's burning the air."

Holt didn't waste time. "Evac sequence?"

"Carrier One loaded and ready. Two at eighty percent. Three at thirty."

"Load three faster. We roll in staggered formation, not one convoy. Too easy to lose everything at once."

Reeves fell into step. "Sir, civilians outside are demanding more seats. We have nothing left to give."

"They get the truth," Holt said. "And the caches. That's all we can do."

They exited the building and were immediately greeted by a blast of hot wind that carried flakes of gray ash. Soldiers worked frantically to secure gear onto the carriers—reinforced trucks designed to handle heat, debris, and rough terrain. Carrier One rumbled with life, the engine loud against the unnatural stillness of the surrounding hills.

Medics guided the last priority civilians toward the bunkers. A woman clutching a newborn. A teenager with severe asthma. An elderly man leaning on his son as they descended the ramp. The bunker doors hissed as they cycled shut behind them.

Reeves checked her watch. "Bunker One sealed. Two sealing now."

The ground vibrated—stronger this time. A deep, rolling shudder that passed beneath their feet like something massive dragging its belly across the Earth. Carrier Two jolted in place.

Malik called from the turret mount. "Commander! Ignition band visible!"

Holt turned.

And across the eastern ridge, he saw it: a towering wave of fire-laced atmospheric discharge, churning like a living storm. It rolled over the far hills, devouring the valley below in a curtain of glowing particles. Trees flashed to ash. Houses ignited without flame. The sky boiled.

Reeves whispered, "My God…"

"No," Holt said. "This is science, not God. And science told us to move."

He looked at her, eyes steady. "Launch Carrier One now."

Reeves ran to relay the order. Carrier One surged forward, the heavy vehicle tearing across the tarmac toward the access road.

Avery shouted, "Carrier Two ready!"

"Launch it."

The second carrier followed, hauling half the engineering corps with it.

Holt turned to Carrier Three, the last evac unit—the one he would command personally. The ignition front was rising behind them like a tidal wave.

Reeves returned, breathless. "Sir—we're out of minutes."

Holt nodded once. "Then we leave now."

As he climbed into Carrier Three, the sky behind Arnold Air Force Base turned fully red.

And the Rain of Fire came for Tennessee.

Carrier Three lurched forward just as the ignition front rolled over the far ridge. The entire eastern sky convulsed with fire-bright haze, swallowing the valley below in a churning wall of heat and glowing particulate. Holt gripped the overhead rail as the carrier accelerated, tires grinding across cracked pavement that buckled from distant shockwaves. Behind them, the lower edge of Arnold Air Force Base lit up in flickering pulses—hangars shimmering, runway concrete glowing faintly, fences rattling like they were being shaken by an unseen hand.

"Keep her steady!" Holt ordered.

"Yes, sir!" the driver shouted, knuckles white on the wheel.

Reeves braced herself at the radio console. "Carrier One reports they've cleared the east access road and are climbing toward the plateau edge. Carrier Two's engine temp is high but stable."

"Tell them not to slow down," Holt said. "If that ignition front catches us on the flat, we're done."

Malik shouted from the rear hatch window, "Sir! The base is starting to ignite around the perimeter! Storage lots, vehicle bays—everything's lighting up!"

Holt didn't turn around. He could feel it in the air—the heat rising, the wind thinning, the way light flickered unnaturally across the road. He kept his eyes on the incline ahead. Elevation was survival.

The carrier hit the rise hard, bouncing as it climbed the sloping road that cut behind the hangars. Trees on the lower slope curled black in seconds. The grass along the drainage ditch combusted into drifting ash. But the higher they climbed, the more the heat spread out, diffusing just enough to give them a fighting chance.

Reeves exhaled sharply as the temperature alarms dropped from critical red to deep orange. "We're outrunning the worst of it."

"Good," Holt said. "Get us to the hangar ridge. We regroup there before the final push out."

Carrier Three climbed one last sharp incline and leveled out behind the upper maintenance hangars—Hangar 2 and Hangar 3. Carrier One and Two were already parked in a staggered line, engines running, soldiers spilling out to gather equipment. Heat shimmered through the air, but the ignition front had slowed as it reached the higher elevation. The base below glowed like a dying furnace—bright, violent, but not rising fast enough to swallow the upper ridge line.

"Ground here is stable," Reeves said, checking thermometers. "Hot, but survivable."

"Unload quick," Holt ordered. "Final checks. We're leaving in five."

Soldiers scrambled, transferring fuel cans, weapons crates, med kits, and remaining gear from the hangars into the carriers. Holt jogged toward Hangar 2, where a line of metal storage boxes still sat under flickering emergency lights. He swept his flashlight over the stacked pallets—MREs, water drums, seeds, tools—all sealed tight, built to withstand heat and time.

"These stay," Holt said quietly. "All of it."

Reeves joined him. "Sir, should we destroy anything to prevent looting?"

"No. If we don't make it… someone else will need this."

He stepped toward a workbench near the hangar wall—the same bench Joe would someday find. Papers lay scattered from earlier inventory work. Holt pulled a blank mission log sheet toward him, the ink already fading at the edges from rising heat, and began to write. His hand moved fast but precise, the way it did on after-action reports. He dated it and labeled the top in block letters: AFB ARNOLD – CHARLIE-NINE SURVIVOR BRIEF. In short, clipped lines he laid out what any future reader would need to know: Rain of Fire onset; the StratCom order shifting from continuity-of-government to continuity-of-survival; confirmation that emergency supply sectors one through five were sealed inside Hangars 2 and 3; the fact that additional caches had been staged farther north along the overland route; and that mobile personnel from Arnold had been ordered to regroup on the high, dry ground of the Cumberland Plateau rather than the low-lying federal complexes at Cheyenne Mountain, Raven Rock, and Mount Weather. At the bottom, he signed his name and added one final line, underlined once: KEEP MOVING NORTH.

He slid the pages into a bright yellow envelope, wrote FOR SURVIVORS on the front, and left it in the open center of the desk, plain as day, meaning for whoever outlived him to find it. Then Malik called out, "Sir! Ignition front rising again!"

"Fall back to the carriers!" Holt shouted.

After Holt walked out, a harried airman rushed through the workspace and dropped a thick maintenance file onto the corner of the bench without looking. The binder slid across the surface and came to rest directly on top of the envelope, hiding it completely. When Holt left the hangar for the last time, he never knew the message he'd written had been buried under paperwork, waiting for someone else's hand to pull it free.

Outside, the heat intensified, but it wasn't the same killing wave now rolling through the valley. Up here, the firestorm had thinned, slowed, barked but couldn't bite through the stone and elevation. It would scorch, crack, and ruin—but not annihilate.

Reeves climbed into Carrier Three. "Convoy ready!"

Holt took one final look at Arnold Air Force Base. The runways glowed like molten scars. The lower barracks smoldered. Ash drifted like sick snowfall. But the upper hangars—Hangar 2 and Hangar 3—still stood. Their thick steel frames held. Their pallet stacks remained untouched. Their sealed supplies waited for whomever would one day come.

"Move!" Holt ordered.

Carrier One roared forward. Carrier Two followed. Holt climbed into Carrier Three and slammed the hatch shut as the ignition front licked the hangar roofs below.

The convoy surged up the final incline toward the northern ridge. And beyond that ridge lay the endless forests and shadowed caverns of the Cumberland Plateau—high, dry, and hidden. A place to rebuild. A place to gather what remained of the U.S. military. A place Joe and his people would someday search for.

Reeves steadied herself beside him. "Do you think anyone will find what we left?"

Holt watched the hangars recede, glowing but strong. "Someone will," he said. "And when they do—they'll have a chance."

The convoy disappeared into the mountains, leaving Arnold Air Force Base scorched, silent, and waiting.

CHAPTER 38
THE WEDDING FIRE

The freeze didn't end. It eased, just enough to let people move without feeling like their bones would crack. The air was still sharp, breath still burned, but the wind dropped from knife to steady slap. It was enough. Ridge Camp woke different that morning—not to panic or orders, but to murmurs and nervous, almost-giddy whispers.

Word moved fast: there was going to be a wedding.

Joe started the day the way he always did—walking the ridge with his slate. Checking fuel. Checking food. Checking the med hut. The numbers still hurt. They'd hurt for a long time. But as he moved through camp, people kept catching his sleeve or nodding at him with something new in their faces.

Hope. Thin. Fragile. But real.

By the main fire pit, Morales and two others were rebuilding the ring of stones, rearranging them like they were setting the stage for something sacred. Donnie was there too, hammering a metal bracket into place to support an extra spit. He pretended not to see Joe at first, then glanced up with a smirk.

"You're supposed to look nervous," Donnie said. "You look like you're planning a refueling op."

"This is logistics," Joe said. "Just with more feelings."

"Yeah, well, you're marrying out of your league," Donnie said. "So I'd suggest you adjust the plan accordingly."

Joe snorted and moved on before his chest did something embarrassing.

On the south side of camp, Mom had conscripted half the available women—and a few willing men—into Decorations Command. Strips of old shirts and blankets were being cut, braided, and tied into garlands. Someone had dug out a jar of dried wildflowers pressed between scavenged books last summer. Those went into glass jars and chipped mugs, set on crate tops like centerpieces for a feast they barely had food for.

"Careful with that," Mom told a girl trying to tug a stubborn knot out of a faded ribbon. "That's not just cloth. That's atmosphere."

The girl giggled. The sound carried.

A cluster of kids followed Mom around, holding up bits of string and fabric. "Can we put some by the med hut?" one boy asked. "So the sick kids can see?"

Mom's face softened. "Yeah," she said. "Yeah, that's a good idea. We'll make sure they get the best ones."

Near the laundry lines, two of the older women from Harrow's group were attacking a pile of clothing like it had insulted them personally. Needles flashed. Thread looped. Jackets got new buttons. Pants got patched. Someone even turned an old curtain into a rough but clean overskirt for Valery.

"Weddings need fuss," one of the women said when Joe passed. "If we can't give fuss, what's the point of surviving?"

Valery herself stood near the entrance to their tent, letting a young girl adjust the ribbon someone had sewn into her collar. Her sling was freshly wrapped. Someone had taken the time to stitch a narrow, dark strip of cloth along its edge—simple, neat, almost elegant.

"You look…" Joe started, then stalled.

"Like I'm about to trip over my own feet?" Valery offered.

"Like I'm out of my league," he said instead.

Her smile flickered, then steadied. "Good. I'd hate to be the only one nervous."

For a few hours, Ridge Camp felt like it belonged to another world. People still worked. Still checked lines. Still hauled wood. But over everything else, there was that hum—the low, excited murmur of a community preparing for something that wasn't dire, or deadly, or necessary to survival.

Something good.

Joe stepped back toward the edge of camp for a moment, letting the noise blur. He watched lanterns being polished, scraps of fabric going up, kids arguing over who got to carry what. The cold was still there. The hunger was still there. The fear was still there.

But so was laughter.

It had been a long time since those two things shared the same space.

For the first time in days, his shoulders eased.

Today wasn't about charts or rations or structural integrity.

Today was about giving his people—his family—one thing winter couldn't touch.

They lit the fire early.

The main pit roared higher than it had since the first weeks on the ridge, fed carefully by extra wood Morales had been hoarding "for an emergency." He'd declared this counted. The flames climbed high enough to throw light across half the camp, turning the snow to gold and the rough-built shelters into something almost beautiful.

People circled wide around the fire, leaving an open space at the front. Someone had laid down two old blankets end-to-end to make an aisle. The braided fabric garlands hung between two poles, framing the spot where Joe and Valery would stand. It was crooked. It was uneven. It was perfect.

Joe stood near the front with Donnie and Morales, trying not to fidget. His hands still smelled faintly of metal and oil from helping fix a lantern that morning. He'd scrubbed them three times. It didn't help. He wiped them on his pants again.

"Stop that," Donnie muttered. "You'll wear a hole in them before she gets here."

"She's late," Joe said.

"She's with your mother," Donnie said. "Of course she's late. That's the law."

Morales straightened the collar of Joe's coat like she was adjusting a uniform before inspection. "You look okay," Morales said. "For a man about to make the worst tactical decision of his life."

Joe gave her a look.

Morales smiled faintly. "Relax. I meant committing to one person when you could have committed to worrying about all of us forever."

"That's still happening," Joe said.

"Yeah," Morales said. "But at least now you'll have someone making sure you sleep occasionally."

The murmurs around the circle softened. A quiet fell over the crowd, the kind that wasn't fear, wasn't tension, wasn't grief. It was anticipation. The good kind.

Mom appeared first, stepping into the open space with a seriousness that almost made Joe laugh. She looked at him, and for a second he saw every version of her overlay—bunker Mom, Ridge Camp Mom, Turkey Point Mom, the Mom who'd stood in a hospital hallway with shaking hands and refused to cry. Her eyes were bright, rimmed in red, but her smile was steady.

"Ready?" she asked.

Joe swallowed. "Yeah."

"Good," she said. "Because here she comes."

Valery walked toward him, arm looped through a makeshift escort —Guido, for reasons no one had fully explained and everyone had accepted. Guido walked with exaggerated care, like he'd been given charge of a priceless artifact. Valery's coat was clean, sling neat, hair pulled back with a strip of pale cloth someone had found in a crate. The firelight caught in her eyes and on the ring already on her finger.

Joe forgot how to breathe for a second.

She met his eyes and didn't look away, not once. Not at the fire. Not at the crowd. Not at the sky. Just him.

When she reached him, Guido released her arm with an overly dramatic bow that made a few people snort. He stepped aside, wiping at his face quickly when he thought no one was looking.

Mom took a breath and lifted her hands slightly, not quite in a formal gesture, but close. "We don't have a book for this," she said, voice carrying clear over the fire. "No official words. No official titles.

We're short on everything except people who've decided to keep living anyway."

A few quiet laughs, a few sniffles.

"But we do have this," she went on. "Two people who have already walked through hell together and still chose each other at the end of it. That's enough."

She nodded at Joe. "Say what you need to say."

Joe pulled a folded scrap of paper from his pocket. His hands shook slightly. He could have spoken without it—he'd rewritten the words in his head fifty times—but he needed something to hold onto.

"Val," he said, and the rest found him. "I don't have much to give you. The world took most of my good options."

A ripple of rueful agreement moved through the crowd.

"But I have this," he said. "I have every day we get from here on out. I have every hard decision. Every stupid argument. Every quiet morning. Every night we're too tired to talk but not too tired to lie next to each other and breathe the same air. I have my hands, my stubborn brain, and whatever passes for a future in this mess."

He swallowed once. "And I want all of it to be with you."

Valery's eyes were wet. She didn't look away.

He crumpled the paper in one fist, done with it. "I love you," he said simply. "In Miami. On the Savannah. In the bunker. On this ridge. In every version of the world we've lost and every one we haven't found yet. I love you when I'm right and when I'm an idiot. I'm not promising to be easy. I'm promising to stay."

He hadn't meant to say that last part, but it landed truer than anything else.

Mom nodded to Valery. "Your turn."

Valery exhaled slowly, a small puff of white. "I didn't write anything down," she said. "Joe stole all the paper this week."

A light chuckle floated through the crowd. Joe ducked his head.

"I don't have fancy vows," she went on. "What I have is this: before the world ended, I thought I knew what love was. It was beaches and dinners and jokes about white clothes and Hawaii. It was future tense. Someday. When."

She shook her head slightly. "That was the draft version. This is the real thing. Real love is you holding a wrench in a flooded hallway, refusing to leave until everyone else is out. It's your voice on the radio when I thought we'd lost you. It's your hand on my shoulder when I wake up at two in the morning sure the roof is going to cave in. It's you letting me see you scared and going anyway."

Her throat tightened. She kept going. "I don't know how long we get. None of us do. But however long it is, I want to spend it being your problem."

A few actual laughs now, wet and heartfelt.

"I love you," she finished. "I'm not promising to be easy either. I'm promising to be here. That's all I have. That's everything I have."

Mom cleared her throat, and even she sounded a little broken around the edges. "Good enough for me," she said. "Rings are already on, vows are said, witnesses are freezing. By the power vested in me by absolutely nobody but common sense and the fact that we all agreed to this, I pronounce you married."

She smiled, sudden and bright. "Now kiss her so we can stop pretending we're not crying."

Joe didn't hesitate. He stepped forward, cupped Valery's face gently, and kissed her. It wasn't dramatic. It wasn't staged. It was slow, sure, steady, full of everything they'd already lived through and everything they were choosing anyway.

The camp erupted—cheers, whistles, clapping, a few loud sobs that no one tried very hard to hide. For one blazing, impossible moment, the ridge felt like the safest place on Earth.

They didn't have enough for a real feast.

They did it anyway.

Tables made from salvaged doors and crates were dragged close to the main fire. Metal pans and mismatched bowls lined the surfaces. The smell of food—thin but real—drifted through camp and pulled people in like gravity.

Mom and the kitchen crew had orchestrated it like a military op. Every family had been asked, quietly, to bring one thing from their stores if they could spare it. A handful of beans. A half jar of pickled something. A broken chocolate bar held back for "someday." A heel of stale bread. A clutch of withered apples that somehow hadn't rotted.

Now it all sat together, transformed.

A pot of stew rich with more vegetables than usual, and even a little meat. A tray of root slices roasted until the edges browned. A precious jar of summer berry preserves opened for the first time, dark and glossy in the firelight. Someone had even managed to make flatbread from hoarded flour, cooked on a salvaged griddle until it blistered.

"It's not much," Mom said as people lined up. "So eat slow and pretend you're full."

The line moved, but this time there was no tight, nervous tension, no eyes calculating the depth of the pot. People smiled as they handed plates down the line, adding a little extra to a kid's bowl, trading a piece of bread for a spoonful of preserves.

Joe moved through the crowd with Valery, accepting congratulations and slaps on the back. Someone handed him a plate loaded heavier than he'd allow on a normal night. He opened his mouth to argue—then shut it again.

Tonight, he could let that go.

At one end of the table, a man from Harrow's original crew cut the last remaining summer sausage into thin coins with careful precision. At the other, Elise sat with two kids from the med hut, their bowls balanced on their knees. They laughed at something she said, and the cough that followed didn't seem as sharp.

A few sailors from the Savannah's skeleton crew had unearthed an ancient guitar from storage. It had five strings and a crack along one side, but Dave tuned it with the concentration of a man defusing a bomb. When he finally strummed a chord that didn't buzz, a small cheer went up.

"Don't expect much," Dave warned.

"We never do," Morales yelled back. "Play anyway."

The music that followed was rough, simple, and desperately needed. Old campfire songs. Bits of half-remembered radio hits. Choruses people mangled together but sang loud anyway. Kids clapped along, off-beat but enthusiastic. Someone tried to harmonize and failed miserably. That made it better.

Laughter rose—not the brittle, strained kind they'd heard in recent weeks, but something looser. Genuine. It rolled through the crowd, cutting the frost like a new kind of fire.

Joe watched as two families who'd barely spoken since arrival pulled their benches closer together to share a blanket. A man from the ridge's original residents handed a slice of bread to a teen from Harrow's group without any visible hesitation. A toddler toddled between tables, collecting head pats and crumbs like treasure.

"This feels wrong," Joe said quietly to Valery as they stood off to the side, plates in hand.

"In what way?" she asked, eyebrow raised.

"Like we're getting away with something," he said. "Like the universe is going to notice and even the score."

Valery took a slow bite of stew, savoring it. "The universe can file a complaint," she said. "We're busy tonight."

He huffed a laugh despite himself.

Across the fire, Mom caught his eye. For a heartbeat, he saw her as she'd looked in the bunker, counting rations with that pinched, tight expression. But now her shoulders were looser. Her face was still lined with worry, but tonight there was something else there too.

Pride. Relief. A little joy.

She lifted her cup toward him in a small toast. He did the same.

"Look at them," Valery whispered.

He did.

Ridge Camp—this patched-together, half-frozen, exhausted collection of souls—was laughing. Eating. Singing badly. Kids were running in circles near the fire. Someone told a joke that made three people wheeze with laughter. Even Guido looked content, leaning back on a crate, mouth full, eyes half closed.

For one night, survival wasn't the only thing on the menu.

Life was.

• • •

The music shifted into something almost slow, more by accident than design. Dave's fingers slipped into a softer pattern, and the voices around the fire dropped to a hum. The flames snapped high, sparks flying into the dark like tiny fireworks.

People started to sway. Not a formal dance—just couples and friends moving in place to chase the chill from their bones.

Manny stood near the edge of the light, hands jammed into his pockets, watching everyone else move. He wasn't dressed any different from usual—same jacket, same boots—but he'd actually run a comb through his hair, which in Ridge Camp counted as formal wear.

Ashley sat on a low crate beside the fire, plate empty, hands wrapped around a chipped mug of something hot. Her shoulders were relaxed, for once. The worst of the hollow look was gone from her eyes. She watched the dancing with a mix of curiosity and uncertainty, as if it were a language she understood but had never spoken.

Manny took a breath that felt like it started somewhere around his boots. Then he walked over before he could talk himself out of it.

"Hey," he said.

Ashley looked up. "Hey."

He shifted his weight, searching for words. "So, um. There's… music."

"Yeah," she said, glancing at the guitar. "That's usually how that works."

Right. Good. Off to a strong start.

He cleared his throat. "I was thinking—since it's a wedding and all, and people are, you know… moving to the noise…"

"Dancing," she supplied, lips twitching.

"Right," he said. "Dancing. I was wondering if maybe you'd want to… do that. With me."

She stared at him, mug halfway to her mouth. The firelight made her eyes look darker, deeper. For a second, she looked almost offended.

"No," she said.

The word landed like a dropped stone.

Manny's face didn't change much, but something behind his eyes shuttered. "Right," he said quickly. "Yeah. No problem. Just figured I'd ask. I'll, uh, go see if Morales wants to embarrass herself instead."

He turned, already giving her space, already pulling back into that careful, respectful distance he always kept unless she moved closer first.

Ashley watched him take two steps away.

Her chest tightened, sudden and sharp.

He'd been there on the worst days. On the boats. In the flood. In the tunnels. At the ridge. He'd handed her tools, water, stupid jokes, steady looks. He'd stood close without crowding her, ready without pushing. And now, when he finally asked for something for himself, she'd shoved a wall up out of old habit.

Fear. Reflex. Not truth.

"Wait," she blurted.

Manny stopped immediately, half turning back. "Yeah?"

Her heart hammered. She set the mug down before she dropped it. "I don't dance," she said, words tumbling out too fast. "I don't know how. I don't like people looking at me. I don't like… all of that." She waved a hand toward the crowd.

Manny raised his hands slightly. "Okay. That's okay. I wasn't trying to—"

"Shut up a second," she said, standing.

He shut up.

She stepped in close, close enough to see the tiny flecks of lighter color in his eyes, the faint scar along his jaw from a cable snap last year. Her hands shook, but not from cold.

"I don't want to dance," she said again, more quietly. "But I do want… this."

Before he could ask what "this" meant, she grabbed a fistful of his jacket, rose onto her toes, and kissed him.

It wasn't practiced or graceful. It was fast, fierce, a little clumsy, driven more by adrenaline than technique. But it was real. Every bit of tight, coiled fear and gratitude and aching almost had a path now, and it went straight through him.

Manny froze for half a heartbeat, then melted into the moment like he'd been waiting on the other side of it for months. His hands hovered at her waist, asking permission even now. When she didn't pull away, he let them rest there, grounding her.

Ashley broke the kiss first, breathless, eyes wide as if surprised by her own courage. "I said no to dancing," she panted. "Not to you."

Manny blinked, then laughed—a low, stunned sound that turned into something bright. "Okay," he said. "That's… that's good data to have."

She huffed, embarrassed, and dropped her forehead briefly against his chest. "Don't make it weird," she muttered.

"I'm pretty sure we just sprinted past weird," he said gently. "But I'm good. I can handle this."

He tipped her chin up with one knuckle, checking. Really checking. "You okay?"

"For right now?" she said. "Yeah. I think so."

"Good," he said. "Because now I really don't care if Morales wants to dance with me."

She snorted, the sound half laugh, half sigh. They stayed close, not quite dancing, just swaying in the heat of the fire while the music wrapped around them and the camp spun on without noticing every detail.

Manny didn't push for more.

Ashley didn't run.

For both of them, that was enough.

Later, when the food was mostly gone and the music had softened into tired strums, Ridge Camp settled into that hazy, golden place between celebration and sleep.

Lanterns swung from hooks and tent poles, their patched glass glinting. The big fire burned lower now, a deep, steady bed of coals licking at a few last logs. Kids dozed against parents' sides. Conversations dropped to murmurs. Laughter still bubbled up here and there, softer now, like the afterglow of a good joke.

Joe and Valery slipped away from the main circle, moving up the slight rise at the edge of camp where the ridge opened to the dark beyond. Snow crunched under their boots. The air was still bitter, but the cold didn't feel as sharp with the fire at their backs and the sound of their people at their feet.

They stopped where the hill started to drop, just far enough to see the whole camp laid out below them.

It looked nothing like the siege scene of the night before.

Lantern light turned the tents into small constellations. The central fire glowed like a heart. Threads of smoke rose into the dark, carrying the smell of wood and stew and something sweet—maybe the last of the preserves. Shadows moved slowly between shelters, not in frantic rushes, but in irregular, easy shapes. A child's laugh pealed out, followed by a stern shush and a softer giggle.

"This looks…" Joe started, searching.

"Alive," Valery said.

He nodded. "Yeah. That."

She slid her hand into his, thumb brushing the edge of his knuckles. The ring on her finger was warm from her skin, a tiny circle of silver catching stray firelight when she moved.

"How's it feel?" he asked quietly. "Being married in the apocalypse."

She considered. "Cold toes," she said. "Decent stew. Excellent kiss. Would recommend."

He huffed a laugh. "That your official review?"

"Four out of five stars," she said. "Docking one because there was no Hawaii and you still refused to wear white."

"Absolutely not," he said. "Camp's been through enough."

They stood in comfortable silence for a moment, watching their people breathe.

"I needed this," Valery admitted, voice softer. "Not just the vows. The whole thing. The decorations. The ridiculous flatbread. The kids running around like it's a festival instead of the end of the world."

"Yeah," Joe said. "Me too."

He let his eyes roam over the camp—the patched tarps, the uneven fences, the smoke-stained tents. It would still be hard tomorrow. Fuel would still be short. Food would still need rationing. Sickness would still need watching. The freeze wasn't done with them yet.

But tonight proved something he'd forgotten he needed to see.

They could build joy on purpose.

They didn't just have to wait for it to surprise them.

"You see it?" Valery asked.

"See what?"

"The future," she said. "Bits of it. Like... if we find land. If we build something more permanent than this. If we have a place that isn't always on the edge of freezing or flooding or falling apart."

He listened to the faint music drifting up, to the low rumble of many voices layered together, to the small, bright bursts of laughter cutting the wind. He pictured walls that didn't leak. Fields instead of ration charts. A room that belonged to them alone. A door they could close without checking the hinges three times.

"Yeah," he said. "I see it."

Valery leaned her head against his shoulder. "Good. Because I want to get there with you."

"You will," he said. "We will. All of us who can."

She was quiet for a beat. "You really believe that?"

He squeezed her hand. "Tonight?" he said. "Yeah. I do."

Below them, someone whooped loud enough to startle a few kids. Manny's laugh followed, bright and surprised. Ashley's voice floated up, mock-annoyed and happy underneath. Mom clapped along to the last, stubborn run of the guitar.

Ridge Camp glowed.

Not just from the fire, not just from the lanterns, but from something warmer. Something that had nothing to do with weather and everything to do with choice.

Joe turned, pressing a kiss to Valery's temple. "Come on," he murmured. "Let's go back down before Guido decides to give a speech."

"Oh God," she said. "Run. Run now."

They walked back toward the light, hand in hand, boots leaving twin tracks in the thin snow. Behind them, the dark waited. Ahead of them, for this one night, the world was small and bright and full of people they loved.

Tomorrow would be hard again.

But tonight, the ridge belonged to the living.

CHAPTER 39
THE LETTER

The morning after the wedding tasted like cold air and woodsmoke. Ridge Camp moved slowly—quiet, softened, as if everyone wanted to hold onto the warmth of last night just a little longer. Frost clung to every line of rope and every roof tarp, glittering under a pale sun that hadn't yet decided whether it wanted to shine. Laughter drifted from a few tents, tired but genuine. Someone had left the flower garland from Valery and Joe's ceremony tied to the center pole of the commons, now stiff with ice but still beautiful.

Joe woke to the weight of Valery curled against his chest, her breath warm on his shirt. Her hand rested over his ribs, fingers relaxed, her injured shoulder cushioned against the pillow they'd doubled up for her. For a moment, nothing existed but her—her scent, her steady breathing, the faint hum of contentment in his bones after years of survival, fear, and endurance.

"Morning," she murmured without opening her eyes.

"Morning," he whispered back, brushing a kiss across her hairline.

The canvas walls glowed dim gold with the early light. Outside, someone chopped wood—slow, methodical thunks. Lenny barked twice. Normal sounds. Human sounds. Sounding like a world that had remembered how to breathe.

Valery shifted, looking up at him. "You're doing the thinking face."

"I am not."

She arched a brow. "You are absolutely doing the thinking face. The one that says you're already planning twelve steps ahead while I'm still half-asleep."

Joe huffed a quiet laugh. "It's been a while since we woke up after something good. I forgot what that feels like."

Her smile softened. "Feels like hope."

"Yeah," he said. "Hope."

But hope had sharp edges. It always had.

A voice called outside their tent. "Joe?" Morales. Firm, clipped, businesslike. Too early for that tone.

Valery sighed and eased up onto one elbow. "Duty calls."

Joe kissed her shoulder and stood, pulling on his jacket. "Come in," he called.

Morales opened the flap partway, breath fogging out in front of her. Her dark hair was pulled into a short ponytail , and she held a leather-bound notebook under one arm—the one she only carried when something serious was brewing.

"Sorry to interrupt your first morning as honeymooners," she said dryly. "But we need the council."

Joe's posture changed instantly. "What happened?"

"Tracks," she said. "And smoke signatures on the ridge north-west. Could be nothing. Could be something. But after what we found with Holt's supplies…" She shook her head. "We're not ignoring anything."

Valery sat up straighter. "Core group or the full council?"

"The full council." Morales said. "In the ops tent in fifteen minutes. Bring coffee if you've got any."

She closed the flap and disappeared into the cold.

Joe turned back to Valery. She was already tying her boots with her good hand, her expression serious but steady.

"Well," she said, "married one day and already called to a council meeting. That tracks."

He knelt beside her, took her hand, and squeezed. "You okay?"

"Always," she said. "As long as we're doing this together."

He kissed her lightly. "Always."

They stepped out into the cold morning, the remnants of the wedding still scattered around camp—lanterns burned low, ribbons frozen in place, footprints in the snow tracing where people had danced long after the music died. Children chased each other around the warming fire. Manny waved from across the clearing with Ashley beside him, both bundled in heavy jackets. Donnie stood near the tool shed, sharpening an axe, nodding when he saw Joe and Valery approach.

Morales waited at the entrance of the operations tent, steam rising from her mug.

"Let's get to it," she said. "We may have something. Or we may have nothing. But after Holt... we're checking twice."

Joe took a slow breath, feeling the weight of leadership settle back where it always lived—between his ribs, right next to the hope he'd woken with.

"Then let's figure it out," he said.

Another chapter of the world—whatever was left of it—was already beginning.

The council meeting started fast. Morales didn't waste time; she never did. The council gathered around the central table—Joe at the head, Valery at his right, Mom at his left, Donnie beside her, with Manny, Guido, and Dr. Elise rounding out the circle. Outside, camp life moved softly, unaware that another turning point had just begun.

Morales set down her notebook. "We've swept the ridge twice. The tracks were faint, half-frozen, but real. And the smoke signatures came from a spot near the old service road." She paused, letting her gaze sweep across the group. "We've gone north, south, east. But the west ridge? That leads straight toward Arnold AFB."

A quiet settled over the tent.

Mom folded her hands. "You think someone was there?"

"I think," Morales said, "we missed something."

Joe nodded slowly. "We brought back everything we could carry. Fuel, medical kits, food caches. But that base..." He looked at Donnie. "It saved us. Or Holt did."

Donnie leaned forward, elbows resting on his knees. "If we missed a piece of his plan, we need it now before winter hits full force."

Valery sipped her tea. "This time we search every file cabinet, every locker, every wall panel. Holt was thorough. He wouldn't leave anything important out in the open."

Morales tapped her pen. "I say we go today. Before the weather shifts. Two helos. Light crew. In and out."

Joe felt the decision forming long before he spoke it. "We go."

Mom exhaled, not quite relieved, not quite surprised. "Be careful."

Guido crooked a tired grin. "When are we not careful?"

Manny elbowed him. "Do you want the list alphabetized or chronological?"

Joe rose. "Gear up. We leave in thirty."

The council dispersed with purpose. Outside, the morning sun had risen higher, still pale, still weak, but enough to cast long shadows across the camp as everyone prepared.

Joe walked with Valery toward the landing zone. Frost crunched under their boots, each step crisp in the morning quiet. She nudged him lightly. "Day after we get married, and you drag me into a burned-down Air Force base. Very romantic."

He smirked. "We can call it a honeymoon if you want."

"We could," she said, looping her good arm through his. "But then we'd have to explain to the children one day why we spent our honeymoon reading classified military files."

"Educational?" he offered.

She laughed—soft, warm, too rare these days. "Sure. That's what we'll call it."

By the time they reached the helos, Manny and Ashley were already loading gear. Manny handed Ashley her headset. She stared at it like it was a snake.

"Why is this thing always cold?" she muttered.

"Because," Manny said, "life hates us."

She huffed. "Fair point."

Guido jogged up with a crate of sample bags. "Joe, I want to hit Holt's storage office first. Last time we barely touched the place."

Joe nodded. "That's the idea. Every drawer, every shelf, every vent."

Mom appeared with two thermoses. She thrust one into Joe's hand, one into Valery's. "Drink. I'm not having you two freeze solid halfway there."

Joe kissed her cheek. "Yes, Mom."

She softened just a touch, smoothing his jacket collar. "Find what we missed."

The pilots signaled readiness. Helo One's rotors spun up, scattering frost in wide spirals. Helo Two rumbled awake behind it.

Joe helped Valery climb in, then strapped himself into the lead seat.

"Council team en route," he radioed. "Heading to Arnold for a full sweep."

"Copy," came the reply. "And Joe?"

"Yeah?"

"Find answers."

Joe tightened the last strap on his pack and slung it into Helo One's storage compartment. The rotors were already spinning, slow and heavy, throwing ribbons of frost across the landing zone. The cold bit straight through his jacket, sharper than the morning before, as if winter had decided to remind them who was in charge.

Morales jogged over with the tablet tucked beneath her arm. "Pilots are spun up. Weather window's holding, but it's narrow. We'll want to be back before dusk."

Joe nodded. "We're not staying long."

"You say that every time," she shot back. But her eyes softened. "She climbed in."

Valery climbed into the cabin behind Joe, settling in with a wince as her shoulder pulled. He reached up to help her adjust the harness straps. "You sure you're good?"

She nudged him with her knee. "You married me. You don't get to question whether I'm good."

He laughed once—quiet, warm—and brushed a piece of hair from her cheek.

Manny and Ashley boarded Helo Two, the second aircraft roaring to life in churns of white air. Manny tapped Ashley's helmet as she buckled in.

"You okay?"

She gave him a look. "This is nothing."

But when Manny held onto the doorframe and leaned closer, she admitted softly, "I'm okay… just don't disappear on me in there."

"Never," Manny said. "You couldn't lose me if you tried."

Ashley swallowed once, then nodded.

The pilot of Helo One gave Joe a thumbs-up. "Strap in. Thirty seconds."

Joe slid into the front seat beside him, headset on.

"Council team departing Ridge Camp," he radioed. "Destination Arnold Air Force Base. Full structural re-check. Beginning search sweep for remaining documentation."

Mom's voice came back instantly, warm but tight. "Be careful, honey. All of you. This feels like one of those days."

"Copy," Joe said gently. "We'll be back before dark."

The rotors surged, the helicopter rising slowly at first, then steadier, climbing above the tree line. Helo Two lifted behind them, banking right, forming a staggered formation as they pushed north.

Ridge Camp fell away beneath them—rows of tents, wisps of smoke from morning fires, kids running between cook lines, the Savannah's silent hull gleaming cold in the distance.

"Strange seeing it like this," Valery murmured. "Quiet. Peaceful."

"It won't stay that way long," Joe said. "Not when we keep finding more traces of Holt's people."

Guido leaned forward between the seats. "Boss—look."

He pointed at the horizon.

A long black scar cut across the snowfields: the old fire line left behind by the atmospheric ignition. Even now, two years later, it looked wrong—as if the land had been peeled open and frozen mid-scream.

Manny's voice crackled over the radio. "Looks worse every time."

Joe keyed his mic. "Eyes open when we land. Last time we were here, we didn't sweep the entire administrative wing. That changes today."

"Copy that," Manny said. "Ashley already made me promise not to fall into any holes."

Ashley muttered in the background, "Because you nearly did last time, genius."

"I was testing it."

"By falling into it?"

"Scientific method," Manny said.

Ashley groaned loudly.

Valery smiled. "We have to adopt them, don't we?"

"Too late," Joe said. "They already think we're responsible adults."

She snorted. "Good thing they don't know the truth."

The helos crossed the river system, following the break in the forest where the old access road ran like a scar. Snow dust kicked up against the skids as they descended toward the remains of the base.

As they approached, Arnold Air Force Base rose into view—blackened hangars, torn fences, collapsed roofs, and stripped concrete still bearing the scorch lines from the firestorm Holt barely outran.

Joe stared out the window. "Every time," he whispered. "Every time it looks worse."

The pilot circled once, scanning for structural hazards, then lowered Helo One toward the cracked pavement outside the administrative complex. Helo Two touched down seconds later. Engines roared, then slowly spooled down as the blades creaked to a stop.

The moment the doors opened, cold swept inside like a living thing.

Joe jumped out first. Boots hit ice. The buildings loomed ahead—silent, skeletal, frost-painted. The place felt abandoned… but watched.

Manny jogged up beside him, adjusting his pack. "Boss?"

"We go straight in," Joe said. "Same sectors as last time. But this time…" He looked at each of them—Valery, Manny, Ashley, Guido, Morales. "…we don't miss a damn thing."

They walked toward the shattered doorway. Joe placed a hand against the frame. The metal was warped, melted, frozen. Burned and rebuilt by nature all at once.

He stepped inside.

And the search began.

The wind funneled through the shattered windows of the old administrative wing, carrying the cold like a blade. Joe stepped inside first, boots crunching frost that formed a brittle skin over the scorched tile. The others filed in behind him—Valery at his shoulder, Manny and Ashley guarding the rear, Guido sweeping to the right, Morales checking structural beams with a quick, practiced eye.

"Same plan as before," Joe said quietly. His voice echoed down the long corridor, thin and hollow. "Spread out. Slow. Nothing gets ignored this time."

Morales nodded. "Copy. Watch for weak flooring. Heat damage still runs deep."

The hallway forked into three branches. Papers littered the floor—burnt edges, half-melted ID cards, scraps of maps that crumbled when touched. Someone had once left this place in a hurry.

"Guido, Manny—south offices," Joe ordered. "Morales, Ashley—check the operations wing. Valery, you're with me. Commander Holt's office is straight ahead."

Everyone moved at once, splitting like spokes from a wheel.

Guido flicked on his flashlight and disappeared down the left corridor. "If I fall through the floor," he called, "tell Mom I wasn't doing anything stupid."

"Lie," Manny said, following him.

Ashley brushed frost off a nameplate on the wall—MAJ. L. CHANDLER—before heading toward the operations wing with Morales. Their flashlights bobbed in the distance as they vanished into darkness.

Joe and Valery stepped deeper into the main hallway. The temperature dropped fast. Frost thickened along the walls in delicate veining, like the cold was growing its own vines across steel.

"Joe," Valery whispered, touching a scorch mark, "this must've been the outer edge of the ignition wave. Look at the metal…"

The aluminum doorframe along the wall had softened and warped, frozen mid-melt. It looked like someone had taken a torch to it and then poured winter over the top.

"Yeah," Joe murmured. "Firestorm passed right through."

They reached the wide archway that once led to the primary administrative suite. A sign hung crooked above them, half-melted:

COMMAND OPERATIONS

AUTHORIZED PERSONNEL ONLY

Valery stepped through first, flashlight sweeping across overturned chairs, a cracked coffee pot welded to the counter by melted plastic, a bulletin board scorched so black the pushpins had fused. A thin dusting of ash lay across everything like gray snowfall.

"Someone's been through here," she said.

Joe slowed. "How recent?"

She pointed at a set of footprints in the frost—wide treads, military issue, but softened by time. "A few days. Maybe less."

Joe didn't like that. Not at all.

They continued down the hall. The deeper they went, the more intact things became. Heat had licked the outer structure but spared parts of the interior—lucky, or deliberate design. Filing cabinets stood in crooked lines. Frost clung to the handles. A fallen ceiling panel split the walkway in two.

"This place feels… haunted," Valery whispered.

"Not haunted," Joe murmured. "Interrupted."

Ahead, a door sat slightly ajar, its glass panel cracked but still in place.

COMMANDER N. HOLT

Joe and Valery exchanged a look.

"Ready?" she asked.

"No," he whispered. "But let's do it."

They pushed the door open slowly. Hinges groaned like something waking from a long sleep.

Inside, Holt's office looked frozen in mid-motion. A mug sat on the desk with a ring of hardened ash inside. A jacket hung on the back of his chair. His wall map—U.S. flight corridors—was warped at the edges but still recognizable. Papers lay scattered across the floor, as if shaken loose by the tremor they barely outran.

And the desk…

The desk held layers. A thick personnel file lay open, half-burnt at the corner. Another stack—training reports, requisition forms, flight logs—sat crooked on the left side, the edge of one folder hanging over the desk like an unfinished thought.

Joe's flashlight passed over a patch of paper peeking from beneath the stack.

Valery saw it too. "Joe. There."

He reached carefully, lifting the top file—a dust-smeared folder marked CLEARANCE: BETA LEVEL. As he moved it…

A sealed envelope slid free.

Thick paper.

Red Air Force band.

Commander Holt's signature across the front.

Valery's breath hitched. "Oh my God…"

Joe stared at the envelope. It was heavier than it should've been. As if whatever was inside carried weight beyond paper and ink.

"I think this is it," he whispered. "The letter we've been missing."

Behind them, down the hall, Manny shouted:

"Boss! You're gonna want to see this!"

But Joe didn't move yet. The envelope sat in his hand like a heartbeat.

Valery touched his arm gently. "Joe… Holt wanted this found."

"Yeah," Joe said softly. "And he wanted it found by someone who'd keep going."

He slid the letter into his jacket.

Then he called back:

"Everyone meet in Holt's office. Now."

They hadn't just come back for scraps.

They came back for answers.

And Holt had left one waiting on the desk the whole damn time—buried under the last file he touched on the day the sky burned.

Manny appeared in the doorway first, breathing hard and holding something in both hands like it was a treasure from a lost civilization. Joe blinked at it. Valery blinked too. It took a full two seconds before Joe realized what he was looking at.

"Is that a… coffee maker?"

Manny nodded proudly. "A Mr. Coffee. Untouched. In the box. Still got the instruction manual."

Ashley stood behind him, arms folded, unimpressed. "He almost slipped on his ass carrying it."

Manny ignored her, hugging the box like a newborn. "It just needs electricity…"

Guido walked in behind them and groaned. "Morales, if he lugs that thing back to camp, I'm not listening to him cry when it doesn't work."

Manny countered, "Oh, it'll work. I'll MAKE it work."

Despite the cold and the weight of the moment, Joe's lips twitched. A coffee pot—found during what might be the most important discovery of their survival—felt ridiculous and perfect. Something human in all the ruin.

Morales stepped in last, brushing frost off her sleeves. "What'd you find?"

Joe tapped the envelope inside his jacket. "The letter. Holt's real message."

The room fell quiet, the kind of quiet that settles when everyone understands they are standing on the edge of something important.

Valery pointed to the desk. "We should check everything. Not just the envelope."

Morales nodded. "Spread out. Papers on the floor, wall files, maps—anything Holt left will matter."

They moved in sync, seasoned from years of working together. Ashley knelt by the lowest drawer, prying it open with Guido's multi-tool. Manny set down the Mr. Coffee on the filing cabinet like a sacred relic and went for the shelf above Holt's desk. Valery began gently stacking the papers on the floor into piles—logs, memos, flight rosters, requisition sheets.

And Joe… Joe stood behind Holt's chair and slowly pulled it back to its full height. He could almost see Holt sitting there—pen in hand, alarms screaming outside, fire rolling toward the valley—deciding what was worth saving and what wasn't.

Joe exhaled, a slow breath that frosted the air.

"Look at this," Manny said from the corner.

He lifted a large folded map pinned beneath a rusted metal clamp. The edges were burned, but the center was intact—three red circles marked in permanent ink.

Joe recognized them immediately. "Those match the ones in the Holt Squadron case."

Morales looked over Manny's shoulder. "He was staring at this before he left."

Ashley found another stack in a side drawer—thin, sealed packets stamped with AIR FORCE LOGISTICS. "These weren't touched," she said. "He meant to take them."

Guido knelt beside her. "Or meant someone else to."

Valery passed Joe a folder she'd just uncovered from under the desk. "Look at the date."

The timestamp on the corner matched the morning of the Rain of Fire. Holt didn't grab it before evacuating. He'd pushed it aside—maybe when the tremors started—maybe when someone called him to the last carrier.

Joe spread it open.

Inside were hand-marked notes, coordinates scrawled in pen, a partially completed memo titled:

CONTINUITY-OF-SURVIVAL — NORTHERN ROUTE

He stared at it.

"He wasn't just leaving caches," Joe murmured. "He was planning a route. A corridor."

Morales stiffened. "A corridor to what?"

Joe didn't answer. Not yet. Not until they opened Holt's letter.

Valery stood, brushing frost off her gloves. Her face was pale, thoughtful. "Joe… everything here points the same direction."

"Yeah," he said softly. "North."

Manny set the Mr. Coffee beside the pile of recovered files and cleared his throat. "Well… if Holt wanted us to find this place…"

Ashley finished for him. "Then he wanted us to see whatever's in that envelope."

Joe felt the weight of it again—the heaviness, the intention. He looked around the office one more time, taking in every detail: the warped map on the wall, the frost-covered mug, the jacket still hanging by the door.

Holt had left in a rush.

But he'd left this on purpose.

"All right," Joe said, voice steady. "Let's get back to camp."

He slid the folder of maps and Holt's notes into his pack beside the envelope. "We open it together."

Morales nodded. "Council session tonight."

Manny picked up the Mr. Coffee again with reverence. "And I'm making this thing a priority."

Guido groaned. "Please don't."

Ashley muttered, "If he wires that to the generators, we're all gonna die."

Valery smiled faintly. "Let him dream."

Joe led them out into the icy corridor. They walked through the cold ruin of Arnold Air Force Base carrying maps, files, a sealed envelope that could rewrite the next years of their lives—

And, thanks to Manny, a Mr. Coffee.

Ahead, the helicopters waited.

Behind them, Holt's last message finally stirred in the dark.

And by nightfall, every answer would be on the table... ready to change everything.

The council gathered in the commons tent as the sun dipped behind the ridge, turning the canvas walls gold. The air smelled like smoke, paper, and cold wool. Everyone had been waiting—Mom, Donnie, Morales, Guido, Dr. Elise—faces drawn tight with expectation. Joe stood at the center table, Valery at his side, the sealed envelope and the stack of maps Holt had been studying spread out in front of him.

Manny and Ashley slipped in last. Manny still carried the Mr. Coffee under one arm like it was a sacred artifact.

Mom blinked. "Is that...?"

Manny lifted it proudly. "It just needs electricity."

Several people actually laughed—a small, tired sound, but real.

Then the tent settled into silence as Joe placed Holt's letter on the table. He didn't open it yet. Not until everyone sat. Not until everyone was ready.

Morales leaned forward. "Whatever's in that letter, it scared Holt enough that he hid it. And important enough that he made sure someone would come back and find it."

"Us," Valery murmured. "He meant us."

Joe opened the letter slowly, careful with the brittle paper. The writing inside was steady, sharp, unmistakably Holt—part mission log, part confession, part set of orders for whoever came next. He skimmed the first lines about the Rain of Fire, the ignition front, the emergency beacons, the caches locked in Hangars Two and Three, and the decision to abandon the big federal bunkers at Cheyenne, Raven Rock, and Mount Weather. Then he found the part that mattered most now.

He read aloud:

"If you found this, you lived. If you lived, then the fire didn't take the last of us. Good. You're going to need what I left. We didn't scatter supplies. We made a path. Follow it. North. Always north. Fort Aster is the end of the line—high, dry, and hidden. If it's standing, you'll have a future. If it's not… keep moving."

Joe lowered the page.

The tent felt suddenly smaller, the air heavier.

Donnie rubbed his jaw. "Fort Aster… I remember that from one of his logs. He called it a 'final post' before the Rain of Fire hit."

Morales tapped the map Joe had brought. "Look. These aren't random drops. These are scheduled cache points." She traced the route with her finger—bold red dots Holt had drawn in thick strokes. "Food. Water. Fuel. Medical crates. Enough for a caravan."

Guido leaned in, brows knitted. "Spacing's too wide for soldiers… but perfect for families. For a community walking or traveling with gear."

"He was planning for civilians," Dr. Elise said. "Survivors."

Mom exhaled shakily. "He must have known he couldn't take everyone. So he left a way for whoever came next."

Joe laid out another page—Holt's original distribution plan. "Every cache has identical contents. Exactly matched. Enough for a group our size."

Manny set the Mr. Coffee down like punctuation. "Holt wasn't surviving. He was bread crumbing."

Ashley frowned. "What's Fort Aster? A shelter?"

Morales flipped to the back map—Holt's northern schematic. "Fort Aster was a Continuity-of-Survival installation," Morales said. "Up on the Cumberland Plateau—high, dry, and hidden. Underground caverns. Military-grade habitat space. Storage, water treatment, fuel reserves. Heavily shielded. It's where personnel and evacuated civilians were supposed to ride out the ignition if the big sites—Cheyenne, Raven Rock, Mount Weather—failed or couldn't be reached."

Joe scanned the annotations—Holt's handwriting tight and urgent:

ASTER HOLD LINE.

LAST SAFE.

CIVILIAN-INTEGRATED.

FINAL NORTH HAVEN.

Valery whispered, "If they made it…"

"Then they might still be there," Dr. Elise finished. "Or… what's left of them."

The possibilities spun through the tent—hope, dread, questions none of them could answer yet.

Joe looked at the letter again—Holt's final line scrawled in the corner:

If you're reading this, you're meant to follow.

Don't stop.

Don't wait.

NORTH.

He folded the paper and slid it into his jacket.

Then he looked around the table, meeting each pair of eyes—his mother's, steady with faith; Donnie's, hard with resolve; Manny's, fiery; Ashley's, wary but ready; Guido's, tense; Morales's, focused; Valery's, warm and unshakable.

"We've hit the limit of what this valley can give," Joe said quietly. "Fort Aster wasn't just a bunker. It was designed to hold hundreds. Maybe thousands. If there's even a chance Holt's people made it… or supplies survived… or anything is still active…"

He paused. The world outside the tent felt impossibly silent.

"We need to know."

Morales nodded. "And if someone else reached Aster first—friend or not—we need to know that too."

Donnie leaned back. "It's a hell of a journey."

"Yeah," Joe said. "It is."

Valery took his hand. "But we don't have to do it alone."

Joe looked around one last time, the decision already solid in his chest.

Then he said the words that would change everything:

"We go north."

End of Book 3

Continued in Book 4 - Ascent

www.ingramcontent.com/pod-product-compliance
Lightning Source LLC
LaVergne TN
LVHW090256210926
848480LV00001B/39
* 9 7 9 8 9 9 4 1 8 5 7 7 3 *